Lecture Notes in Computational Science and Engineering

35

Editors

Timothy J. Barth, Moffett Field, CA
Michael Griebel, Bonn
David E. Keyes, New York
Risto M. Nieminen, Espoo
Dirk Roose, Leuven
Tamar Schlick, New York

Springer-Verlag Berlin Heidelberg GmbH

Eberhard Bänsch
Editor

Challenges in Scientific Computing – CISC 2002

Proceedings of the Conference
Challenges in Scientific Computing
Berlin, October 2-5, 2002

Springer

Editor

Eberhard Bänsch

Weierstrass-Institut
für Angewandte Analysis und Stochastik
Mohrenstr. 39
10117 Berlin, Germany
e-mail: baensch@wias-berlin.de

Cover figure by Jürgen Fuhrmann, Weierstrass-Institut Berlin

Cataloging-in-Publication Data applied for

A catalog record for this book is available from the Library of Congress.

Bibliographic information published by Die Deutsche Bibliothek
Die Deutsche Bibliothek lists this publication in the Deutsche Nationalbibliografie;
detailed bibliographic data is available in the Internet at <http://dnb.ddb.de>.

Mathematics Subject Classification (2000): 65-06, 65M06, 65M55, 65M60, 65N06, 65N30, 65N50, 65N55, 65Y05, 68-06, 76D05, 76S05, 82D37

ISSN 1439-7358
ISBN 978-3-642-62406-3 ISBN 978-3-642-19014-8 (eBook)
DOI 10.1007/978-3-642-19014-8

springeronline.com

Originally published by Springer-Verlag Berlin Heidelberg New York in 2003
Softcover reprint of the hardcover 1st edition 2003

Cover design: Friedhelm Steinen-Broo, Estudio Calamar, Spain
Cover production: *design & production*
Typeset by the authors using a Springer LaTeX macro package

Printed on acid-free paper 46/3142/LK - 5 4 3 2 1 0

Preface

The conference ***Challenges In Scientific Computing*** (CISC 2002) took place from October, 2 to 5, 2002. The hosting institution was the Weierstrass Institute for Applied Analysis and Stochastics (WIAS) in Berlin, Germany.

The main purpose of this meeting was to draw together researchers working in the fields of numerical analysis and scientific computing with a common interest in the numerical treatment and the computational solution of systems of nonlinear partial differential equations arising from applications of physical and engineering problems.

The main focus of the conference was on the problem class of nonlinear transport/diffusion/reaction systems, chief amongst these being: the Navier-Stokes equations, semiconductor-device equations and porous media-flow problems. The emphasis was on *unsolved* problems, challenging open questions from applications and assessing the various numerical methods used to handle them, rather than concentrate on accurate results from "solved" problems.

Thanks to the participants it was an interesting meeting. The presentations stimulated exchanging ideas and lively discussions.

This proceedings comprises 13 papers form the conference, ranging from numerical methods for flow problems, multigrid methods, semiconductor and microwave simulation, solution methods, finite element analysis to software aspects.

This interesting conference would not have been possible without the help of the staff of the WIAS. I thank all participants, and all our supporters, especially those not onstage, for making the conference a success.

Berlin, September 2003 Eberhard Bänsch

Organization

Scientific Committee

Eberhard Bänsch (Chair)	Freie Universität Berlin and Weierstraß–Institut für Angewandte Analysis und Stochastik, Berlin, Germany
Michael Griebel	Rheinische Friedrich–Wilhelms–Universität Bonn, Germany
Ralf Kornhuber	Freie Universität Berlin, Germany
Ulrich Langer	Johannes Kepler Universität Linz, Austria
Ricardo Nochetto	University of Maryland, College Park, Maryland, United States of America
Alfio Quarteroni	Ecole Polytéchnique Fédérale de Lausanne, Switzerland and Politecnico di Milano, Italy

Organising Committee

Eberhard Bänsch (Chair), Dominic Davis, Jürgen Fuhrmann, Friedrich Grund, Georg Hebermehl, Hartmut Langmach, Marion Lawrenz, Rainer Schlundt, Manfred Uhle (all from WIAS).

Sponsoring Organization

The financial support provided by the following institutions is gratefully acknowledged:

German Research Foundation (DFG, Deutsche Forschungsgemeinschaft), Bonn, Germany

Hewlett-Packard Company, Munich, Germany

Weierstrass Institute for Applied Analysis and Stochastics, Berlin, Germany

Contents

Higher Order Discontinuous Galerkin Methods for Flow and Transport in Porous Media
Peter Bastian 1

Macroscopic Models of Fluids with Microstructure
Noel J. Walkington 23

The Fictitious Boundary Method for the Implicit Treatment of Dirichlet Boundary Conditions with Applications to Incompressible Flow Simulations
Stefan Turek, Decheng Wan, Liudmila S. Rivkind 37

Numerical Simulation of Three Dimensional Free Surface Flows with Bubbles
Alexandre Caboussat, Vincent Maronnier, Marco Picasso, Jacques Rappaz 69

A Computational Comparison of Two FEM Solvers for Nonlinear Incompressible Flow
Jaroslav Hron, Abderrahim Ouazzi, Stefan Turek 87

Algebraic Multigrid for Industrial Semiconductor Device Simulation
Tanja Clees, Klaus Stüben 110

Simulation of Microwave and Semiconductor Laser Structures Including Absorbing Boundary Conditions
Georg Hebermehl, Friedrich-Karl Hübner, Rainer Schlundt, Thorsten Tischler, Horst Zscheile, Wolfgang Heinrich 131

Finite Element Methods for Coupled Problems in Ferrohydrodynamics
Olga Lavrova, Gunar Matthies, Teodora Mitkova, Viktor Polevikov, Lutz Tobiska 160

Iterative Substructuring Methods for Advection – Diffusion Problems in Heterogeneous Media
Paolo Zunino 184

Efficient Preconditioning of Linear Systems Arising from the Discretization of Radiative Transfer Equation
Mohammed Seaïd, Axel Klar 211

On Using Common Lisp for Scientific Computing
Nicolas Neuss 237

The Discontinuous Galerkin Finite Element Method for Singularly Perturbed Problems
Hans-Görg Roos, Helena Zarin 246

Black-Box Preconditioning for Mixed Formulation of Self-Adjoint Elliptic PDEs
Catherine Powell, David Silvester 268

Author Index 287

Higher Order Discontinuous Galerkin Methods for Flow and Transport in Porous Media

Peter Bastian

Interdisziplinäres Zentrum für Wissenschaftliches Rechnen, Universität Heidelberg, Im Neuenheimer Feld 348, 69120 Heidelberg, Germany

Summary. In this paper we describe higher order discontinuous Galerkin (DG) methods for application in subsurface transport. Based on a scheme developed by Oden, Babuška and Baumann we present a method for the solution of the elliptic flow equation and describe a multigrid method for the fast solution of the arising algebraic equations. For the solution of the transport equation we combine the DG space discretization with higher order explicit Runge-Kutta schemes in the convection-dominated case and with diagonally implicit Runge-Kutta schemes in the diffusion-dominated case. Numerical results are presented for single-phase flow in heterogeneous media, solute transport and two-phase flow.

1 Introduction

In this paper we are interested in the accurate numerical solution of the equations describing the flow of fluids and dissolved components in the subsurface.

Let Ω be a domain in $\mathbb{R}^d$, $d = 2, 3$, with outward unit normal n. The equation for groundwater flow in head-based formulation is given by

$$\nabla \cdot u = f \quad \text{in } \Omega, \quad u = -K\nabla H, \tag{1}$$

with Dirichlet boundary conditions $H = H_0$ on Γ_H and flux boundary conditions $u \cdot n = U$ on boundary Γ_U. H is the hydraulic head, u is the Darcy velocity and K is the permeability tensor.

The challenge for numerical methods solving (1) is to achieve high accuracy for the flow velocity u subsequently entering the transport equation. Moreover, the permeability tensor may vary over many orders of magnitude.

The generic transport equation including convective and dispersive transport, radioactive decay and a reaction term reads

$$R\Phi\left(\frac{\partial C}{\partial t} + \lambda C\right) + \nabla \cdot j = q(C) \quad \text{in } \Omega, \quad j = uC - D(u)\nabla C \tag{2}$$

with Dirichlet boundary conditions $C = C_0$ on Γ_C, flux boundary conditions $j \cdot n = J$ on boundary Γ_J and outflow boundary conditions $j \cdot n = (uC - D(u)\nabla C) \cdot n$ on Γ_O. We assume that $u \cdot n \geq 0$ on Γ_O. R is the retardation factor, Φ the effective porosity, $\lambda = \log 2/T$ with T the half life time of the element and D the diffusion/dispersion tensor.

The flow equation (1) is of elliptic type while the transport equation (2) is of hyperbolic type if $D = 0$ and parabolic if D is a symmetric positive definite tensor.

A nonlinear extension of the equations above describes the flow of two incompressible and immiscible fluids, e. g. water (the wetting phase w) and oil (the non-wetting phase n), in a porous medium. The primary unknowns are the pressure of the water phase p_w and the saturation of the non-wetting phase S_n. For details of the modeling we refer to [28, 4, 24].
The phase pressure equation is given by

$$\nabla \cdot u = q \quad \text{in } \Omega, \quad u = -\lambda K(\nabla p_w - G) - \lambda_n K \nabla p_c \tag{3}$$

with boundary conditions

$$p_w = p_{wd} \text{ on } \Gamma_{wd}, \quad u \cdot n = U \text{ on } \Gamma_{wn}, \tag{4}$$

and the total velocity $u = u_w + u_n$ being the sum of the phase velocities, K the absolute permeability, q the source term, $k_{rw}(1-S_n), k_{rn}(S_n)$ the relative permeabilities, $p_c(1-S_n)$ the capillary pressure function, μ_w, μ_n the dynamic viscosities of the two fluids, $\lambda_w = k_{rw}(1-S_n)/\mu_w$, $\lambda_n = k_{rn}(S_n)/\mu_n$ the phase mobilities, $\lambda = \lambda_w + \lambda_n$ the total mobility, ϱ_w, ϱ_n the densities of the two fluids and $G = g(\lambda_w \varrho_w + \lambda_n \varrho_n)/\lambda$, where g is a vector that points in the direction of gravity and has the length of the gravitational acceleration.

The phase pressure equation (3) is coupled to the saturation equation

$$\Phi \frac{\partial S_n}{\partial t} + \nabla \cdot j_n = q_n \quad \text{in } \Omega, \quad j_n = f_n(S_n) w(u, S_n) - hK\nabla p_c \tag{5}$$

with boundary and initial conditions

$$S_n = S_{nd} \text{ on } \Gamma_{nd}, \quad j_n \cdot n = \phi_n/\varrho_n \text{ on } \Gamma_{nn}, \quad S_n(t=0) = S_{n0}, \tag{6}$$

and coefficients

$$f_n = \lambda_n/\lambda, \quad w(u, S_n) = u - \lambda_w(\varrho_w - \varrho_n)Kg, \quad h = \lambda_w \lambda_n/\lambda. \tag{7}$$

For a given function $S_n(x,t)$ the phase pressure equation is elliptic, while for a given total velocity field $u(x,t)$ the saturation equation is either nonlinear hyperbolic if $p_c \equiv 0$ and nonlinear parabolic else.

In this paper we present the application of higher order discontinuous Galerkin (DG) methods to the equations given above. Due to their flexibility, DG methods have been popular among the finite element community and they have been applied to a wide range of computational fluid problems. Since the first DG method introduced in [29] the methods have been developed for hyperbolic problems known as the Runge-Kutta DG method [15, 14, 11, 12, 17] and for elliptic problems in [38, 27, 16, 33, 34]. A unified analysis for many DG methods for elliptic problems has been given recently in [3]. A general overview is available in [13].

Advantages of DG methods are their higher order convergence property, local conservation of mass and flexibility with respect to meshing and *hp*-adaptive refinement. Their uniform applicability to hyperbolic, elliptic and parabolic problems as well as their robustness with respect to strongly discontinuous coefficients renders them very attractive for porous medium flow and transport calculations [32, 1]. The application of DG methods to a benchmark problem in atomic waste repository simulation is given in [8]. DG methods for elliptic problems are comparable in quality with mixed finite element methods, as will be shown below.

In this work we use a DG formulation due to Oden, Babuška, and Baumann [27] for elliptic problems and combine it with either the explicit Runge-Kutta DG methods from [12, 17] or with diagonally implicit Runge-Kutta time discretizations.

This paper is organized as follows: Section 2 introduces the necessary notation, Section 3 presents the DG scheme for the eliptic flow equation while Section 4 describes the DG scheme for hyperbolic and parabolic transport equations. Numerical results for single and two-phase flow problems are presented in Section 5. Finally, conclusions are given in Section 6.

2 Notation

Let $E_h = \{e_1, \dots, e_{n_h}\}$ be a non-degenerate quasi-uniform subdivision of Ω where $e \in E_h$ is a triangle or quadrilateral if $d = 2$ and e is a tetrahedron, pyramid, prism or hexahedron with planar faces if $d = 3$. Let h denote the maximum diameter of the elements in E_h. The domain covered by $e \in E_h$ is denoted by Ω_e and the outward unit normal to Ω_e is n_e. The subdivision does not have to match up at element boundaries (although this is not yet implemented in the code).

The space of polynomial functions of degree r on element $e \in E_h$ is defined by

$$P_r(\Omega_e) = \{w : \Omega_e \to \mathbb{R} \mid w(x,y) = \sum_{0 \le a+b \le r} c_{ab}\, x^a y^b\}. \tag{8}$$

The extension to three space dimensions is obvious. Note that P_r can be used on triangles (tetrahedra) *and* quadrilaterals (hexahedra). In the implementation P_r is generated from basis polynomials on the reference element. Moreover, we use basis polynomials that are L^2-orthogonal on the reference elements. This improves the conditioning of the arising matrices and leads to diagonal mass matrices.

The finite element space used in the DG method is defined as

$$V^r(E_h) = \prod_{e \in E_h} P_r(\Omega_e). \tag{9}$$

Note that functions in $V^r(E_h)$ are discontinuous on the so-called internal skeleton Γ_{int} which is defined as

$$\Gamma_{int} = \{\gamma_{e,f} \mid \gamma_{e,f} = \partial\Omega_e \cap \partial\Omega_f \ \forall e, f \in E_h, e \neq f\}. \tag{10}$$

Correspondingly, the external skeleton is defined as

$$\Gamma_{ext} = \{\gamma_e \mid \gamma_e = \partial\Omega_e \cap \partial\Omega \ \forall e \in E_h\}. \tag{11}$$

With each $\gamma_{e,f} \in \Gamma_{int}$ we associate a unit normal n. The orientation can be selected arbitrarily. With any $\gamma_e \in \Gamma_{ext}$ we associate the unit normal n oriented outward to Ω.
For any $x \in \gamma \in \Gamma_{int}$ we denote the jump of a function $v \in V^r(E_h)$ by

$$[v](x) = \lim_{\epsilon \to 0+} v(x + \epsilon n) - \lim_{\epsilon \to 0+} v(x - \epsilon n). \tag{12}$$

In addition to the jump we also define the average of a function $v \in V^r(E_h)$ at $x \in \gamma \in \Gamma_{int}$:

$$\langle v \rangle(x) = \frac{1}{2}\left(\lim_{\varepsilon \to 0+} v(x + \varepsilon n) + \lim_{\varepsilon \to 0+} v(x - \varepsilon n)\right). \tag{13}$$

3 DG Method for Flow Equation

3.1 Scheme

The DG method due to Oden, Babuška, and Baumann [27] for solving the elliptic problem (1) is given as follows: Find $H_h \in V^r(E_h)$ such that for all $v \in V^r(E_h)$

$$\begin{aligned}
&\sum_{e \in E_h} \int_{\Omega_e} (K \nabla H_h) \cdot \nabla v \, dx \\
&\quad + \sum_{\gamma \in \Gamma_{int}} \int_{\gamma} \langle K \nabla v \cdot n \rangle [H_h] - [v] \langle K \nabla H_h \cdot n \rangle \, ds \\
&\quad + \sum_{\gamma \in \Gamma_{ext} \cap \Gamma_H} \int_{\gamma} (K \nabla v \cdot n) H_h - v \, K \nabla H_h \cdot n \, ds \qquad (14) \\
&= \sum_{e \in E_h} \int_{\Omega_e} f v \, dx - \sum_{\gamma \in \Gamma_{ext} \cap \Gamma_U} \int_{\gamma} U v \, ds + \sum_{\gamma \in \Gamma_{ext} \cap \Gamma_H} \int_{\gamma} (K \nabla v \cdot n) H_0 \, ds
\end{aligned}$$

which will be abbreviated as

$$a_h(H_h, v) = f_h(v) \quad \forall v \in V^r(E_h). \tag{15}$$

Note that the Dirichlet boundary condition is approximated weakly. Assuming that the solution is sufficiently regular the convergence rate of the scheme in the energy norm (and thus for the velocity $u = -K\nabla H$) is $O(h^r)$ and the convergence rate in L^2 is $O(h^r)$ if r is even and $O(h^{r+1})$ if r is odd. This anomaly can can be remedied with other stabilizations such as the nonsymmetric interior penalty DG method [30, 34] or the local DG method [16, 3].

The local conservation property of the DG scheme becomes obvious when a test function $v \in V^r(E_h)$ is inserted which is constant on each element. Then the scheme reduces to

$$\sum_{\gamma \in \Gamma_{int}} \int_{\gamma} [v] \, \langle u \cdot n \rangle ds + \sum_{\gamma \in \partial\Omega} \int_{\gamma} u \cdot n \; v \; ds = \sum_{e \in E_h} \int_{\Omega_e} f v \; dx \tag{16}$$

which shows that the conserved flux is the average $\langle u \cdot n \rangle$.

3.2 Variational Multigrid Formulation

Insertion of a basis representation into (14) results into a large system of linear equations. Since the underlying problem is elliptic it should be possible to construct optimal order multigrid solvers. Such an algorithm has been presented in [9]. For an introduction to multigrid methods we refer to [21, 22].

Here, we will derive the coarse grid correction of a two-grid method in variational form. Let $E_0, E_1, \ldots, E_J$ be a nested mesh hierarchy. Let $H_l^{old} \in V^r(E_l)$ be an approximate solution of (14) on mesh level l. We seek a coarse grid correction $c_{l-1} \in V^r(E_{l-1})$ that removes the low frequency errors from H_l^{old}. The subspace correction approach [39] results in the following procedure for the coarse grid correction:
Solve

$$a_{l-1}(c_{l-1}, v_{l-1}) = f_h(v_{l-1}) - a_{l-1}(H_l^{old}, v_{l-1}) \quad \forall v_{l-1} \in V^r(E_{l-1}) \tag{17}$$

and correct

$$H_l^{new} = H_l^{old} + c_{l-1}. \tag{18}$$

Note that polynomials of degree r are used also on the coarse grid. This should not be necessary since low frequency errors can be represented with low-order polynomials but on the other hand it does not harm the overall complexity and eases implementation.

In the multigrid algorithm, the exact coarse grid correction (17) is replaced by a recursive application of the same procedure. The entries of the restriction and prolongation matrices are obtained from the equation $\varphi_{i,l-1} = \sum_j \omega_{i,j} \varphi_{j,l}$, stating that every coarse grid basis function $\varphi_{i,l-1}$ can be represented by a linear combination of fine grid basis functions $\varphi_{j,l}$. This follows from the fact that the DG spaces are nested: $V^r(E_{l-1}) \subset V^r(E_l)$. The factors $\omega_{i,j}$ are the entries of the prolongation matrix, see [9] for details.

It remains to specify the smoothing iteration. There, we use a zero order incomplete LU decomposition applied to a blocked matrix where all degrees of freedom associated with one element are collected in a block. Blockwise Jacobi and Gauß-Seidel methods (with respect to the same blocking) are not effective as smoothers since errors on the skeleton $\Gamma_{int} \cup \Gamma_{ext}$ are not removed. Numerical results for this multigrid algorithm are presented below.

3.3 BDM Projection

The Darcy velocity $u^{\text{DG}} = -K\nabla H_h$ computed from $H_h \in V^r(E_h)$ is discontinuous at element boundaries and does not have continuous normal component $u^{\text{DG}} \cdot n$. Thus, the average flux $\langle u \cdot n \rangle$ is inconsistent with the fluxes evaluated from left and right. Mathematically, we have $u^{\text{DG}} \notin H(\text{div}; \Omega)$. A velocity field with continuous normal component is, however, required by most transport simulations such as the scheme described below. In [10] we describe a simple projection scheme $\Pi : (V^r(E_h))^d \to H(\text{div}; \Omega)$ and prove that this projection does not reduce the accuracy of the DG scheme. This projected velocity $u^* = \Pi(u^{\text{DG}})$ will be used in the transport simulation.

4 DG for Transport Equation

4.1 Scheme

The semi-discrete DG scheme for solving the transport equation (2) in either its hyperbolic or parabolic form is given as follows: Find $C_h : [0, T] \to V^r(E_h)$ such that for all $v \in V^r(E_h)$

$$
\begin{aligned}
&\frac{\partial}{\partial t}\sum_{e\in E_h}\int_{\Omega_e} R\Phi C_h v\,dx + \sum_{e\in E_h}\int_{\Omega_e} R\Phi\lambda C_h v\,dx \\
&\quad -\sum_{e\in E_h}\int_{\Omega_e}(uC_h - D\nabla C_h)\cdot\nabla v\,dx + \sum_{\gamma\in\Gamma_{int}}\int_\gamma [v]C_h^*\langle u\cdot n\rangle\,ds \\
&\quad +\sum_{\gamma\in\Gamma_{int}}\int_\gamma \langle D\nabla v\cdot n\rangle[C_h] - [v]\langle D\nabla C_h\cdot n\rangle\,ds \\
&\quad +\sum_{\gamma\in\Gamma_{ext}\cap\Gamma_C}\int_\gamma D\nabla v\cdot nC_h - vD\nabla C_h\cdot n\,ds + \sum_{\gamma\in\Gamma_{ext}\cap\Gamma_C^{out}}\int_\gamma vC_h u\cdot n\,ds \\
&\quad +\sum_{\gamma\in\Gamma_{ext}\cap\Gamma_O}\int_\gamma vC_h u\cdot n\ - vD\nabla C_h\cdot n\,ds + J_0^{\sigma,\beta}(C_h,v) \qquad (19)\\
&= \sum_{e\in E_h}\int_{\Omega_e} qv\,dx - \sum_{\gamma\in\Gamma_{ext}\cap\Gamma_J}\int_\gamma Jv\,ds - \sum_{\gamma\in\Gamma_{ext}\cap\Gamma_C^{in}}\int_\gamma vC_0 u\cdot n\,ds \\
&\quad +\sum_{\gamma\in\Gamma_{ext}\cap\Gamma_C}\int_\gamma D\nabla v\cdot n\,C_0\,ds + \sum_{\gamma\in\Gamma_{ext}\cap\Gamma_C}\frac{\sigma}{|\gamma|^\beta}\int_\gamma C_0 v\,ds
\end{aligned}
$$

where we have used the refined decomposition of the boundary into Dirichlet outflow

$$\Gamma_C^{out} = \{x\in\Gamma_C \mid u(x)\cdot n > 0\} \tag{20}$$

and Dirichlet inflow

$$\Gamma_C^{in} = \{x\in\Gamma_C \mid u(x)\cdot n \le 0\} \tag{21}$$

and the interior penalty term [38, 30]

$$J_0^{\sigma,\beta}(C,v) = \sum_{\gamma\in\Gamma_{int}}\frac{\sigma}{|\gamma|^\beta}\int_\gamma [C][v]\,ds + \sum_{\gamma\in\Gamma_{ext}\cap\Gamma_C}\frac{\sigma}{|\gamma|^\beta}\int_\gamma Cv\,ds \tag{22}$$

with user-defined parameters σ and β. This additional penalty term can be used to improve continuity of the solution at internal boundaries.

The concentration in the convective term for $x\in\gamma\in\Gamma_{int}$ is upwinded via

$$C_h^*(x) = \begin{cases} \lim\limits_{\epsilon\to 0+} C_h(x-\epsilon n) & \text{if } \langle u\cdot n\rangle \ge 0 \\ \lim\limits_{\epsilon\to 0+} C_h(x+\epsilon n) & \text{else}\end{cases}. \tag{23}$$

In the scheme (19) we use the projected velocity u^* defined in Subs. 3.3. The spatial error of this formulation is $O(h^{r+1})$ in L^2 in the hyperbolic case ($D=0$) for a sufficiently regular solution. Error estimates are provided in [31].

A nice property of the DG method is the correct treatment of Dirichlet boundary conditions in the transition from the hyperbolic to the parabolic case at outflow boundaries. In the hyperbolic case no boundary condition can be prescribed at outflow boundaries and the numerical method treats the value at the outflow boundary as unknown. If a diffusion term is present a Dirichlet boundary condition is obeyed for $h \to 0$.

4.2 Runge-Kutta Time Discretizations

Eq. (19) can be rewritten in ODE form after inserting a basis and inverting the mass matrix. The mass matrix is diagonal if an orthogonal basis is chosen (otherwise it is block diagonal). We denote the ODE system by

$$\frac{d}{dt} C_h = L_h(t, C_h(t)). \tag{24}$$

This ODE system is discretized by Runge-Kutta methods. The time interval $(0, T)$ is subdivided into $0 = t^0 < t^1 < \ldots < t^M = T$ with $\Delta t^n = t^{n+1} - t^n$. The approximation of $C_h(t)$ to be computed is denoted by C_h^n.

All Runge-Kutta methods used here can be written in the following form which computes C_h^{n+1} from C_h^n:

1. $C_h^{(0)} = C_h^n$;
2. $C_h^{(i)} = \sum_{k=0}^{i} \left[a_{ik} C_h^{(k)} + b_{ik} \Delta t^n L_h(t^n + d_k \Delta t^n, C_h^{(k)}) \right] \quad i = 1(1)s$;
3. $C_h^{n+1} = C_h^{(s)}$;

The number of stages of the scheme is s. Note that schemes with $b_{ii} = 0$ are explicit. Otherwise they are implicit and a large system of nonlinear algebraic equations has to be solved per stage.

For hyperbolic and convection-dominated parabolic problems [35, 15] develop explicit schemes with the total variation diminishing (TVD) property. Diagonally implicit Runge-Kutta schemes with favourable stability characteristics are presented in [2, 23].

Table 1 displays the coefficients of all schemes used in this work. These are the second and third order explicit TVD methods of [35], the second and third order strongly S-stable methods of [2] and the fourth order L-stable method of [23]. The coefficients are written in the form

$$\begin{matrix} a_{10} \ldots a_{1s} & b_{10} \ldots b_{1s} & d_1 \\ \vdots \quad\quad \vdots & \vdots \quad\quad \vdots & \vdots \\ a_{s0} \ldots a_{ss} & b_{s0} \ldots b_{ss} & d_s \end{matrix}, \quad d_0 = 0.$$

The CFL constraint for the explicit schemes is 1/3 for the second order scheme and 1/5 for the third order scheme.

Table 1. Coefficients for various explicit and diagonally implicit Runge-Kutta methods.

Reference	Ord. /s	Coefficients
[35]	2/2	$\begin{array}{ccccccc} 1 & 0 & 0 & 1 & 0 & 0 & 1 \\ \frac{1}{2} & \frac{1}{2} & 0 & 0 & \frac{1}{2} & 0 & 1 \end{array}$
[35]	3/3	$\begin{array}{ccccccccc} 1 & 0 & 0 & 0 & 1 & 0 & 0 & 0 & 1 \\ \frac{3}{4} & \frac{1}{4} & 0 & 0 & 0 & \frac{1}{2} & 0 & 0 & \frac{1}{2} \\ \frac{1}{3} & 0 & \frac{2}{3} & 0 & 0 & 0 & \frac{2}{3} & 0 & 1 \end{array}$
[2]	2/2	$\begin{array}{ccccccc} 1 & 0 & 0 & 0 & \alpha & 0 & \alpha \\ 1 & 0 & 0 & 0 & 1-\alpha & \alpha & 1 \end{array}, \alpha = 1 - \frac{\sqrt{2}}{2}$
[2]	3/3	$\begin{array}{ccccccccc} 1 & 0 & 0 & 0 & 0 & \alpha & 0 & 0 & \alpha \\ 1 & 0 & 0 & 0 & 0 & \tau_2 - \alpha & \alpha & 0 & \tau_2 \\ 1 & 0 & 0 & 0 & 0 & \beta_1 & \beta_2 & \alpha & 1 \end{array}, \begin{array}{l} \alpha = 0.4358665215 \\ \tau_2 = (1+\alpha)/2 \\ \beta_1 = -\frac{6\alpha^2 - 16\alpha + 1}{4} \\ \beta_2 = \frac{6\alpha^2 - 20\alpha + 5}{4} \end{array}$
[23]	4/5	$\begin{array}{cccccccccccc} 1 & 0 & 0 & 0 & 0 & 0 & 0 & \frac{1}{4} & 0 & 0 & 0 & 0 & \frac{1}{4} \\ 1 & 0 & 0 & 0 & 0 & 0 & 0 & \frac{1}{2} & \frac{1}{4} & 0 & 0 & 0 & \frac{3}{4} \\ 1 & 0 & 0 & 0 & 0 & 0 & 0 & \frac{17}{50} & -\frac{1}{25} & \frac{1}{4} & 0 & 0 & \frac{11}{20} \\ 1 & 0 & 0 & 0 & 0 & 0 & 0 & \frac{371}{1360} & -\frac{137}{2720} & \frac{15}{544} & \frac{1}{4} & 0 & \frac{1}{2} \\ 1 & 0 & 0 & 0 & 0 & 0 & 0 & \frac{25}{24} & -\frac{49}{48} & \frac{125}{16} & -\frac{85}{12} & \frac{1}{4} & 1 \end{array}$

4.3 Slope Limiters

According to Godunov's theorem [25] there are no monotone linear schemes of order greater than one. Unphysical oscillations in higher order schemes are suppressed by slope limiters. Slope limiters applicable to discontinuous Galerkin space discretizations have been presented in [12, 17]. They are incorporated into the Runge-Kutta time-stepping procedure as follows:

1. $C_h^{(0)} = C_h^n$;
2. $C_h^{(i)} = \Lambda\Pi \left(\sum_{k=0}^{i} \left[a_{ik} C_h^{(k)} + b_{ik} \Delta t^n L_h(t^n + d_k \Delta t^n, C_h^{(k)}) \right] \right) \quad i = 1(1)s$;
3. $C_h^{n+1} = C_h^{(s)}$;

This scheme is called the RK$\Lambda\Pi$P time-stepping method. $\Lambda\Pi : V^r(E_h) \to V^r(E_h)$ is a projection that is used to postprocess the solution after each stage such that the TVD property is maintained, for details we refer to [12, 17]. $\Lambda\Pi$ is local in the sense that on an element it only uses the solution on that element and in neighboring elements.

We now describe the construction of $\Lambda\Pi$ applied to a function $v \in V^r(E_h)$. Fix a particular element $e \in E_h$. The first step is to extract the linear part v_e^1 on Ω_e through L_2-projection, i. e.

$$(v_e^1, w)_{\Omega_e} = (v, w)_{\Omega_e} \qquad \forall w \in P_1(\Omega_e). \tag{25}$$

$\Lambda\Pi$ relies on the assumption that spurious oscillations in v already show up in its linear part. This is theoretically not proven [17].

In the following we restrict the presentation to the one-dimensional case. The elements are then given by $\Omega_{e_i} = (x_{i-1/2}, x_{i+1/2})$, the center of element e_i is at x_i, $i = 1, \dots, K$. The functions v_{e_i} in $P_1(\Omega_{e_i})$ can be written in the orthonormal basis

$$v^1_{e_i}(x) = \bar{v}_i \phi_{i,0}(x) + \tilde{v}_i \phi_{i,1}(x)$$

with

$$\phi_{i,0}(x) = 1 \quad \text{and} \quad \phi_{i,1}(x) = \frac{x - x_i}{\Delta x_i/2} \quad \text{where } \Delta x_i = x_{i+1/2} - x_{i-1/2}.$$

The two coefficients $\bar{v}_i$ and $\tilde{v}_i$ can be computed from a given $v^1_{e_i}$ by

$$\bar{v}_i = v^1_{e_i}(x_i), \qquad \tilde{v}_i = v^1_{e_i}(x_{i+1/2}) - v^1_{e_i}(x_i). \tag{26}$$

Obviously, $\bar{v}_i$ is the average cell value and $\tilde{v}_i$ measures the slope of the linear function. The slope limiter, as the name implies, will possibliy modify the slope of the function but not its average value (conservation of mass). The limited slope $\tilde{v}^*_i$ on element e_i is computed by

$$\tilde{v}^*_i = m\left(\tilde{v}_i, q(\bar{v}_{i+1} - \bar{v}_i), q(\bar{v}_i - \bar{v}_{i-1})\right) \tag{27}$$

where m is the so-called minmod function

$$m(a_1, a_2, a_3) = \begin{cases} s \min_i |a_i| & \text{if } s = \text{sign}(a_1) = \text{sign}(a_2) = \text{sign}(a_3), \\ 0 & \text{otherwise,} \end{cases}$$

and $q \in (0, 1]$. In (27) the slope on element e_i is compared with the slopes computed from neighboring cell averages. If all slopes have the same sign the slope with the smallest absolute value is selected. Otherwise a local minimum or maximum has been detected and the slope is set to zero. The parameter q is taken as 1 for linear convection and $1/2$ for nonlinear convective terms.

In quadrilateral elements two opposite edges define a coordinate direction and the same procedure is applied accordingly. The limiting in case of triangular elements is more complicated and is described in [12, 17].
It remains to specify the $\Lambda\Pi$-projection:

$$\Lambda\Pi(v)|_{\Omega_{e_i}} = \begin{cases} \bar{v}_i \phi_{i,0}(x) + \tilde{v}^*_i \phi_{i,1}(x) & \text{if } \tilde{v}^*_i \neq \tilde{v}_i, \\ v|_{\Omega_{e_i}} & \text{else.} \end{cases} \tag{28}$$

5 Numerical Results

The schemes described in this paper have been implemented in the software framework “Unstructured Grids” [6] for the numerical solution of partial differential equations. The implementation covers two- and three-dimensional unstructured meshes and polynomials up to degree 6.

5.1 Multigrid Performance

The convergence of the multigrid solver is illustrated in Tables 2 and 3. For comparison we also list the iteration numbers for standard multigrid applied to a vertex centered finite volume discretization. We use a multigrid V-cycle with one ILU pre- and postsmoothing step and list iteration numbers for a 10^{-8} reduction of the initial residual.

Table 2 shows the iteration numbers for a full regularity model problem $-\Delta p = f$ in $(0,1)^2$, $p = p_0$ on $\partial\Omega$. A structured quadrilateral mesh with coarse mesh size $h_0 = 1/2$ has been used. Obviously the method is robust in mesh size h and the polynomial degree r.

Multigrid convergence is independent of the regularity of the problem as is shown by Table 3 where a reentrant corner problem with solution in H^s, $s = 1 + 4/7$, has been solved.

Table 2. Number of multigrid iterations for full regularity model problem.

h^{-1}	FV	$r=2$	$r=3$	$r=4$	$r=5$	$r=6$
4	3	5	5	5	5	4
8	4	7	6	6	5	6
16	4	7	6	6	5	6
32	4	7	6	6	5	6
64	4	7	6	6	5	6
128	4	6	6	6		
256	4					
512	4					

Table 3. Number of multigrid iterations for reentrant corner problem.

Triangles	FV	$r=2$	$r=3$	$r=4$	$r=5$	$r=6$
20	4	7	8	7	7	8
80	6	8	7	7	8	8
320	6	8	8	7	8	8
1280	6	9	8	7	8	8
5120	7	9	8	7		
20480	7	9				
81920	7					

5.2 Single Phase Flow in Heterogeneous Media

The second example explores the quality of the DG solutions for elliptic problems with highly discontinuous coefficients. We solve $-\nabla \cdot \{K\nabla p\}$ in the unit square with $p = 1$ for $x = 0$, $p = 0$ for $x = 1$ and no flow boundary conditions

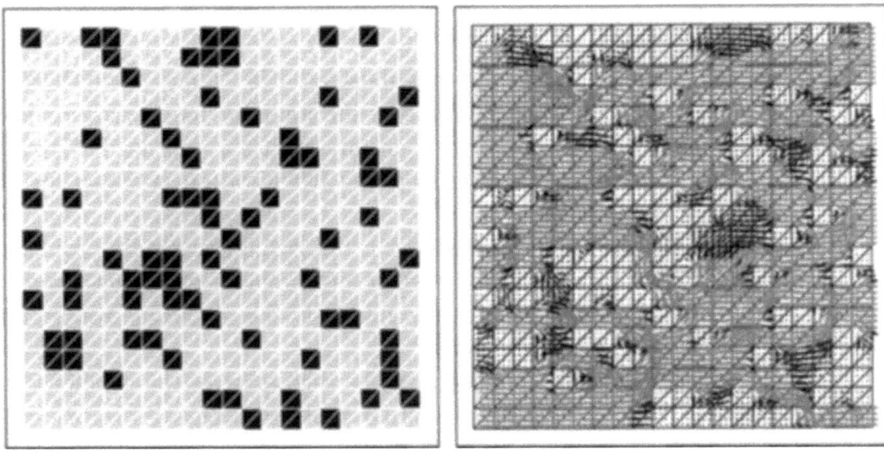

Fig. 1. Permeability and flow field for the discontinuous coefficient example computed with DG(3). Permeability 1 shown in light gray and 10^{-6} in black. Vectors not drawn to scale are indicated by gray color in the vector plot.

for $y = 0$ and $y = 1$. The permeability field is defined on a regular 20×20 mesh and is shown in Fig. 1 on the left. In dark areas the permeability is $K = 10^{-6} \cdot I$, elsewhere it is $K = I$. The example is taken from [19].

The unit square is discretized with $20 \times 20 \times 2$ triangular elements such that the permeability field is resolved with coarse grid elements. Finer grids are obtained through regular refinement. The right plot in Fig. 1 shows the flow field computed with degree $r = 3$ on the coarsest mesh.

Table 4. Total flux through the system for discontinuous coefficent example.

h^{-1}	FV	$r = 2$	$r = 3$	$r = 4$	$r = 5$	$r = 6$	MFE
20	0.6991	0.5094	0.5152	0.5174	0.5232	0.5152	0.4508
40	0.6466	0.5179	0.5181	0.5208	0.5206		
80	0.6170	0.5194	0.5192	0.5201			
160	0.5998	0.5199	0.5198				
320	0.5890						
640	0.5816						

In Table 4 we show results for the unknown total flux through the system. We compare the vertex centered finite volume method (which in this case is identical to P1 conforming finite elements), DG with $r = 2$ up to $r = 6$ and the lowest order mixed finite element method. The value for the mixed method is taken from [19]. The "exact" value has been given in [19] as 0.5205 which was obtained by computing approximations on a sequence of meshes up to 200×200 with a cell centered finite volume scheme and extrapolation to $h = 0$. The results clearly show the unsuitability of the standard finite element

Table 5. Number of multigrid cycles in the discontinuous coefficient example.

l	h^{-1}	FV	$r=2$	$r=3$	$r=4$
1	40	6	14	14	16
2	80	7	14	12	15
3	160	7	13	12	
4	320	8			
5	640	9			

method for this type of problem. Moreover, the error in the mixed finite element solution on the coarsest mesh is about a factor six larger than the error in the DG result on the same mesh. However, the number of unknowns is also about 2.5 times larger for DG (4800 vs. 2000 in the non-hybridized version). This result clearly shows the suitability of the DG method since the MFE is considered optimal for this type of problem.

The performance of the multigrid method is shown in Table 5. For the discontinuous coefficent example we used a multigrid V-cycle with $\nu_1 = \nu_2 = 2$ ILU smoothing steps as a preconditioner in the BiCGSTAB-method [36]. The table shows the number of preconditioner evaluations needed to reduce the norm of the defect by 10^{-8}. Again the iteration numbers seem to be independent of h and r.

5.3 Rotating Pulse Problem

In this example we consider the transport of a Gaussian pulse in a rotating flow field. We solve

$$\frac{\partial C}{\partial t} + \nabla \cdot \{uC - D\nabla C\} = 0 \qquad \text{in } \Omega = (-0.5, 0.5)^2 \tag{29}$$

with Dirichlet boundary and initial condition taken from the exact solution

$$C(x,y,t) = \frac{2\sigma^2}{2\sigma^2 + 4Dt} \exp\left(- \frac{(\bar{x} - x_c)^2 + (\bar{y} - y_c)^2}{2\sigma^2 + 4Dt} \right) \tag{30}$$

which is available for $u = (-4y, 4x)^T$ and $\bar{x} = x\cos(4t) + y\sin(4t)$, $\bar{y} = -x\sin(4t) + y\cos(4t)$. The exact solution is taken from [37] where many methods are compared for this model problem (unfortunately not in the L_2 norm, only maxima and minima are listed). We use the same parameters as in their paper: $D = 10^{-4}$, $x_c = -0.25$, $y_c = 0$ and $2\sigma^2 = 0.004$. The time interval for the simulation is $[0, \pi/4]$, which is the time for a half rotation. Fig. 2 shows the initial condition and final solution.

For comparison we list the L_2 error of the solution at $t = \pi/4$ for different methods in Tables 6 and 7. T denotes the number of elements and TS denotes the number of time steps used. Table 6 containes the rates for several schemes using continuous and piecewise linear trial functions. Spatial and

Fig. 2. Rotating pulse problem. Initial solution (left) and solution after half rotation (right).

temporal mesh are refined such that $h/\Delta t$ is constant. The backward Euler / full upwinding vertex centered finite volume scheme (column FV BE/FU) is formally of first order. However, the results show that this asymptotic rate is achieved only for very fine meshes requiring huge computational effort. Comparison with the other schemes shows that the asymptotic convergence rate is reached for an absolute error smaller than about $5 \cdot 10^{-3}$. Above that absolute error the observed convergence rate may be much worse than the expected rate. The modified method of characteristic (MMOC) [18] is formally first order convergent but has a small error constant. Asymptotic convergence is achieved already on coarse meshes. The finite volume scheme with Crank-Nicolson time-stepping and central evaluation of convective terms is second-order accurate which is confirmed by the table. It should be noted that the solution is smooth but convection-dominated. The second order scheme shows negligible oscillations on the finer meshes.

Table 7 reports the results for discontinuous Galerkin methods using polynomials of degree $1, 2$ and 3 and the corresponding diagonally implicit Runge-Kutta time integrators of order $2, 3$ and 4 (see Table 1). The theoretical asymptotic convergence rates are fully confirmed by the table. Since the problem is smooth, no slope limiters are necessary. With respect to efficiency one can state the following: For an accuracy of $5 \cdot 10^{-4}$ even the second order DG scheme is faster than the second order finite volume scheme. The third order DG scheme is about five times faster than the second order scheme, the fourth order scheme is about 12 times faster for a more accurate solution. Computation times in the table are always for the finest mesh using a Pentium III/850 MHz computer.

Table 8 shows results where the DG space discretization is combined with a DG time discretization resulting in a space-time finite element formulation. In this scheme time is treated as a fourth dimension. Upwinding in time naturally leads to a sequential solution of the equation on time slabs, see [20]. This scheme is very competitive with respect to accuracy per computation time. However, it requires more computer memory than the higher order Runge-Kutta schemes which may be prohibitive in practice.

Table 6. Rotating pulse problem. L_2 error after half rotation for some simpler schemes.

		MMOC				FV BE/FU		FV CN/CEN	
h^{-1}	T	TS	Rate	L_2-Norm	TS	Rate	L_2-Norm	Rate	L_2-Norm
8	256				10		$7.24 \cdot 10^{-2}$		$8.87 \cdot 10^{-2}$
16	1K	2		$5.13 \cdot 10^{-2}$	20	0.07	$6.90 \cdot 10^{-2}$	0.22	$7.60 \cdot 10^{-2}$
32	4K	4	0.77	$3.01 \cdot 10^{-2}$	40	0.12	$6.35 \cdot 10^{-2}$	0.47	$5.47 \cdot 10^{-2}$
64	16K	8	0.90	$1.61 \cdot 10^{-2}$	80	0.19	$5.55 \cdot 10^{-2}$	1.11	$2.53 \cdot 10^{-2}$
128	64K	16	0.92	$8.51 \cdot 10^{-3}$	160	0.30	$4.52 \cdot 10^{-2}$	1.74	$7.56 \cdot 10^{-3}$
256	256K	32	0.86	$4.69 \cdot 10^{-3}$	320	0.43	$3.36 \cdot 10^{-2}$	1.96	$1.94 \cdot 10^{-3}$
512	1M				640	0.57	$2.26 \cdot 10^{-2}$	2.00	$4.86 \cdot 10^{-4}$
comp. time						565 min		384 min	

Table 7. Rotating pulse problem. L_2 error after half rotation for the DG schemes.

			Order 2		Order 3		Order 4	
h^{-1}	T	TS	Rate	L_2-Norm	Rate	L_2-Norm	Rate	L_2-Norm
8	256	10		$4.37 \cdot 10^{-2}$		$3.56 \cdot 10^{-2}$		$1.73 \cdot 10^{-2}$
16	1K	20	0.75	$2.59 \cdot 10^{-2}$	1.09	$1.67 \cdot 10^{-2}$	2.69	$2.68 \cdot 10^{-3}$
32	4K	40	1.31	$1.04 \cdot 10^{-2}$	1.86	$4.60 \cdot 10^{-3}$	3.70	$2.05 \cdot 10^{-4}$
64	16K	80	1.83	$2.92 \cdot 10^{-3}$	2.53	$7.91 \cdot 10^{-4}$	3.92	$1.35 \cdot 10^{-5}$
comp. time			21 min		67 min		265 min	

Table 8. Rotating pulse problem. L_2 error after half rotation for space-time DG schemes.

			$p = q = 1$		$p = q = 2$		$p = q = 3$	
h^{-1}	T	TS	Rate	L_2-Norm	Rate	L_2-Norm	Rate	L_2-Norm
8	256	10		$5.53 \cdot 10^{-2}$		$3.29 \cdot 10^{-2}$		$1.61 \cdot 10^{-2}$
16	1K	20	0.65	$3.52 \cdot 10^{-2}$	1.90	$8.77 \cdot 10^{-3}$	3.52	$1.40 \cdot 10^{-3}$
32	4K	40	1.37	$1.36 \cdot 10^{-2}$	3.40	$8.27 \cdot 10^{-4}$	4.80	$5.02 \cdot 10^{-5}$
64	16K	80	2.23	$2.88 \cdot 10^{-3}$	2.51	$7.22 \cdot 10^{-5}$	4.09	$2.94 \cdot 10^{-6}$
comp. time		.	14 min		42 min		232 min	

5.4 Buckley-Leverett Problem

The Buckley-Leverett problem is an often used test example for two-phase flow without capillary pressure effects in a one-dimensional porous medium. The corresponding equation

$$\frac{\partial S_w}{\partial t} + \frac{u}{\Phi}\frac{\partial}{\partial x} f_w(S_w) = 0 \quad \text{in } (0, 300[m]) \tag{31}$$

for the wetting phase saturation is nonlinear hyperbolic. The fractional flow function

$$f_w(S_w) = \frac{k_{rw}(S_w)}{k_{rw}(S_w) + \frac{\mu_w}{\mu_n} k_{rn}(1 - S_w)} \tag{32}$$

is S-shaped for typical applications. Here we use Brooks-Corey relative permeabilities [24]

$$k_{rw}(S_w) = S_w^{\frac{2+3\lambda}{\lambda}}, \quad k_{rn}(S_n) = S_n^2 \left(1 - (1 - S_n)^{\frac{2+\lambda}{\lambda}}\right) \tag{33}$$

with $\lambda = 2$. The velocity was set to $u = 3 \cdot 10^{-7}[m/s]$, the porosity to $\Phi = 1/5$ and the viscosity ratio was $\mu_w/\mu_n = 1$. The following boundary and initial conditions are imposed:

$$S_w(0, t) = 1, \qquad S_w(x, 0) = 0. \tag{34}$$

The analytical solution of this problem can be constructed with the method of characteristics [25, 24].

Figure 3 shows the results at the final time $T = 1500[d]$ for polynomial degree 1 and 2 using the explicit TVD Runge-Kutta methods of order 2 and 3 (see Table 1). The CFL number was 0.3 for the second order scheme and 0.18 for the third order scheme. The shock is resolved sharply within about three mesh cells. Experimental orders of convergence in the L_1 and L_2 norm are shown in Table 9. Due to the low regularity of the problem the higher order does not pay off in the shock region. A comparison with standard finite volume methods from [5, 7] exhibits that two levels of mesh refinement are needed with the finite volume scheme to match the accuracy of the DG solution.

Table 9. Error and experimental order of convergence of RKDG applied to the Buckley-Leverett problem.

Method	elements	L_1-error	L_1-rate	L_2-error	L_2-rate
DG(1)	32	$4.41 \cdot 10^{+0}$		$1.06 \cdot 10^{+0}$	
TVD RK(2)	64	$2.40 \cdot 10^{+0}$	0.88	$6.48 \cdot 10^{-1}$	0.71
$Cr = 0.3$	128	$1.37 \cdot 10^{+0}$	0.81	$5.29 \cdot 10^{-1}$	0.29
minmod	256	$7.76 \cdot 10^{-1}$	0.82	$4.38 \cdot 10^{-1}$	0.27
	512	$4.06 \cdot 10^{-1}$	0.93	$2.98 \cdot 10^{-1}$	0.56
DG(2)	32	$3.96 \cdot 10^{+0}$		$9.83 \cdot 10^{-1}$	
TVD RK(3)	64	$2.03 \cdot 10^{+0}$	0.96	$5.89 \cdot 10^{-1}$	0.74
$Cr = 0.18$	128	$1.19 \cdot 10^{+0}$	0.77	$5.02 \cdot 10^{-1}$	0.23
minmod	256	$6.94 \cdot 10^{-1}$	0.78	$4.27 \cdot 10^{-1}$	0.23
	512	$3.58 \cdot 10^{-1}$	0.95	$2.68 \cdot 10^{-1}$	0.67

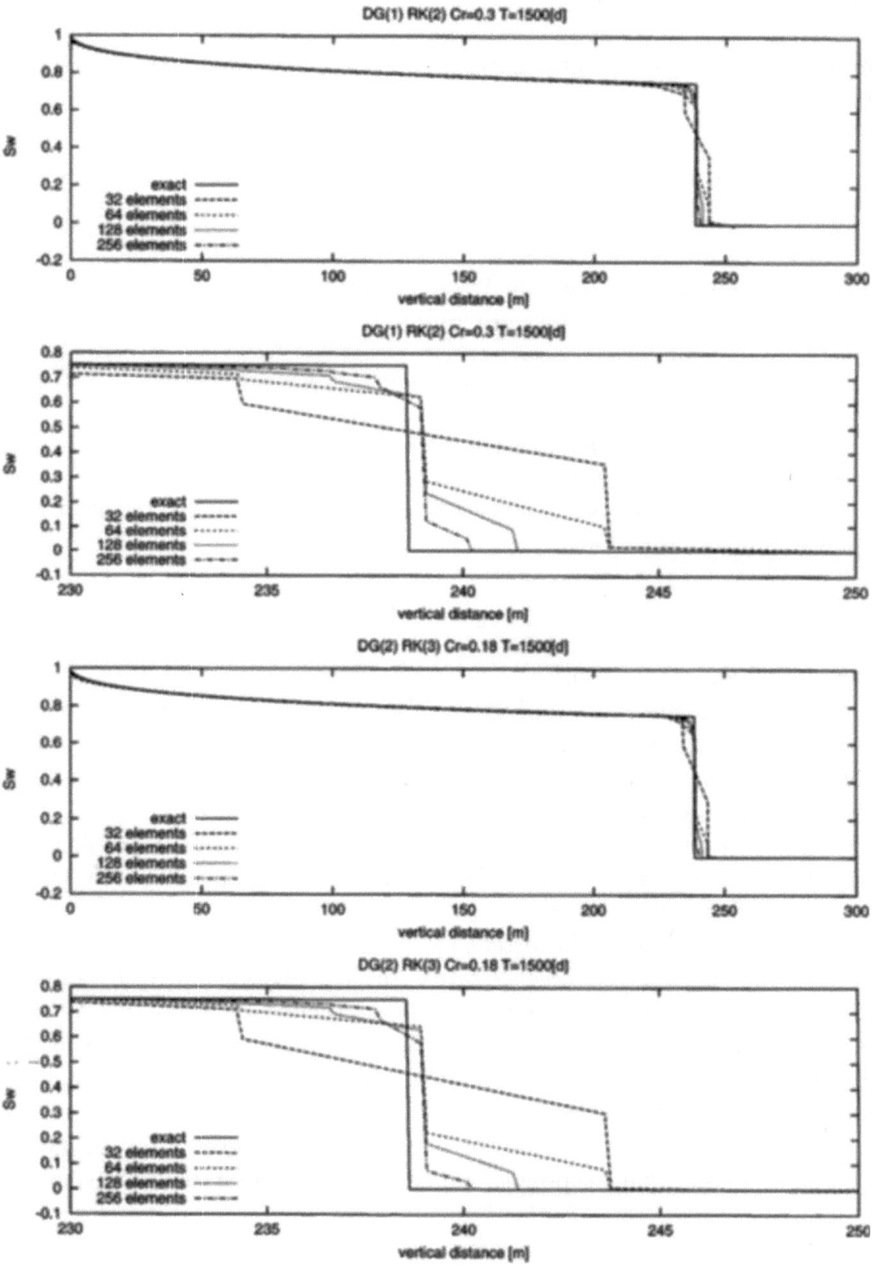

Fig. 3. DG for the Buckley-Leverett problem.

5.5 McWhorter Problem

For a one-dimensional counter–current two-phase flow the wetting phase saturation S_w can be described by the following doubly degenerate parabolic equation:

$$\Phi \frac{\partial S_w}{\partial t} + \frac{\partial}{\partial x}\left(\frac{\lambda_n \lambda_w}{\lambda_w + \lambda_n} p_c' K \frac{\partial S_w}{\partial x}\right) = 0 \quad \text{in } (0, 1.6[m]) \tag{35}$$

where we recall $\lambda_w(S_w) = k_{rw}(S_w)/\mu_w$, $\lambda_n(S_n) = k_{rn}(S_n)/\mu_n$. The following boundary and initial conditions are imposed:

$$S_w(0,t) = 1, \quad S_w(1.6,t) = 0, \qquad S_w(x,0) = 0. \tag{36}$$

A quasi-analytical solution for this problem has been presented in [26]. Here we use the following parameters: $\Phi = 0.3$, $K = 10^{-10}[m^2]$, $\mu_n = \mu_w = 10^{-3}[Pa\ s]$, Brooks–Corey relative permeability functions with $\lambda = 2$ and Brooks-Corey capillary pressure function

$$p_c(S_w) = p_d S_w^{-\frac{1}{\lambda}} \tag{37}$$

where $\lambda = 2$ and $p_d = 5000[Pa]$.

Solution plots are shown in Figure 4. Because the problem is parabolic we use diagonally implicit Runge-Kutta methods of corresponding order. This problem requires the use of the J_0 penalty term in the saturation equation, otherwise the method does not converge with the initial condition $S_w = 0$ in Ω. The reason for this is that the Oden-Babuška-Baumann stabilization term vanishes for extreme values of saturation. In the J_0 penalty term we use $\sigma = 10^{-3}$ and $\beta = 1$.

The solution of the McWhorter problem does not have enough regularity to realize higher order convergence rates. Nevertheless the much smaller error constant of the higher order schemes renders them very attractive. A comparison of Table 10 with the finite volume method used in [7] shows that the mesh used for the finite volume scheme must be refined three times in space and time to match the accuracy of the DG method.

We observe that the solution for polynomials of degree 2 lies below the exact solution and that for polynomial degree 3 lies above the exact solution.

6 Conclusion

In this paper we formulated higher order discontinuous Galerkin methods for elliptic, hyperbolic and parabolic equations which describe single-phase and two-phase flow in porous media. For smooth problems the higher-order convergence is fully confirmed. For problems with low regularity it is shown consistently over several examples that the solutions of the presented schemes

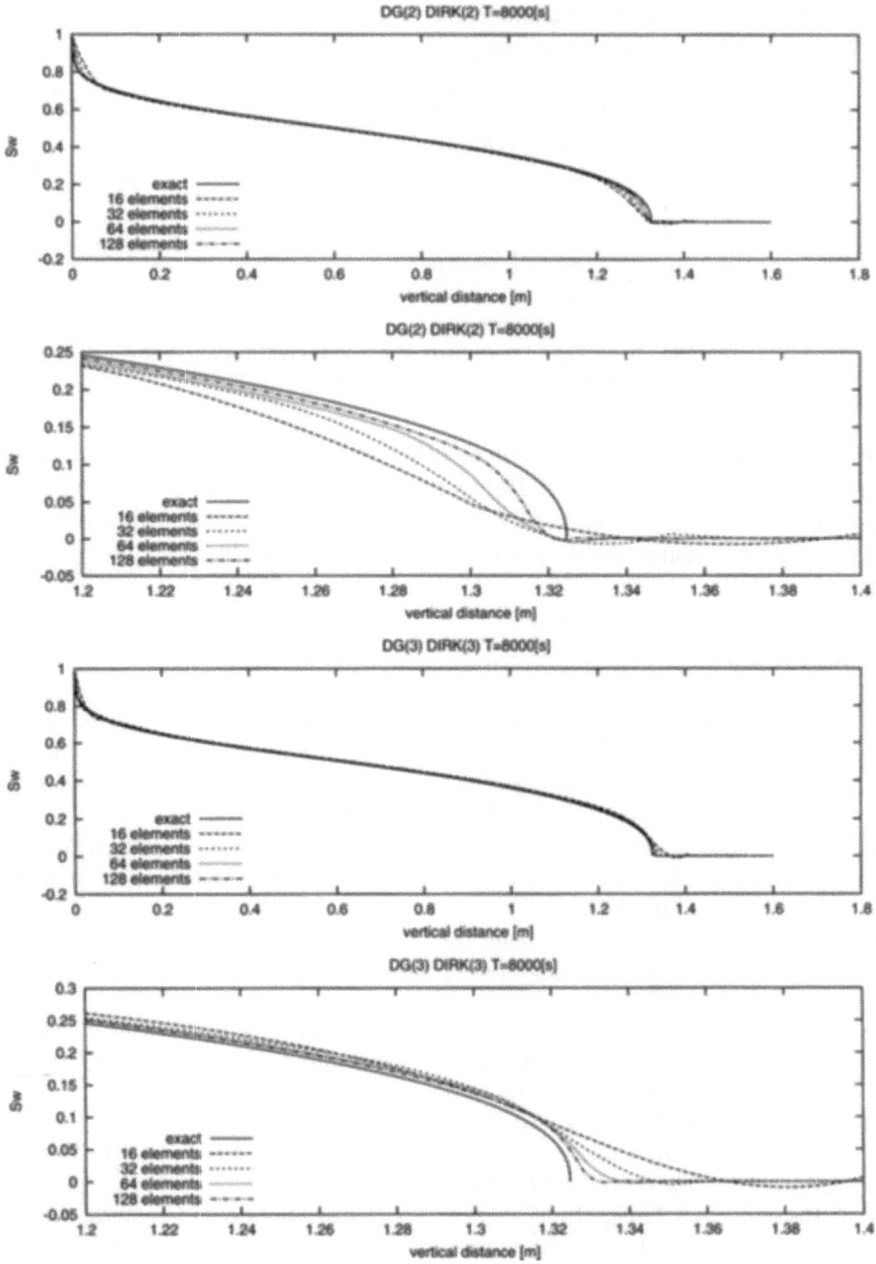

Fig. 4. DG for the McWhorter problem.

Table 10. Error and experimental order of convergence of DG applied to the McWhorter problem using implicit Runge-Kutta time discretizations.

Method	elements	steps	L_1-error	L_1-rate	L_2-error	L_2-rate
DG(2)	16	6	$1.96 \cdot 10^{-2}$		$2.63 \cdot 10^{-2}$	
DIRK(2)	32	12	$1.42 \cdot 10^{-2}$	0.46	$1.98 \cdot 10^{-2}$	0.41
	64	24	$8.98 \cdot 10^{-3}$	0.66	$1.37 \cdot 10^{-2}$	0.53
	128	48	$5.02 \cdot 10^{-3}$	0.84	$8.78 \cdot 10^{-3}$	0.64
DG(3)	16	6	$1.22 \cdot 10^{-2}$		$1.49 \cdot 10^{-2}$	
DIRK(3)	32	12	$7.66 \cdot 10^{-3}$	0.67	$9.37 \cdot 10^{-3}$	0.67
	64	24	$4.79 \cdot 10^{-3}$	0.68	$5.67 \cdot 10^{-3}$	0.72
	128	48	$3.15 \cdot 10^{-3}$	0.60	$3.43 \cdot 10^{-3}$	0.73

are very competitive in comparison to other well established methods such as the mixed finite element and finite volume methods. In contrast to the previous approaches with discontinuous elements the new methods allow efficient solution of the arising linear systems with multigrid methods.

References

1. V. Aizinger, C. Dawson, B. Cockburn, and P. Castillo, *The local discontinuous Galerkin method for contaminant transport*, Adv. Wat. Res. **24** (2001), 73–87.
2. R. Alexander, *Diagonally implicit Runge-Kutta methods for stiff O.D.E.'s*, SIAM Journal Numer. Anal. **14** (1977), 1006–1021.
3. D. N. Arnold, F. Brezzi, B. Cockburn, and L. D. Marini, *Unified analysis of discontinuous Galerkin methods for elliptic problems*, SIAM J. Numer. Anal. **39** (2002), no. 5, 1749–1779.
4. K. Aziz and A. Settari, *Petroleum reservoir simulation*, Elsevier, 1979.
5. P. Bastian, *Numerical computation of multiphase flow in porous media*, Habilitationsschrift, 1999.
6. P. Bastian, K. Birken, S. Lang, K. Johannsen, N. Neuß, H. Rentz-Reichert, and C. Wieners, *UG: A flexible software toolbox for solving partial differential equations*, Computing and Visualization in Science **1** (1997), 27–40.
7. P. Bastian and R. Helmig, *Efficient fully-coupled solution techniques for two-phase flow in porous media. Parallel multigrid solution and large scale computations*, Adv. Water Res. **23** (1999), 199–216.
8. P. Bastian and S. Lang, *Couplex benchmark computations with UG*, Tech. Report 2002–31, IWR (SFB 359), Universität Heidelberg, 2002, submitted to Computational Geosciences.
9. P. Bastian and V. Reichenberger, *Multigrid for higher order discontinuous Galerkin finite elements applied to groundwater flow*, Tech. Report 2000-37, SFB 359, 2000.
10. P. Bastian and B. Rivière, *Superconvergence and H(div)-projection for discontinuous Galerkin methods*, Tech. Report 2002-23, IWR (SFB 359), Universität Heidelberg, 2002, to appear in Int. J. Numer. Meth. Fluids.

11. B. Cockburn, S. Hou, and C.-W. Shu, *TVB Runge-Kutta local projection discontinuous Galerkin finite element method for conservation laws III: One-dimensional systems*, J. Comput. Phys. **84** (1989), 90.
12. ———, *TVB Runge-Kutta local projection discontinuous Galerkin finite element method for conservation laws IV: The multidimensional case*, Math. Comput. **54** (1990), 545.
13. B. Cockburn, S. Y. Lin, and C.-W. Shu (eds.), *Discontinuous Galerkin methods. Theory, computation and applications*, Lecture Notes in Computational Science and Engineering, vol. 11, Springer-Verlag, 2000.
14. B. Cockburn and C.-W. Shu, *TVB Runge-Kutta local projection discontinuous Galerkin finite element method for conservation laws II: General framework*, Math. Comput. **52** (1989), no. 186, 411–435.
15. ———, *The Runge-Kutta local projection P^1-discontinuous Galerkin method for scalar conservation laws*, M^2 *AN* **25** (1991), 337.
16. ———, *The local discontinuous Galerkin finite element method for convection-diffusion systems*, SIAM J. Numer. Anal. **35** (1998), 2440–2463.
17. ———, *The Runge-Kutta discontinuous Galerkin method for conservation laws V: Multidimensional systems*, J. Comput. Phys. **141** (1998), 199–224.
18. J. Douglas Jr. and T. F. Russel, *Numerical methods for convection dominated diffusion problems based on combining the method of characteristics with finite element or finite difference procedures*, SIAM J. Numer. Anal. **19(5)** (1982), 871–885.
19. L. J. Durlofsky, *Accuracy of mixed and control volume finite element approximations to Darcy velocity and related quantities*, Water Resources Research **30** (1994), no. 4, 965–973.
20. K. Eriksson, D. Estep, P. Hansbo, and C. Johnson, *Introduction to adaptive methods for differential equations*, Acta Numerica, 1995.
21. W. Hackbusch, *Multi–grid methods and applications*, Springer–Verlag, 1985.
22. ———, *Iterative solution of large sparse systems of linear equations*, Springer, 1994.
23. E. Hairer and G. Wanner, *Solving ordinary differential equations II*, Springer, Berlin, 1991.
24. R. Helmig, *Multiphase flow and transport processes in the subsurface – A contribution to the modeling of hydrosystems*, Springer–Verlag, 1997.
25. R. J. LeVeque, *Numerical methods for conservation laws*, Birkhäuser, 1992.
26. D. B. McWhorter and D. K. Sunada, *Exact integral Solutions for two–phase flow*, Water Resources Research **26(3)** (1990), 399–413.
27. J. T. Oden, I. Babuška, and C. E. Baumann, *A discontinuous* hp *finite element method for diffusion problems*, Journal of Computational Physics **146** (1998), 491–519.
28. D. W. Peaceman, *Fundamentals of numerical reservoir simulation*, Elsevier, 1977.
29. W. H. Reed and T. R. Hill, *Triangular mesh methods for the neutron transport equation*, Tech. report, Los Alamos Scientific Laboratory, 1973.
30. B. Rivière, *Discontinuous Galerkin methods for solving the miscible displacement problem in porous media*, Ph.D. thesis, The University of Texas at Austin, 2000.
31. B. Rivière and M. F. Wheeler, *Nonconforming methods for transport with nonlinear reaction*, Fluid Flow and Transport in Porous Media: Mathematical and Numerical (Z. Chen and R. E. Ewing, eds.), vol. 295, American Mathematical Society, 2002, pp. 421–432.

32. B. Rivière, M. F. Wheeler, and K. Banaś, *Part II. Discontinuous Galerkin method applied to a single phase flow in porous media*, Comput. Geosci. **4** (2000), 337–349.
33. B. Rivière, M. F. Wheeler, and V. Girault, *Improved energy estimates for interior penalty, constrained and discontinuous Galerkin methods for elliptic problems I*, Comput. Geosci. **3** (1999), 337–360.
34. ———, *A priori error estimates for finite element emthods based on discontinuous approximation spaces for elliptic problem*, SIAM Journal on Numerical Analysis **39** (2001), no. 3, 902–931.
35. C.-W. Shu, *Total-variation-diminishing time discretizations*, SIAM J. Sci. Stat. Comput. **9** (1988), no. 6, 1073–1084.
36. H. A. Van der Vorst, *BiCGSTAB: A fast and smoothly converging variant of Bi–CG for the solution of non–symmetric linear systems*, SIAM J. Sci. Stat. Comput. **13** (1992), 631–644.
37. H. Wang, H. K. Dahle, R. E. Ewing, M. S. Espedal, R. C. Sharpley, and S. Man, *An ELLAM scheme dor advection-diffusion equations in two dimensions*, SIAM J. Sci. Stat. Comput. **20** (1999), no. 6, 2160–2194.
38. M. F. Wheeler, *An elliptic collocation finite element method with interior penalties*, SIAM J. Numer. Anal. **15** (1978), no. 1, 152–161.
39. J. Xu, *Iterative methods by space decomposition and subspace correction*, SIAM Review **34** (1992), 581–613.

Macroscopic Models of Fluids with Microstructure

Noel J. Walkington

Department of Mathematics, Carnegie Mellon University, Pittsburgh, PA 15213*

Summary. Liquid crystals, fluids containing elastic particles, and polymer fluids, all exhibit non-trivial macroscopic behavior due to interactions occurring at micro/mesoscopic scales. Frequently the particles are small enough to be influenced by Brownian motion so the classical equations of mechanics must be coupled to appropriate Fokker Planck equation(s). Currently the coupling between the Fokker Planck equations modeling the microstructure and the macroscopic equations of mechanics is poorly understood. In this talk I'll present some of the microscopic models appearing in the physics literature and the systems of pde's that arise when they are coupled to the equations of mechanics.

1 Introduction

One distinguished property of fluids is that they readily form mixtures. Fluid mixtures can take on many states depending upon their physical properties of the constituents. Certain fluids may dissolve within each other, while immiscible fluids, such as oil and water, typically form fine mixtures containing droplets. Fluids are also used to transport solid materials either as particles or as a solute. Macroscopically such mixtures appear homogeneous; however, physical effects, such as surface tension, elasticity of the particles, etc, play an important role in determining their macroscopic properties. Frequently the interaction or "competition" between the elastic and kinetic energies at small scales gives rise to viscoelastic phenomena at the macroscopic scale. Common examples of such fluids include:

- MIXTURES OF IMMISCIBLE FLUIDS: In the absence of surface tension determining macroscopic properties can be viewed as a homogenization problem. If surface tension is present this will give rise to additional elastic effects.

* Supported in part by National Science Foundation Grants DMS-0208586 and ITR-0086903.

- FLUIDS CONTAINING PARTICLES: Examples include slurries and fluidized beds. Macroscopic properties will depend upon particulate concentration, geometry, elastic properties etc.
- BLOOD: Human blood is essentially a mixture consisting of a Newtonian plasma (60%) containing red blood cells (40%). In small cappilaries red blood cells undergo large deformations, and their elastic properties give rise to the nonlinear variation of apparent viscosity verses shear rate shown in Figure 1.

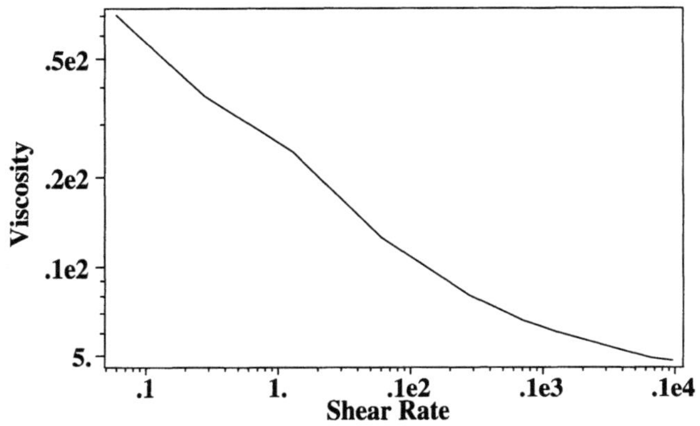

Fig. 1. Blood Viscosity vs. Shear Rate (from M. Kameneva, University of Pittsburgh).

Red blood cells consist of a membrane enclosing hemoglobin. The mechanical properties of the membrane and hemoglobin are not well understood; the membrane exhibits elastic properties and the hemoglobin appears to be fluid-like. In order to design prosthetic devices macroscopic models of such mixtures are required. To understand blood cell damage and transport in small capillaries models at the micro scale are required.

- POLYMER FLUIDS AND LIQUID CRYSTALS: Polymer and liquid crystal molecules are long and flexible and interact in various ways. This gives rise to many macroscopic properties ranging from visco-elasticity to the well known optical properties of liquid crystals as illustrated in Figure 2. The polymer and liquid crystal molecules are several orders of magnitude smaller than red blood cells so may be subject to Brownian motion.

The complexity of these fluid systems gives rise to many computational challenges. Brownian motion, particle elasticity and interaction etc. are very complex, and in the physics literature are frequently considered separately. Often the initial challenge for the computational scientist is to assemble a consistent model and well posed set of equations from possibly incompatible sub-models which may arise due to inconsistent assumptions and simplifications. The

Fig. 2. Alignment of long molecules and defect structures in a nematic liquid crystals gives rise to macroscopic optical properties.

complexity of the equations results in complicated codes which take a long time to develop and require significant computational resources to resolve the physics. For this reason it is important to start with a sound model and to understand the structure of the underlying partial differential equations. For example, very few attempts to compute solutions of the Ericksen Leslie equations, discussed in Section 3.1 below, were made prior to the development of a suitable existence and uniqueness theory. Below we discuss issues ranging from the validity of models, to analysis of the underlying equations, to approximation theory, all of which impact the accuracy, robustness, and validity of various computational models.

2 Models with Macroscopic Particles

2.1 Fluid-Fluid Mixtures

One of the simplest mixtures to model is the flow of two immiscible fluids having densities and viscosities (ρ_f, μ_f) and (ρ_s, μ_s) as illustrated in Figure 3. If $\Omega_f(t)$ and $\Omega_s(t)$ denote regions occupied by each fluid it is convenient to introduce a phase function ϕ satisfying $\Omega_f(t) = \{(x,t) \in \Omega \mid \phi(x,t) = 1\}$ and $\Omega_s(t) = \{(x,t) \in \Omega \mid \phi(x,t) = -1\}$. An appropriate weak form of the Navier Stokes equations modeling such a mixture takes the form

$$\int_\Omega \rho(\phi) \left(\mathbf{v}_t + (\mathbf{v}.\nabla)\mathbf{v}\right).\mathbf{w} - p\nabla.\mathbf{w} + \mu(\phi)D(\mathbf{v}) \cdot D(\mathbf{w}) = \int_\Omega \rho(\phi) f.\mathbf{w}$$

$$div(\mathbf{v}) = 0$$

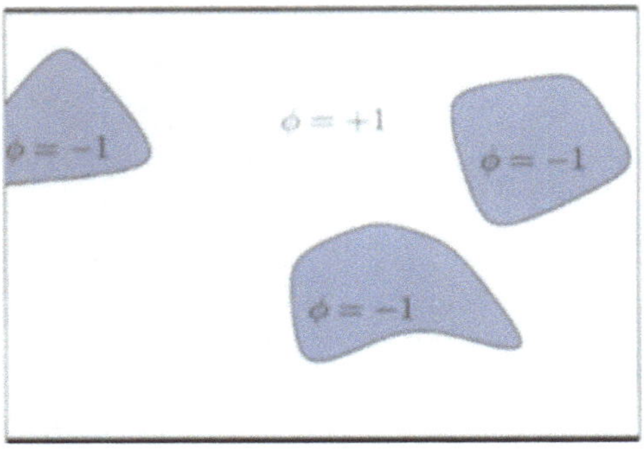

Fig. 3. Immiscible fluids.

$$\phi_t + \mathbf{v}.\nabla\phi = 0.$$

The density and viscosity are affine functions of ϕ,

$$\rho(\phi) = (1/2)(1+\phi)\rho_f + (1/2)(1-\phi)\rho_s,$$

$$\mu(\phi) = (1/2)(1+\phi)\mu_f + (1/2)(1-\phi)\mu_s.$$

In the above, $D(\mathbf{v}) = (1/2)(\nabla\mathbf{v}+(\nabla\mathbf{v})^T)$ is the symmetric part of the velocity gradient.

Existence of solutions to this system of equation was established by DiPerna and Lions [7, 22]; however, many issues pertinent to their numerical approximation remain unresolved. Such issues include:

- Regularity of Solutions to the Transport Equation: While solutions exist when $\mathbf{v} \in L^2[0,T; H^1_0(\Omega)]$, characteristics may not be well defined. In particular, solutions will *not* have bounded variation so that schemes developed to solve conservation laws which bound the variation (e.g. ENO and TVD schemes) will not converge. Convergence results for the discontinuous Galerkin method under minimal regularity have recently been obtained by the author in [30].
- Regularity of Solutions for the Coupled Problem: Currently there are no regularity results for solutions $(\phi, \mathbf{v})$ of the coupled system. In order to guarantee that ϕ has bounded variation it appears necessary to establish $W^{1,\infty}$ estimates for $\mathbf{v}$.
- Uniqueness: While solutions of the convection equation with $\mathbf{v} \in L^2[0,T; H^1_0(\Omega)]$ are unique, only uniqueness results for smooth solutions of the coupled system are available. Flows of mixtures have discontinuous densities and viscosities, so even piecewise smooth classical solutions will only have $\mathbf{v} \in W^{1,\infty}$.
- Numerical Approximation: It is difficult to design numerical schemes in the absence of uniqueness and regularity of solutions. A minimal require-

ment would be convergence of a sub-sequence of the discrete solutions to a solution of the continuous problem.

In order to predict the morphology of immiscible fluid droplets it is necessary to include surface tension. Classical surface tension models introduce singular terms on the (possibly very irregular) interface between the fluids. An alternative is to consider "phase field" models which approximate the interface by a finite transition layer [4, 5, 11]. In the present context this is equivalent to allowing ϕ to smoothly change from -1 to $+1$. This approach was considered in [24] and leads to equations similar in form to Ericksen's equations for liquid crystals discussed below.

2.2 Fluid-Particle Mixtures

It is possible to extend the ideas above to develop relatively concise equations modeling the flow of elastic particles in a Newtonian fluid [24]. However, these equations contain the equations of nonlinear elasticity about which very little is known. One of the major issues concerns properties of solutions to the equation governing the evolution of the deformation gradient F.

Given a velocity field $\mathbf{v}$ recall that the flow map $x = x(t, X)$ satisfies $\dot{x}(t, X) = \mathbf{v}(t, x(t, X))$ with initial condition $x(0, X) = X$. The deformation gradient is defined to be tensor with components $F_{i\alpha} = \partial x_i / \partial X_\alpha$, and an application of the chain rule shows that

$$F_t + (\mathbf{v}.\nabla)F = (\nabla\mathbf{v})F, \qquad F(0) = I.$$

Currently very little is known about the compactness properties of solutions of this equation. The Cauchy stress for a visco-elastic material is a nonlinear function of F, and in the absence of any compactness results, existence of solutions to the equations of motion for visco-elastic materials can not be established. Specifically the Cauchy stress of the simplest visco-elastic material is

$$T = \mu D(\mathbf{v}) + (1/J) D\mathcal{W}(F) F^T,$$

where $J = det(F)$ is equal to one if $div(\mathbf{v}) = 0$, and $\mathcal{W}$ is the elastic strain energy function. Notice that T is not linear in F even when $\mathcal{W}$ is quadratic. Nonlinearities persist when the equations are written in Lagrangian coordinates; the Piola Kirchoff stress becomes

$$S = \mu J \left((\nabla_X \mathbf{v}) F^{-1} \right)_{sym} F^{-T} + D\mathcal{W}(F).$$

While the elastic part of S is linear if $\mathcal{W}$ is quadratic, the viscous term is nonlinear in F.

The lack of compactness of the deformation gradient for elasticity problems is well known and is persists in the equations modeling mixtures containing macroscopic elastic particles. In the following section we show that certain models of flows containing microscopic particles can circumvent this issue.

3 Macroscopic Equations Modeling Flows with Microscopic Particles

The problem encountered in the previous section arises due to the property that the elastic stresses can not be determined by examining the present configuration of an elastic body; it is necessary to know the relative displacements between material particles. Since small particles in a flow can freely slide past each other the elastic stresses are independent of the reference configuration representing their relative positions. In certain situations this allows the elastic stresses to be determined by examining the present configuration which circumvents the problems encountered in the previous section.

3.1 Ericksen Leslie Equations

Fig. 4. Nematic liquid crystals.

A classical example of a macroscopic model of a flow containing microscopic particles is Ericksen's model of nematic liquid crystals. Nematic liquid crystals are modeled as fluids containing rod-like molecules which are free to flow past each other (see Figure 4). These long molecules tend to align and the resulting "tangling" resists misalignment. Introducing a unit vector field $\mathbf{n}$ to represent the orientation of the molecules, the balance of linear and angular momentum derived by Ericksen [9, 10] take the form

$$\rho\dot{\mathbf{v}} - div(T) = \rho\mathbf{f},$$

$$(\rho\bar{r}^2)\ddot{\mathbf{n}} - div(C) - \mathbf{g} = \rho\mathbf{m}.$$

Here T is the Cauchy stress tensor, C is a couple stress tensor, and the vector $\mathbf{g}$ arises due to the constrains that (i) $|\mathbf{n}| = 1$ and (ii) the molecules can not sustain couples parallel to their axis (they have no inertia about this axis). The stress and couple tensors derived by Ericksen and Leslie [18, 19] take the form

$$T = -pI - \left[\frac{\partial \mathcal{W}}{\partial n_{k,j}} n_{k,i}\right] + \cdots + \alpha_2(\dot{\mathbf{n}} - W(\mathbf{v})\mathbf{n}) \otimes \mathbf{n} + \mu D(\mathbf{v}) + \cdots$$

$$C = \left[\frac{\partial \mathcal{W}}{\partial n_{i,j}}\right] \qquad \mathbf{g} = -\theta \mathbf{n} - \frac{\partial \mathcal{W}}{\partial \mathbf{n}} + \cdots$$

Here p and θ are Lagrange variables dual to the constraints $div(\mathbf{v}) = 0$ and $|\mathbf{n}| = 1$ respectively, and many of the viscous terms have been omitted. The elastic stresses appear as derivatives of the Oseen-Frank strain energy function[2] [14, 26]:

$$\mathcal{W}(\mathbf{n}, \nabla \mathbf{n}) = (1/2)\left(k_1\, div(\mathbf{n})^2 + k_2\,(\mathbf{n}.curl(\mathbf{n}))^2 + k_3\,|\mathbf{n} \times curl(\mathbf{n})|^2\right)$$

The three terms represent the elastic energy stored by virtue of bending, splaying and twisting of the molecules, see Figure 5. Each of these deformations can be determined from the current configuration of the molecules which eliminates the need to introduce a reference configuration and the corresponding deformation gradient.

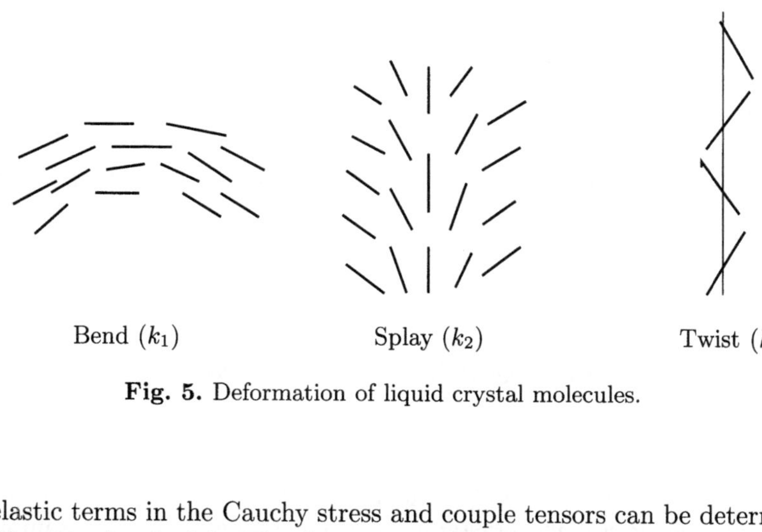

Bend (k_1) Splay (k_2) Twist (k_3)

Fig. 5. Deformation of liquid crystal molecules.

The elastic terms in the Cauchy stress and couple tensors can be determined from Hamilton's principle. The elastic part of the Cauchy stress is the variation of $\mathcal{W}$ under perturbations of the flow map, $x(X,t) \to x(X,t) + \delta x(X,t)$. For the incompressible fluids under consideration $div(\delta x) = 0$. Then the identities

$$\mathbf{n}_r(X,t) = \mathbf{n}(x(X,t),t), \qquad \text{and} \qquad (\nabla_x \mathbf{n}) = (\nabla_X \mathbf{n}_r)F^{-1},$$

allow the variation can be computed as:

$$\delta \int_0^T \int_\Omega \mathcal{W}(\mathbf{n}, \nabla \mathbf{n})\, dx\, dt = \int_0^T \int_\Omega \left[\frac{\partial \mathcal{W}}{\partial n_{k,j}} n_{k,i}\right] \cdot (\nabla \delta x)\, dx\, dt.$$

[2] We have omitted the chiral parameter and a null Lagrangian from the elastic energy.

The couple stresses are determined from the variations of the director, $\mathbf{n}(x,t) \to \mathbf{n}(x,t) + \delta\mathbf{n}(x,t)$,

$$\delta \int_0^T \int_\Omega \mathcal{W}(\mathbf{n}, \nabla\mathbf{n})\, dx\, dt = \int_0^T \int_\Omega \left[\frac{\partial \mathcal{W}}{\partial n_{k,j}}\right] \cdot (\nabla\delta\mathbf{n}) + \left(\frac{\partial \mathcal{W}}{\partial \mathbf{n}}\right) . \delta\mathbf{n}\, dx\, dt.$$

The pioneering work of Lin and Liu [20, 21] developed the existence and uniqueness theory for the Ericksen equations. This work facilitated the development and analysis of numerical schemes to approximate the flow of nematic liquid crystals [23, 25].

3.2 Tensor Models of Liquid Crystals (de Gennes)

While the Ericksen Leslie model is very successful at modeling many liquid crystal systems, some configurations can not be modeled. In certain experiments degree 1/2 singularities, which can not be modeled by a continuous vector field, have been observed, see Figure 6.

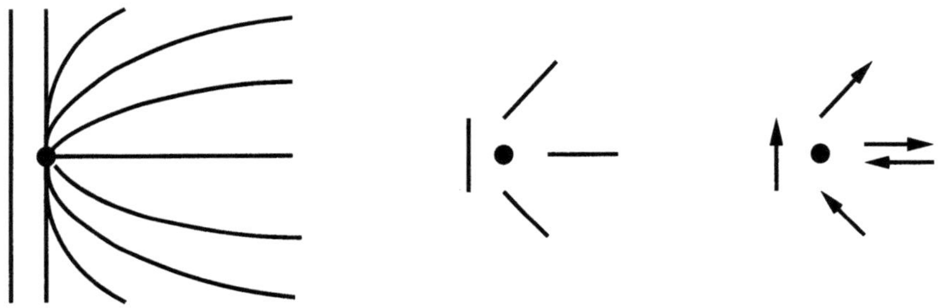

Fig. 6. Degree half singularities can not be modeled with vector fields.

Refinements of Ericksen's model of liquid crystals are based upon the premise that molecules in a liquid crystal are not aligned perfectly; for example, the liquid crystal molecules are small enough to be subject to Brownian motion. To model such a system, de Gennes [6] introduced a probability distribution $f(\mathbf{n}) = f(x,t,\mathbf{n})$ to characterize the alignment:

$$P\{\mathbf{n} \in A \subset S^2\} = \int_A f(\mathbf{n})\, d\sigma(\mathbf{n}).$$

If the molecules are symmetric from end-to-end, then $f(\mathbf{n}) = f(-\mathbf{n})$ and the first non-trivial moment of f is the second:

$$\int_{S^2} f(\mathbf{n})\, d\sigma(\mathbf{n}) = 1, \quad \int_{S^2} \mathbf{n} f(\mathbf{n})\, d\sigma(\mathbf{n}) = 0, \quad Q \equiv \int_{S^2} (\mathbf{n} \otimes \mathbf{n}) f(\mathbf{n})\, d\sigma(\mathbf{n}).$$

Systems of equations involving probability distributions have been introduced to study a wide variety of polymer fluids in addition to liquid crystals [3, 8].

Several approaches have been developed to determine equations governing the evolution of f. The classical approach pursued by the continuum mechanics community is to consider the "most general" frame indifferent equations [29]. This leads to models containing many physical constants which need to be determined experimentally. An alternative is to consider a "mesoscale" approach where various properties of the embedded particles or molecules are considered [3, 8]. This later approach will typically have fewer unknown parameters. In each instance, the evolution of f is typically governed by a Fokker Planck equation, and the theory for the (rather complicated) coupled systems of equations that arise is in its infancy. This gives rise to many mathematical and computational challenges.

3.3 Fokker Planck Equations

Models of liquid crystals, and more generally polymer fluids, frequently introduce a probability distributions to characterize the orientation and positions of various molecules or (microscopic) particles. In this section we recall the structure of the Fokker Planck equations used ubiquitously to model the evolution of these distributions.

It is well known that the distribution function of the solution to the stochastic differential equation

$$dx(t) = \mathbf{a}(x,t)\, dt + B(x,t)\, d\mathbf{w}(t), \qquad x(0) = X,$$

($X = X(\omega)$ is a random variable) will satisfy the Fokker Planck equation

$$f_t = div\left((1/2)\nabla(BB^T f) + \mathbf{a} f\right).$$

Recall that the distribution f is characterized by

$$\int_A f(x,t)\, dx = meas\{\omega \mid x(t;\omega) \in A\}, \qquad A \subset \mathbb{R}^d.$$

The general form of the Fokker Planck equation encountered in the polymer dynamics literature takes the form

$$f_t = div\left(f\,\nabla\left(\frac{\delta\mathcal{A}}{\delta f}\right)\right), \tag{1}$$

where $\mathcal{A} = \mathcal{A}(f, x)$ is the "free energy" energy of the system.

Example 1 (Einstein). If $f(t,x)$ is the distribution function for the density of of spherical particles in a dilute solution then

$$f_t = div\left(\frac{1}{\xi}\left(k_b \nabla f + f\nabla\psi\right)\right).$$

- k_b is the product of a Boltzmann constant with the temperature.
- ψ is a potential for body forces (e.g. gravity).
- $\mathcal{A} = (1/\xi) \int_{\mathbb{R}^d} (k_b \ln(f) + \psi) f \, dx$ is the "Gibbs Thompson entropy".
- $\xi = 6\pi\nu a$ is the constant relating force and velocity in steady state motion of particles of radius a in a viscous fluid (Stokes flow).

This example illustrates the mesoscopic nature of such models: The particles are assumed to be so small that Brownian motion of the fluid molecules is important; however, classical hydrodynamics is used to characterize the motion of a particle. This mixing of scales characterizes these models.

3.4 Variational Principles and the Wasserstein Metric

The evolution of a probability distribution governed by equation (1) can be characterized as a gradient flow in the space of probability measures [16, 27, 28]. Recall that natural numerical schemes for gradient flows enjoy many desirable stability and monotonicity properties. In [16] Jordan, Kinderlehrer, and Otto showed that solutions of (1) could be approximated by an implicit scheme of the form

$$I(f) = \frac{1}{2\tau} d_w^2(f, f^{n-1}) + \int_\Omega \mathcal{A}(f) \to \min .$$

Here τ is a time step and $f^n \sim f(n\tau)$ is the minimizer of the indicated functional over the space

$$X = \{f \in L^1(\Omega) \mid f(x) \geq 0, \int_\Omega f = 1\}.$$

In the above $d_w(.,.)$ is the Wasserstein metric. The following characterization, due to Benamou and Brenier [2], illustrates the natural mechanical interpretation of this quantity:

$$d_w^2(f, f^{n-1}) = \tau \inf_{(\rho, \mathbf{v})} \int_0^\tau \int_\Omega \rho \, |\mathbf{v}|^2,$$

where the infimum is taken over all pairs $(\rho, \mathbf{v})$ for which

$$\rho_t + div(\rho \mathbf{v}) = 0, \qquad \rho(0) = f^{n-1} \text{ and } \rho(\tau) = f.$$

This states that the Wasserstein metric is the minimal kinetic energy of all mass conserving flows which carry f^{n-1} to f (c.f. Kelvin's 1849 variation principle for incompressible irrotational flows).

An explicit formula for the Wasserstein distance in one dimension was given by Frechet in [15], and this was used by Kinderlehrer and Walkington [17] to construct natural numerical approximations of the one dimensional Fokker Planck equation. One of the computational challenges in this area is the development of algorithms that will quickly compute the Wasserstein distance in multiple dimensions. Such algorithms would facilitate the development of robust schemes to approximate solutions of the Fokker Plank equation.

3.5 Mixture Models and Distribution Functions

The complex fluids models discussed above introduce a probability distribution *at each point* to model microscopic interactions. The distribution function at each $(x,t) \in \Omega \times (0,T)$ then takes the form $f(x,t,y)$ where $y \in Y$ is the variable required to characterize the microstructure.

Example 2 (Distribution Functions on $\mathbb{R}^d$). If the microstructure is characterized by a vector $y \in \mathbb{R}^d$, and it is assumed that the mixture is "dilute" in the sense that adjacent distributions do not interact, then the Fokker Planck equation takes the form

$$f_t + \mathbf{v}.\nabla_x f + (\nabla_x \mathbf{v}) y.\nabla_y f = div_y \left((\delta \mathcal{A}/\delta f) \right).$$

The expression on the left is the convected time derivative of f and may be derived by introducing the motion $x = x(X,t)$ and $y = FY$ from a reference vector bundle as shown in Figure 7 (recall that $F = [\partial x/\partial X]$ is the deformation gradient). If $f_r(t,X,Y) = f\left(t, x(X,t), F(X,t)Y\right)$ then

$$\frac{d}{dt} f_r(t,X,Y) = f_t + \mathbf{v}.\nabla_x f + (\nabla_x \mathbf{v}) y.\nabla_y f.$$

The microstructure gives rise to additional stresses which couple with the

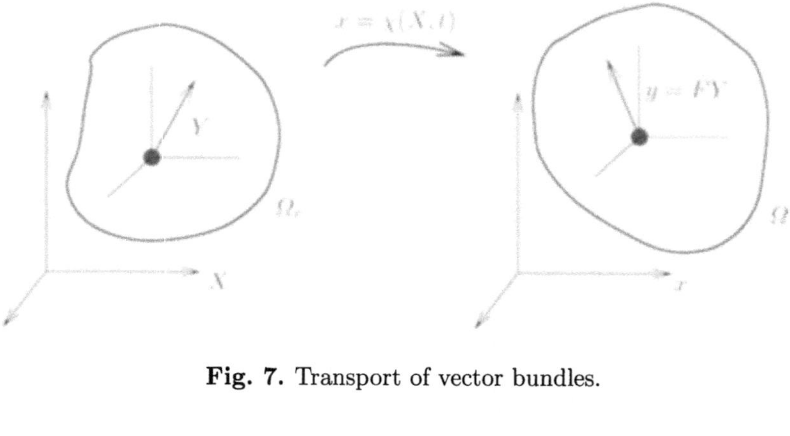

Fig. 7. Transport of vector bundles.

linear momentum equation. The form of this coupling can be determined from Hamilton's principle. If the flow map is varied according to

$$x(X,t) \to x(X,t) + \delta x(X,t), \qquad div_x(\delta x) = 0,$$

(the second condition corresponding to an incompressible fluid) then

$$\delta \int_0^T \int_{\Omega_r} \int_{\mathbb{R}^3} \mathcal{A}(f(t, x(X,t), F(X,t)Y) = \int_0^T \int_{\Omega_r} \int_{\mathbb{R}^3} -div_x(y \otimes \nabla_y \psi\, f).\delta x.$$

The momentum equation then takes the form

$$\rho(\mathbf{v} + (\mathbf{v}.\nabla)\mathbf{v}) - div\left(-pI + \mu D(\mathbf{v}) - \int_{\mathbb{R}^3} (y \otimes \nabla_y \psi) f\, dy\right) = \rho \mathbf{f}.$$

This model satisfies the natural energy estimate

$$\frac{d}{dt} \int_\Omega \left((\rho/2)|\mathbf{v}|^2 + \int_{\mathbb{R}^3} \mathcal{A}(f)\, dy\right) dx = - \int_\Omega \left(\mu |D(\mathbf{v})|^2 + \int_{\mathbb{R}^3} \left|\frac{\delta \mathcal{A}}{\delta f}\right|^2 f\, dy\right) dx.$$

This example is typical in the sense that $Y \subset \mathbb{R}^m$ for some $m > 1$ so that $(x,y) \in \Omega \times Y \subset \mathbb{R}^{d+m}$. Typically $(d+m) \geq 6$ and direct approximation of f on a mesh of $\Omega \times Y$ is impossible on todays computers. One of the challenges in this area is to derive reduced models, or clever algorithms, to circumvent the problem of high dimensionality so that distribution functions defined on "bundles" of the form $\Omega \times Y$ can be efficiently computed.

A classical approach used to construct reduced models is to derive a moment expansion for f. Equations for the evolution of the moments are derived by multiplying the Fokker Planck equation by products of y (e.g. $y \otimes y$ for the second moment) and integrating over Y.

Example 3 (Dumbbell Molecules and Oldroyd-B Fluids:). We continue the previous example where $Y = \mathbb{R}^3$. If we assume that the entropy is $\mathcal{A}(f) = (k_b \ln(f) + (K/2)|y|^2) f$, then an exact reduced model, corresponding to an Oldroyd-B fluid, can be derived. This form of $\mathcal{A}$ models spring-like molecules subject to Brownian perturbations in length. Upon extension by y, the stored energy is $\psi(y) = (K/2)|y|^2$.

If we write the second moment (which corresponds to the microstress) as

$$\hat{T} \equiv \int_{\mathbb{R}^3} (y \otimes y) f\, dy,$$

then substituting $\nabla_y \psi = Ky$ into the linear momentum equation gives

$$\rho \dot{\mathbf{v}} - div\left(-pI + \mu D(\mathbf{v}) - K\hat{T}\right) = \rho \mathbf{f}.$$

For the specific form of the entropy introduced here a closed form expression for $\hat{T}$ can be deduced by multiplying the Fokker Planck equation by $y \otimes y$ and integrating over $Y = \mathbb{R}^3$. Upon assuming that $div(\mathbf{v}) = 0$ it follows that

$$\hat{T}_t + (\mathbf{v}.\nabla)\hat{T} + (\nabla \mathbf{v})\hat{T} + \hat{T}(\nabla \mathbf{v})^T = -k_b I - K\hat{T}.$$

This equation for the "extra stress" corresponds to a classical Oldroyd-B fluid.

This example is typical in the sense that $f(.,.,y) = f(.,.,-y)$ so that all of the odd moments vanish; however, the existence of an exact reduced model is exceptional. The equation for a particular moment almost always involves higher moments so a "closure rule" must be adopted [12, 13]. A common closure rule is to write fourth and higher moments as products of second moments to obtain a single equation in the second moment $Q = Q(x,t)$. One of the challenging open problems in this area is the development of a convergence analysis to determine the accuracy of such approximations.

4 Concluding Remarks

The exposition here only introduced the simplest examples to both motivate and highlight some of the fascinating challenges encountered when modeling complex fluids. In particular, no examples of the "non-dilute" case have been considered where adjacent distributions interact. One reason for this is that these interactions are poorly understood, and characterizing these is an open challenge in itself. We have also omitted any discussion of classical mixture theory [1] which has a lot in common with the above.

Acknowledgment: The author learned of many of the issues and ideas discussed in this monograph through many years of collaboration, conversation, and companionship with his colleagues David Kinderlehrer and Chun Liu.

References

1. A. Bedford. Theories of immiscible and structured mixtures. *Int. J. Engng. Sci.*, 2(8):863–960, 1983.
2. J. Benamou and Y. Brenier. A computational fluid mechanics solution to the Monge–Kantorovich mass transfer problem. *preprint*, November, 1998.
3. A. N. Beris and B. J. Edwards. *Thermodynamics of Flowing Systems.* Number 36 in Oxford Engineering Science Series. Oxford, 1994.
4. G. Caginalp. An analysis of a phase field model of a free boundary. *Archive for Rational Mechanics and Analysis*, 92:205–245, 1986.
5. G. Caginalp. Stefan and Hele–Shaw type models a asymptotic limits of phase field equations. *Physics Review A*, 39:887–896, 1989.
6. P. G. de Gennes. *The Physics Of Liquid Crystals.* Oxford, 1974.
7. R. J. DiPerna and P. L. Lions. Ordinary differential equations, transport theory and Sobolev spaces. *Inventiones Mathematicae*, 98:511–547, 1989.
8. M. Doi and S. F. Edwards. *The Theory of Polymer Dynamics.* Number 73 in Internatiional Series of Mongraphs on Physics. Oxford, 1986.
9. J. Ericksen. Conservation laws for liquid crystals. *Trans. Soc. Rheol.*, 5:22—34, 1961.
10. J. Ericksen. Nilpotent energies in liquid crystal theory. *Archive for Rational Mechanics and Analysis*, 10:189—196, 1962.

11. G. Fix. Phase field methods for free boundary problems. In B. Fasano and M. Primicerio, editors, *Free Boundary Problems*, pages 580–589. Pitman, London, 1983.
12. M. G. Forest, Q. Wang, and H. Zhou. Exact banded patterns from a doi-marrucci-greco model of nematic liquid crystal polymers. *Phys. Fluids*, 12(3):490–498, 2000.
13. M. G. Forest, Q. Wang, and H. Zhou. Homogeneous pattern selection and director instabilities of nematic liquid crystal polymers induced by elongational flows. *Phys. Fluids*, 12(3):490–498, 2000.
14. F. C. Frank. On the theory of liquid crystals. *Discuss. Faraday Soc.*, 28:19–28, 1958.
15. M. Frechet. Sur la distance de deux lois de probabilite. *Comptes Rendus Acad. Sci*, 244:689–692, 1957.
16. R. Jordan, D. Kinderlehrer, and F. Otto. Dynamics of the Fokker-Planck equation. *Phase Transitions*, to appear, 1999.
17. D. Kinderlehrer and N. J. Walkington. Approximation of parabolic equations using the Wasserstein metric. *MMAN Math. Model. Numer. Anal.*, 33(4):837–852, 1999.
18. F. Leslie. Some constitutive equations for liquid crystals. *Archive for Rational Mechanics and Analysis*, 28:265–283, 1968.
19. F. Leslie. Theory of flow phenomenum in liquid crystals. In W. Brown, editor, *The Theory of Liquid Crystals*, volume 4, pages 1–81, New York, 1979. Academic Press.
20. F. H. Lin and C. Liu. Nonparabolic dissipative systems, modeling the flow of liquid crystals. *Comm. Pure Appl. Math.*, XLVIII(5):501–537, 1995.
21. F. H. Lin and C. Liu. Global existence of solutions for the Ericksen Leslie–system. *Archive for Rational Mechanics and Analysis*, accepted, 1998.
22. P. L. Lions. *Mathemaitcial Topics in Fluid Mechanics, Volume 1: Incompressible Models.* Oxford Press, Oxford, U.K., 1996.
23. C. Liu and N. J. Walkington. Approximation of liquid crystal flows. *SIAM Journal on Numerical Analysis*, 37(3):725–741, Feb. 2000.
24. C. Liu and N. J. Walkington. An Eulerian description of fluids containing visco-elastic particles. *Arch. Ration. Mech. Anal.*, 159(3):229–252, 2001.
25. C. Liu and N. J. Walkington. Mixed methods for the approximation of liquid crystal flows. *Math. Modelling and Numer. Anal.*, 36(2):205–222, March/April 2002.
26. C. W. Oseen. The theory of liquid crystals. *Trans. Faraday Soc.*, 29:883–889, 1933.
27. F. Otto. Dynamics of labyrinthine pattern formantion in magnetic fluids. *Archive for Rational Mechanics and Analysis*, 141:63–103, 1998.
28. F. Otto. The geometry of dissipative evolution equations: the porous medium equation. *Comm. Partial Differential Equations*, 26(1-2):101–174, 2001.
29. A. M. Sonnet and E. G. Virga. Dynamics of dissipative ordered fluids. *Phys. Rev. E*, 64(031705):1–10, 2001.
30. N. J. Walkington. Convergence of the discontinuous Galerkin method for discontinuous solutions. *SIAM Journal on Numerical Analysis*, submitted, June 2002.

The Fictitious Boundary Method for the Implicit Treatment of Dirichlet Boundary Conditions with Applications to Incompressible Flow Simulations

Stefan Turek, Decheng Wan, and Liudmila S. Rivkind

Institute of Applied Mathematics LS III, University of Dortmund,
Vogelpothsweg 87, 44227 Dortmund, Germany

Summary. A 'fictitious boundary method' for computing incompressible flows with complicated small-scale and/or time-dependent geometric details is presented. The underlying technique is based on a special treatment of Dirichlet boundary conditions, particularly for FEM discretizations, together with so-called 'iterative filtering techniques' in the context of hierarchical multigrid approaches such that the flow can be efficienctly computed on a fixed computational mesh while the solid boundaries are allowed to move freely through the given mesh. The presented method provides an easy way of incorporating geometrically complicated objects and time-dependent boundary components into standard CFD codes to simulate (at least) the qualitative flow behaviour of complex configurations. Furthermore, higher accuracy can be reached via local mesh adaptation techniques which might be based on local (coarse) mesh adaptation or mesh deformation techniques to avoid expensive (global) grid reconstruction.

We explain the mathematical and algorithmic details and provide numerical examples based on the FeatFlow [13] software for incompressible flow to illustrate qualitatively and to examine quantitatively the presented fictitious boundary method, for various stationary and time-dependent configurations. In particular, we compare with standard approaches which use geometrically adapted meshes, and with a 'viscosity-density blockage' method which describes internal objects via appropriate settings of the density and viscosity parameters in the Navier-Stokes equations. Moreover, we also discuss implementation details and software techniques which can provide very high MFLOP/s rates for such techniques in combination with special hierarchical data, matrix and solver structures.

1 Introduction

Flow configurations with numerous complex geometrical details and/or moving interfaces and boundaries have important applications in a variety of physical and engineering areas such as flows around objects, fluid-structure interaction, multiphase flows with chemical reactions, stratified flows, bubble dynamics, melting and solidification, crystal growth, etc. The numerical investigation of these physical problems has to take into account the effect of the complicated stationary and/or moving boundaries, and it can require a huge amount of time for the regeneration or deformation of the computational grid when the corresponding interfaces or boundaries are changing. Developing an accurate, effective and robust approach for tackling these problems is necessary: The overall aim is to deal successfully with the moving interfaces or boundaries such that the accuracy of the numerical approximation is increased while at the same time also the computational cost is decreased.

Generally speaking, there are (at least) two major approaches to simulate fluid flows with complex stationary and/or moving boundaries. One of them is a 'body-conformal approach' which always keeps the computational mesh in accordance to the geometrical details. Another one is a 'fixed grid approach' in which case the mesh is (arbitrarily) fixed and internal objects are allowed to move freely through the mesh. One big advantage of such 'fixed grid' approaches over the conventional 'body-conformal' approach is that the computational mesh remains unchanged such that the CPU cost can be significantly decreased - less computational effort due to saving the expensive mesh generation - and that such techniques can be easily incorporated into standard CFD codes which mostly allow fixed computational grids without local adaptivity only.

In this paper, we describe a 'fictitious boundary method' which works on an arbitrarily fixed grid. To be more specific, we concentrate on the problem of how to manage time-dependent domains with complicated geometrical structures inside of 'standard' CFD software. To be even more precise, we additionally assume that (geometrical) multigrid solvers are applied such that appropriate techniques for a sequence of meshes have to be designed. If 'classical' multigrid solvers are applied, one central problem is the sufficiently accurate approximation of complex geometries and their associated boundary curves or surfaces in a hierarchical way. While there is "no" problem to design (very) fine meshes which provide all necessary information, particularly with professional mesh generation tools, the construction of sequences of "coarser" meshes may be more crucial; especially if one keeps in mind that sequences of more or less nested meshes are necessary for multigrid.

One possible approach is to start with a coarse mesh which contains already most of geometrical fine-scale details. This technique may work efficiently in 2D cases, but for analogous 3D applications the resulting "coarsest" mesh will be in general very large (in the range of more than 10,000 - 50,000 mesh cells) such that the typical multigrid efficiency is lost due to a dominating coarse grid solver. In contrast, completely different - since non–nested - grids in comparison to the finest mesh may be used, but the corresponding intergrid transfer routines, which interpolate from one mesh to another, are difficult to handle. Consequently, the resulting multigrid solver spends most of its time with grid transfer routines on lower levels. Further, the convergence rates may deteriorate since they massively depend on the choice of subgrids and corresponding transfer operators as it is well-known from algebraic multigrid solvers.

Taking these problems into account, we propose another approach which is embedded into a general framework for implementing boundary conditions in iterative solution techniques: the *iterative filtering technique* in combination with *fictitious boundary conditions.*

'Employ a (rough) boundary parametrization which sufficiently describes all large-scale structures with regard to the boundary conditions. Treat all fine-scale features as interior objects such that the corresponding components in all matrices and vectors are unknown degrees of freedom which are implicitely incorporated into all iterative solution steps. Hence, standard tools for grid refinement in interior regions are easily applicable and highly accurate approximations can be obtained. Further, utilize filtering techniques to project the corresponding vector components onto the subspace of "correct" boundary conditions, before and directly after each iterative substep. This additionally ensures the typical performance of standard multigrid solvers without requiring additional modifications in the software components.'

Another important aspect is the corresponding modification of standard (geometrical) multigrid components and the resulting multigrid convergence, in particular if we apply the *filtering process* depending on the mesh level. So, we might apply mesh-dependent filters in such a way that the number of 'internal' objects is decreased when the mesh is coarsened, or the shape may become coarser due to less grid points, or it might even happen that the coarsest configurations for flow problems around internal objects do not contain any object: On the coarse levels, we even could perform a pure channel simulation while the internal objects are present only on the finer meshes! Such a technique - assuming that the typical multigrid convergence behaviour is preserved - of modifying the geometrical details on coarser meshes could be a very powerful tool in the context of multilevel approaches.

It is quite understandable that such an approach might work well even for the nonlinear Navier-Stokes equations, since the essential background idea for our applied nonlinear solution schemes of quasi-Newton type is the strong separation between **nonlinear treatment** and **linear solution tools**. Our approach approximates nonlinear effects on the finest mesh only, while multigrid is exclusively applied to solve linear subproblems in a preconditioning step. Consequently, we have to take care that at least the finest mesh is sufficiently accurate to describe all important physical effects. However, the coarser meshes are – roughly spoken – only involved to "provide some spectral approximation for the acceleration of matrix problems" for the performed linear multigrid. In this context, multigrid is only applied as very efficient linear **pre-**

conditioning tool, separated from the discretization aspects regarding the nonlinearity. This is a very essential difference to *direct nonlinear multigrid* approaches which require the approximation of analogous nonlinear problems on the coarser meshes, too. Then, due to the separation of both tasks, we can apply standard multigrid tools without any modification and we can guarantee at least the fast treatment of the linear subproblems in each nonlinear iteration step.

This approach may not be the best for some other problems in which case nonlinear multigrid approaches can lead to better results. However, our experience in the case of the incompressible Navier–Stokes equations shows explicitely that this "separation approach" seems to be superior. One reason is that typically the nonlinear convergence behaviour is only determined by the strength of the nonlinearity on the continuous level, that means the Reynolds number, while mesh dependent effects are of (almost) no importance. Hence, quasi-Newton schemes might be superior in comparison to nonlinear multigrid approaches for such problem configurations.

Before we describe and analyse the corresponding numerical details in the next section, we provide an example for applying this technique in the case of a prototypical simulation of moving flaps in a channel. Figure 1 shows the norm of the velocity as well as corresponding 'Particle Tracing' snapshots for simulating the incompressible Navier-Stokes equations in a channel configuration while the movement of the 'flaps' is analytically prescribed. All simulations are computed on a fixed mesh such that the moving flaps are described via the fictitious boundary conditions only and their position is incorporated via the iterative filtering technique (from [12]).

2 Concepts for fictitious boundary conditions

2.1 The incompressible Navier-Stokes resp. Boussinesq equations

All following concepts for integrating boundary conditions into CFD software for incompressible flow problems are realized in the FeatFlow software [13] which is based on (nonconforming) FEM discretizations, adaptive implicit time-stepping, nonlinear Newton-like methods, (geometrical) multigrid solvers (for velocity, temperature and pressure separately as well as for all physical quantities simultanously) on quite arbitrary domains. There exist special solvers in the framework of the so-called 'pressure Schur complement' methodology (see [8]) which can be directly applied as stationary approaches ('CCnD') as well as in fully nonstationary configurations ('PPnD') following operator splitting ideas. The software is realized for $n = 2$ as well as $n = 3$ dimensions; however, in this paper we mainly restrict to the 2D variants which already contain all important numerical features (Remark: The corresponding 3D realizations in FeatFlow do exist, too [13]).

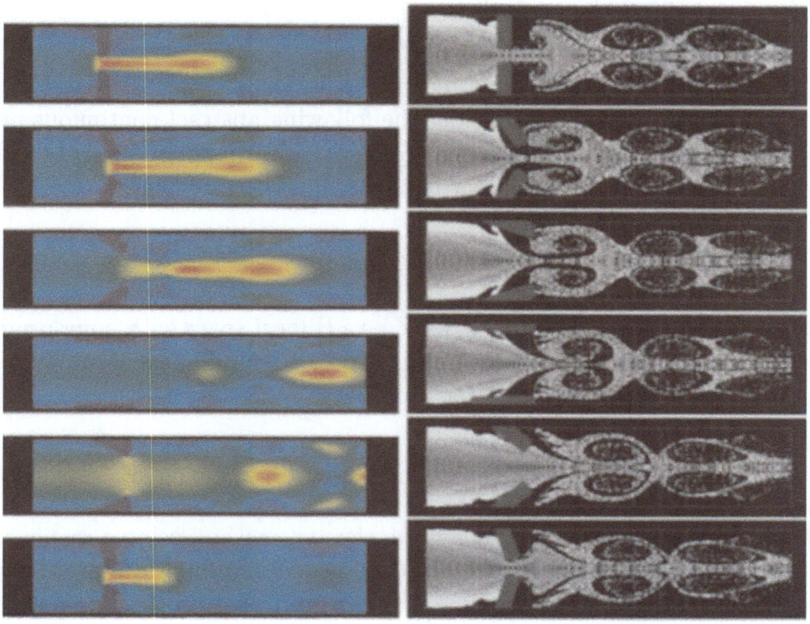

Fig. 1. Prototypical application of the fictitious boundary conditions for 'moving flaps in a channel' (from [12])

For the following considerations, let Ω be a bounded domain with a piecewise smooth boundary Γ. The equations to be solved are the incompressible Navier-Stokes, resp., Boussinesq equations

$$\rho \frac{\partial \mathbf{u}}{\partial t} + \rho \mathbf{u} \cdot \nabla \mathbf{u} = \mathbf{f} - \nabla p + \nu \nabla^2 \mathbf{u} + \rho \mathbf{j} T \quad , \quad \nabla \cdot \mathbf{u} = 0$$

$$\frac{\partial T}{\partial t} + \mathbf{u} \cdot \nabla T = \nu_{\mathbf{T}} \nabla^2 T$$

where $\mathbf{u}$ is the velocity, p the pressure, ν the kinematic viscosity coefficient, ρ the density, $\mathbf{f}$ source term which may include the gravitational force, $\nu_{\mathbf{T}}$ the diffusion coefficient for T which is the prototype for a transported quantity such as temperature, density, viscosity, etc., $\mathbf{j}$ is the unit vector in the y-direction. The Boussinesq approximation for the buoyancy forces is assumed to be valid; i.e., only small temperature excursions from the mean temperature are admitted. If we neglect the quantity T, resp., if $\mathbf{j} = 0$ such that $\mathbf{u}$ and p are independent of T, we obtain the incompressible Navier-Stokes equations. The above equations are to be solved with $\mathbf{u}(\mathbf{x}, t) = \mathbf{u}_\partial(\mathbf{x}, t)$ on parts of the boundaries of the flow domain where $\mathbf{u}_\partial(\mathbf{x}, t)$ is the prescribed boundary velocity, including time-dependent moving boundaries.

2.2 Iterative filtering techniques

For a better illustration of these techniques which are not restricted to the Navier-Stokes equations, we consider the following abstract continuous problem given in operator notation which - for reasons of simplicity - may be linked to a typical second-order PDE (for instance, Poisson problem),

$$Au = f\,. \tag{1}$$

For a general variational approach, let V be a Hilbert space with inner product $(\cdot,\cdot)$ and corresponding norm $\|\cdot\|$, and let $a(\cdot,\cdot)$ be a bilinear form. We seek a solution to the following variational problem:

$$\text{Find } u \in V, \text{ such that:} \quad a(u,\varphi) = (f,\varphi) \quad \forall \varphi \in V\,. \tag{2}$$

This problem is approximated by a finite element method using a sequence of finite dimensional subspaces "$V_h \subset V$" (also nonconforming FEM are allowed) parameterized by a discretization parameter h, such that the discrete problems read as usually:

$$\text{Find } u_h \in V_h, \text{ such that:} \quad a_h(u_h,\varphi) = (f_h,\varphi) \quad \forall \varphi \in V_h\,. \tag{3}$$

In matrix-vector notation, we derive the discrete linear system

$$A_h U_h = F_h\,. \tag{4}$$

Further, we assume that the continuous problem is defined with corresponding boundary conditions $B(u) = g$ on a boundary part Γ. Let be B a boundary operator which involves a combination of function values and partial derivatives of u on Γ. In the following we imagine for simplicity that we perform Dirichlet boundary values $u = g$ on Γ. Then, besides the possiblity of applying a typical penalty approach, there are the following three possiblities after the discretization process to involve these boundary conditions into the solution process of the matrix-vector problem (4). In all cases, we assume that the matrix A_h (and also U_h and F_h) has not yet incorporated any boundary condition which in fact corresponds to the corresponding natural boundary condition due to the partial integration involved, for instance

$$\partial_n u = 0 \quad \text{on } \Gamma = \partial\Omega\,. \tag{5}$$

Further, let $S_h(\Gamma)$ denote all degrees of freedom, resp., all components of the solution vector U_h, which are related to the boundary Γ. For example, in the case of bilinear finite elements these are the nodes on Γ (or better: on Γ_h as an approximation to Γ), while for the nonconforming rotated bilinear elements (see [6]) the edges on Γ, resp., on Γ_h are associated. Then, we can proceed as follows to apply Dirichlet boundary conditions:

1) Fully explicit treatment:
We eliminate in A_h all rows and columns belonging to $S_h(\Gamma)$. Additionally, we modify all components of the right hand side vector F_h not belonging to $S_h(\Gamma)$ according to this elimination process. Further, we may prescribe the correspondingly prescribed Dirichlet values for all these components of the vectors U_h and F_h which belong to $S_h(\Gamma)$. But this step is not necessary since all components of $S_h(\Gamma)$ are treated as being well-known and hence do not longer belong to the set of unknowns. This process is often performed if direct solvers (Gaussian elimination) are applied. However, problems arise if the right hand sides change since corresponding modifications due to (then already) eliminated matrix elements of A_h have to be repeated. As a further consequence, two different matrices may be needed if the same operator A is required with two different boundary conditions, for instance Dirichlet values for the momentum equations and natural settings for associated Poisson problems in the incompressible Navier-Stokes equations.

2) Semi-implicit treatment:
We replace only those rows of A_h by rows of the identity matrix which correspond to $S_h(\Gamma)$. The other rows and columns remain unchanged. Further, during an initialization process only, we have to prescribe the given Dirichlet values for all these components of the vectors U_h and F_h which belong to $S_h(\Gamma)$. This setting guarantees in combination with iterative solvers that the resulting solution vectors always satisfy the prescribed boundary conditions. This approach seems to be the most common in the framework of iterative solution tools. Although the components of $S_h(\Gamma)$ cannot change their value, they belong explicitly to the set of unknowns and are treated by each component of the iterative solver (e.g., smoothing, prolongation, restriction, defect calculation, preconditioning). Again, two different matrices may be needed when two different boundary conditions, for instance Dirichlet and natural conditions, are prescribed for the same operator A.

3) Fully implicit treatment:
We do not modify A_h which consequently is the Neumann matrix due to natural boundary conditions (see [3]). Again, we have to prescribe the given Dirichlet values for all these components of the vectors U_h and F_h which belong to $S_h(\Gamma)$, but before and after each iteration step (*filtering*). All components of the solution vectors have to be treated as unknowns by all components of the iterative solvers. This approach is the most general for iterative solution techniques. Even if different boundary conditions are involved with the same operator A, we can always work with only one matrix A_h. Only the performed *filtering operator* has to be changed.

While the use of direct solution tools often requires the application of the *fully explicit treatment*, all three approaches are equivalent if iterative solvers are applied, and lead to exactly the same results. We prefer the *fully implicit treatment* since this is the most general approach for multigrid solvers and finite element approaches. One important component inside is the *projection filter* which administrates the boundary setting for all vector components of $S_h(\Gamma)$ during the iterative procedure. This tool will be shown to be the essential trick for easy implementations of complicated boundary value settings, in particular for small-scale details due to complicated geometrical boundaries and time-dependent moving boundary parts, and also with respect to several boundary components for discretely divergence-free FEM [11].

Before we examine more carefully the numerical and algorithmic properties of this approach w.r.t. accuracy and robustness, hereby analysing particularly the corresponding multigrid convergence behaviour, we shortly present another (well-known) 'fictitious boundary condition' method which does not necessarily capture (internal) geometrical objects via the computational mesh, too. Moreover, it does not require (complicated) modifications of the source code due to changing bilinear forms for incorporating a typical penalty approach or for applying boundary conditions: it only prescribes problem-adapted input parameters. In the following, we use specific values for density and viscosity inside of the incompressible Navier-Stokes equations which act as the corresponding filter to satisfy - approximately - the corresponding Dirichlet boundary conditions, at least for homogeneous no-slip conditions.

2.3 Density-viscosity blocking techniques

Since the value of the density or viscosity inside of the Navier-Stokes equations determines the state of fluid, the fluid changes into the state of solid when the values of density or viscosity become infinity. Motivated by this idea, we can use 'huge' values for the density or viscosity parameters to mark a solid body which is positioned in the flow field. So, we can view it as another fictitious boundary method. The difference is that in the previous 'mesh blockage' method, the mesh in the area occupied by a solid body is blocked, while in the 'density-viscosity' blockage method, the density or viscosity parameter inside of the stationary and/or moving solid body is 'blocked'.

It is obvious that this method is the easiest way to incorporate complicated fine-scale structures into the code, since no geometrical adaptations have to be performed and no changes in the code due to implementing boundary conditions are necessary: Only the parameter values for density and viscosity have to be prescribed in an appropriate way! However, it is also well-known that multigrid-like solvers have problems with strongly discontinuous coefficients. Moreover, the influence of the 'strength' of the discontinuity, that means the size of the values for the 'solid' viscosity, resp., density, onto the resulting solution, in particular w.r.t. physically interesting values directly on the surface of the interior objects, is not clear. And, probably the most important criticism, choosing finite parameters for density and viscosity as an approximation for the 'infinite' values of a real solid leads to penetration from the surrounding flow into this object since the corresponding - implicitely given - natural boundary conditions on the interface between the solid and the surrounding liquid do not enforce that $\mathbf{u} \cdot \mathbf{n} = 0$ is valid. However, this internal diffusion is mainly related to the actual size of viscosity and density, and the time scale for observing such perturbations may be much larger than the actual time interval for the calculation. These considerations will be subject of our subsequent numerical analysis.

3 Numerical tests: Qualitative analysis

In this section we will show several applications of the fictitious boundary methods in order to illustrate their flexibility and their qualitative behaviour for quite complex configurations; a precise comparative study and quantitative examination will be part of the next section. Here, we concentrate on the main characteristics w.r.t. flexibility and applicability in the context of incompressible flow simulations.

3.1 Numerical Examples for the 'Iterative Filtering' technique

Flow generated by an oscillating heated plate in an enclosure box.

In the first example, we consider a unit square with an interior heated plate which is fixed at the center of the box (see Fig. 2). The length of the plate is 0.6, thickness is 0.04. The enclosure boundary conditions consist of no-slip walls, and constant cold temperature ($T = 0$) at horizontal and vertical walls while the surface of the plate is held at a constant hot temperature ($T = 15$). The fluid is initially at rest.

The time-dependent position of the plate is prescribed by a sinusoidally oscillating rotation with angle velocity $\omega = 0.05\pi sin(0.1\pi t)$ (t is the time). We use the fictitious boundary method which does not take care about the employed mesh. Instead, we just need to generate a simple mesh for the whole box, and then use a corresponding filter procedure to represent those elements which are occupied by the plate. Fig. 3, Fig. 4 and Fig. 5 give snapshots for the contour plots of the norm of velocity, temperature and pressure distributions, respectively. From these pictures, we can see that the fluid motion in the box is initiated by the oscillating rotation of the plate, and the vortex shedding and circulation is generated periodically in the box.

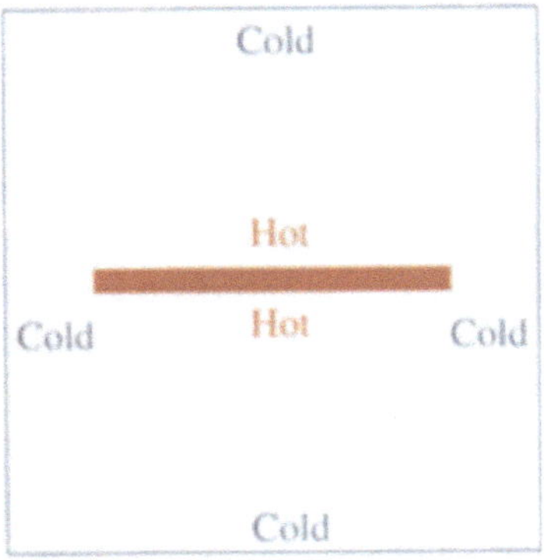

Fig. 2. Description of the 'heated plate in a box' problem

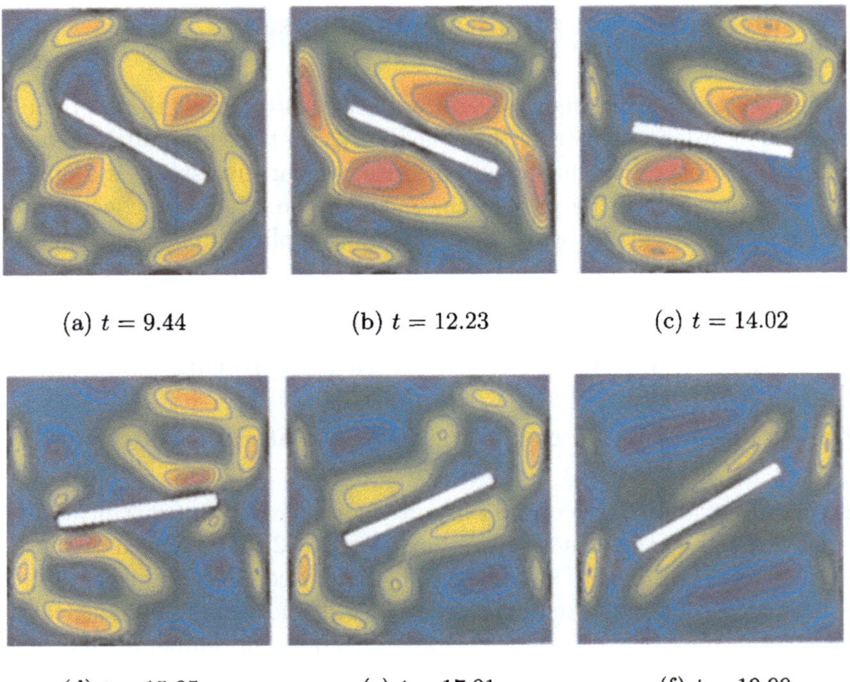

(a) $t = 9.44$ (b) $t = 12.23$ (c) $t = 14.02$

(d) $t = 15.85$ (e) $t = 17.91$ (f) $t = 19.99$

Fig. 3. Oscillating rotation of a heated plate in a box: norm of velocity

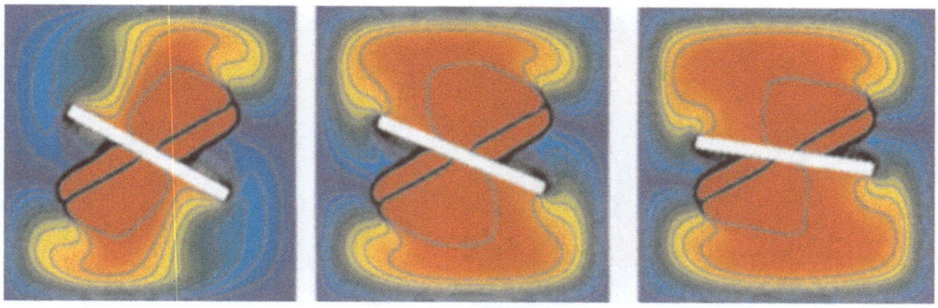

(a) $t = 9.44$ (b) $t = 12.23$ (c) $t = 14.02$

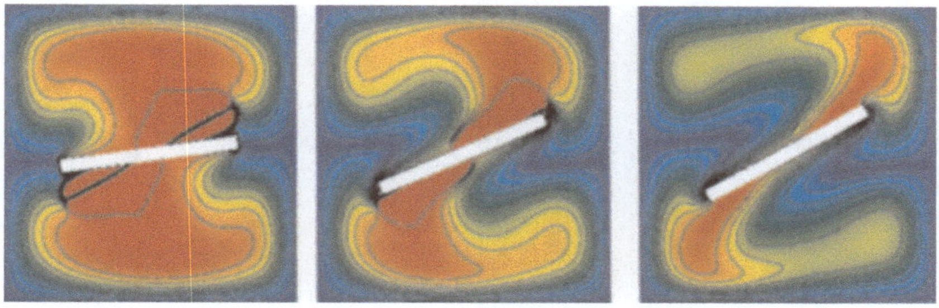

(d) $t = 15.85$ (e) $t = 17.91$ (f) $t = 19.99$

Fig. 4. Oscillating rotation of a heated plate in a box: temperature

Flow generated by an oscillating cylinder in a channel.

The following flow configuration has been chosen to demonstrate the fictitious boundary method for a configuration in which case the flow is disturbed by a transversely oscillating circular cylinder such that vortex shedding is observed. Here, we simulate a cylinder undergoing a sinusoidally transverse oscillation with explicitely specified amplitudes and frequencies. The computational domain size is $(L \times H)$=(2.2×0.41), L being the length of the channel, and H is the width of the channel. The location of the cylinder center (x_0, y_0) is $(0.8, 0.2)$ relative to the left bottom corner of the domain. The cylinder diameter D is equal to 0.2. No-slip condition is prescribed on the walls. The fluid in the channel is initially at rest. The cylinder oscillates sinusoidally such that the location of its center (x_c, y_c) is given by $x_c(t) = x_0 + Asin(2\pi ft)$; $y_c(t) = y_0$, where t is the time, and $A = 0.6$ and $f = 0.5$ are the amplitude and frequency of the oscillation, respectively. A nonuniform mesh is used in the simulation such that enhanced resolution is provided in the cylinder vicinity and in the wake. In the horizontal direction, improved resolution is provided up to three diameters on either side of the cylinder location, which is adequate to cover the near wake for all the oscillation amplitudes (see Fig. 6).

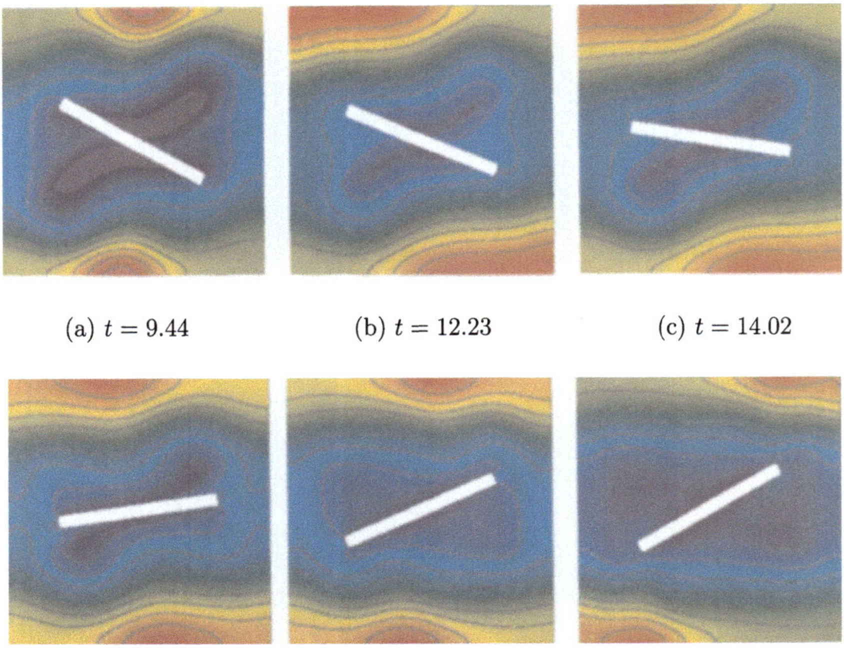

(a) $t = 9.44$ (b) $t = 12.23$ (c) $t = 14.02$

(d) $t = 15.85$ (e) $t = 17.91$ (f) $t = 19.99$

Fig. 5. Oscillating rotation of a heated plate in a box: pressure

Fig. 7 gives a sequence of contour plots for velocity distributions which show the shedding of vortices when the cylinder is oscillating in the channel. Fig. 8 presents corresponding contour plots for the pressure distribution. From these pictures, we can see that the flow in the channel is significantly disturbed by the oscillating cylinder, and vortex shedding is generated periodically in the wake of the cylinder. The range of the wake becomes longest when the cylinder is at the end of the moving direction ($t = 3.59, 4.44, 5.50$), while the flow is seriously perturbed and more complex ($t = 4.01, 5.08$) when the cylinder is in the middle position of its oscillation.

3.2 Numerical Examples for 'Density-Viscosity' blocking

In this subsection, numerical examples for the flow around single or multiple circular cylinders are presented by using the described density-viscosity blockage method. We can see that the density-viscosity blockage method seems to be a flexible and efficient tool to compute the flow with complex immersed stationary boundaries. However, the numerical tests in the following Section will show that this approach can be quite 'dangerous' since penetration into the 'solid' cannot be avoided by this simple approach. Nevertheless, understanding the different time scales due to the simulation process and due to the perturbations which are introduced by the size of the parameters, a qualitative flow prediction with this methodology seems to be possible.

In the following simulations, the computational domain is $L = 6.0$ and $H = 1.2$, with L and H being the length and the height of the channel. The cylinder diameter D is equal to 0.2. No-slip is prescribed on the walls while natural 'do-nothing' boundary conditions are employed at the right boundary. A parabolic velocity $U = U_\infty(H - y)y$ is prescribed at the inflow (the left part of the channel). The Reynolds number is defined as $Re = U_\infty D/\nu$. Again, a fixed nonuniform mesh is used in the time-dependent simulation (see Fig. 6).

Fig. 9 and Fig. 10 show the numerical results for flow around a cylinder with $Re = 100$. In this case, we set the value of density within the cylinder equal to $10^5\rho$, ρ is the normal density of the fluid. Fig. 9 (a) and (b) are the velocity and vorticity in the region nearby the cylinder area. Fig. 10 (a) and (b) are the corresponding pictures in the whole computational domain. We can see that vortex shedding is initiated behind the cylinder which is imitated by the high viscosity and density values. In Fig. 10, we set the value of the viscosity within the two circular cylinders into $10^9\nu$, ν is the normal viscosity of the fluid. Fig. 10 (c) and (d) show the velocity and vorticity for the case of Reynolds number $Re = 100$, respectively. We can see that the interaction of vortices behind the wake of two cylinders is obvious and strong. Fig. 10 shows the corresponding flow around three circular cylinder, again with Reynolds number $Re = 100$. In this case, we set both the values of density and viscosity into 'infinity' ($10^5\rho$ and $10^9\nu$, respectively) within the three cylinder areas. Fig. 10 (e) and (f) present the velocity and vorticity, respectively.

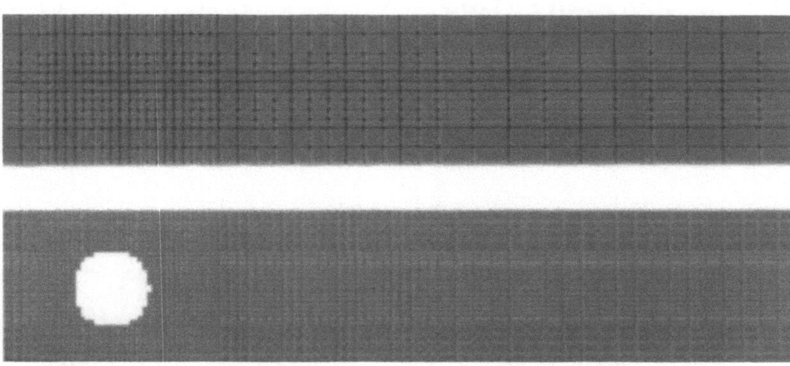

Fig. 6. Complete computational mesh and snapshot for actual 'active elements' on the (globally) refined mesh (Remark: The filtering procedure (de-)activates edges due to the nonconforming FEM approach for the velocity, but not pressure cells!)

(a) $t = 3.59$

(b) $t = 4.01$

(c) $t = 4.44$

(d) $t = 5.08$

(e) $t = 5.50$

Fig. 7. Sequential contour plots for velocity showing the shedding of vortices

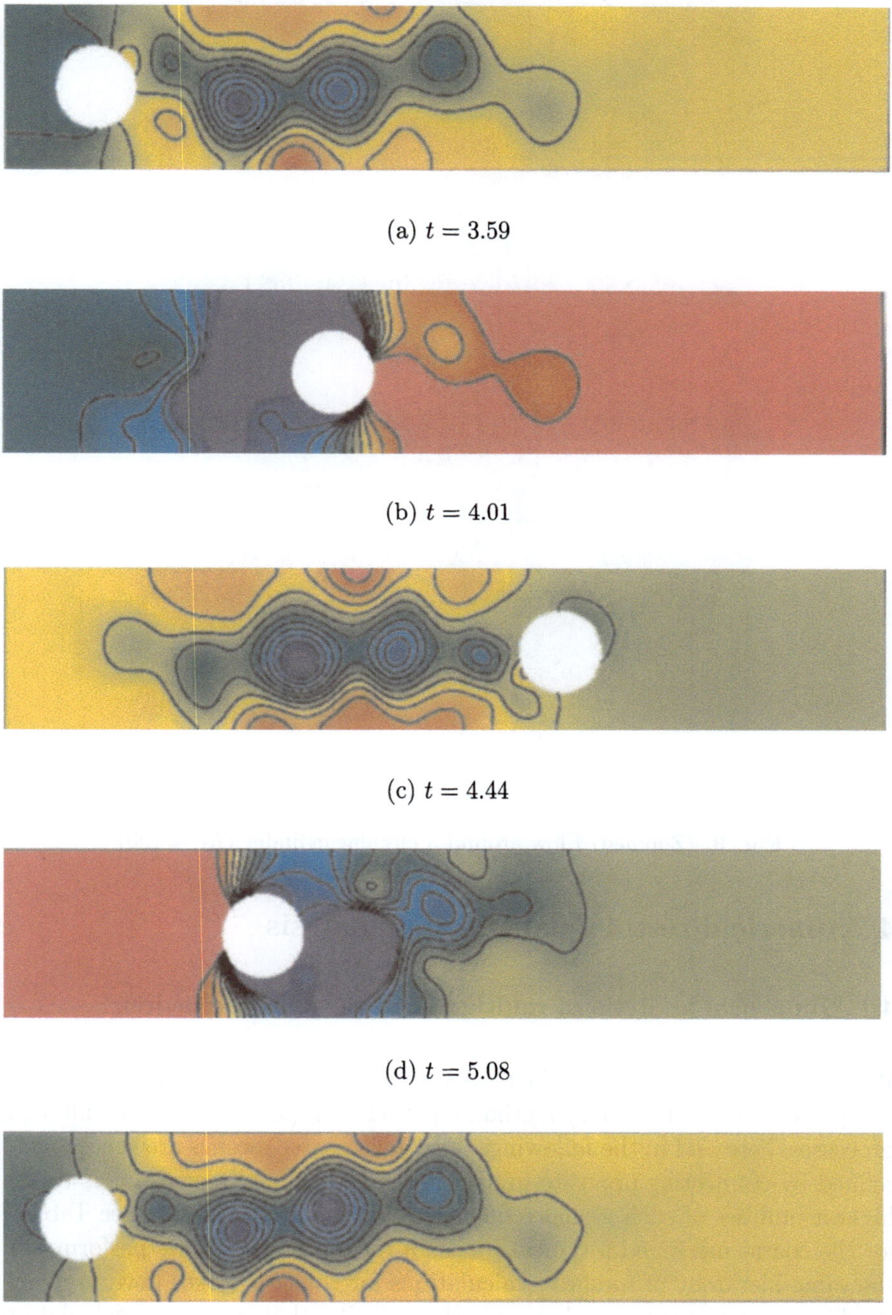

(a) $t = 3.59$

(b) $t = 4.01$

(c) $t = 4.44$

(d) $t = 5.08$

(e) $t = 5.50$

Fig. 8. Sequential contour plots for the pressure distribution

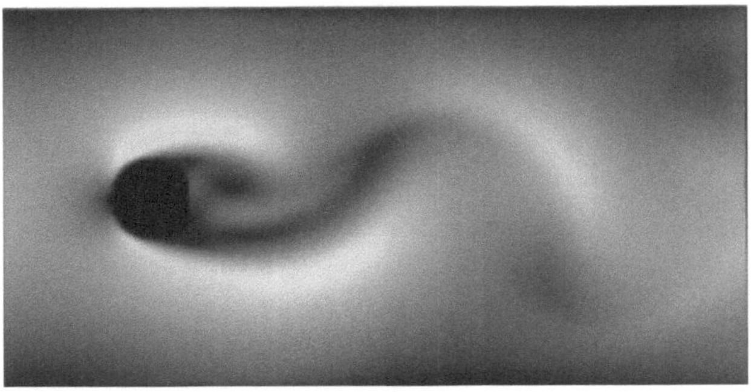

(a) velocity

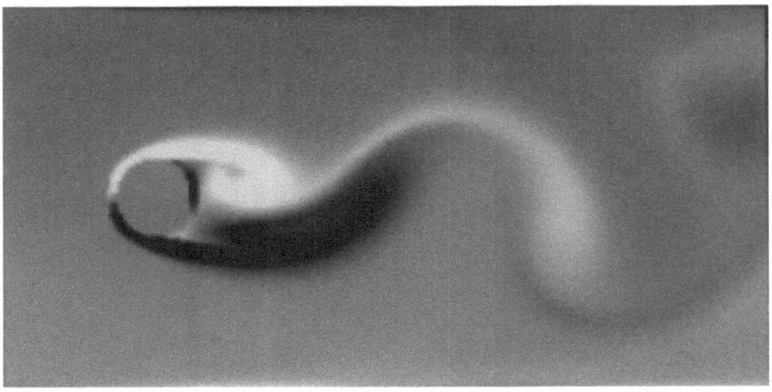

(b) vorticity

Fig. 9. (Zoomed) Flow around a circular cylinder ($Re = 100$)

4 Numerical tests: Quantitative analysis

4.1 Numerical analysis of multigrid convergence behaviour

We start with demonstrating the related (linear) multigrid behaviour when we apply the fictitious boundary method together with (level-dependent) filtering processes: here and in the following, the shown coarse meshes are successively refined by connecting opposite midpoints, and LEVMAX corresponds to the highest number of such global refinements. In the first example (see Table 1 for the coarse mesh and a typical snapshot for the pressure), we perform - on the same hierarchy of meshes - calculations for (linear) Stokes flow. We vary the filtering process by applying no filter at all - that means we calculate a simple parabolic channel flow - in comparison to prescribing the corresponding filter for flow around a cylinder on all mesh levels. In addition, we show results for modified configurations which apply this filtering process on highest mesh levels only, starting from the finest level LEVMAX down to level LEVMAX - `LEVFILT`: Choosing `LEVFILT`=1 means that only the two finest meshes corre-

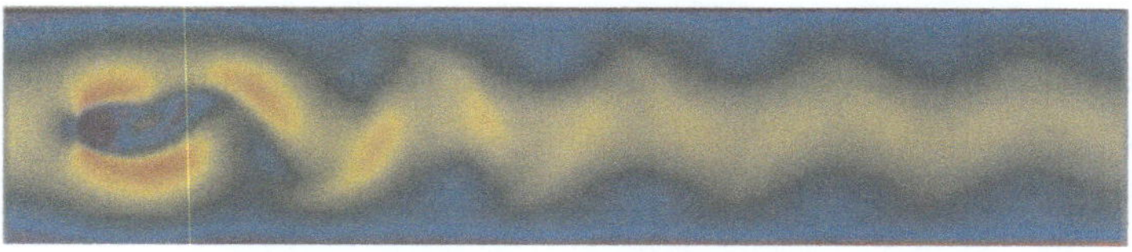

(a) velocity

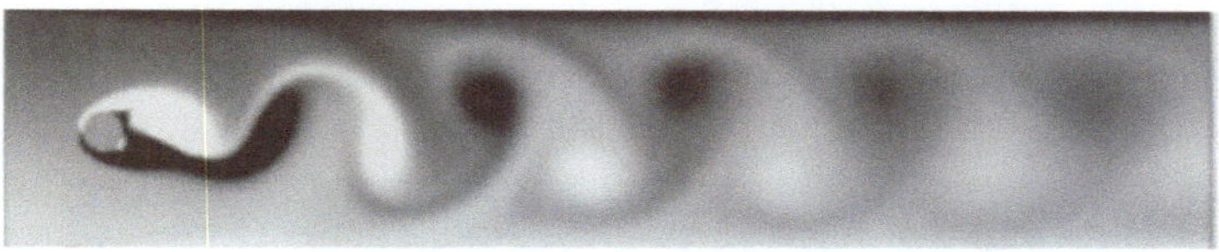

(b) vorticity

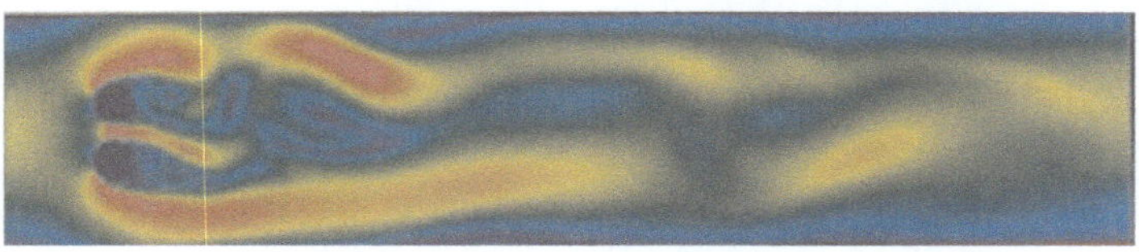

(c) velocity

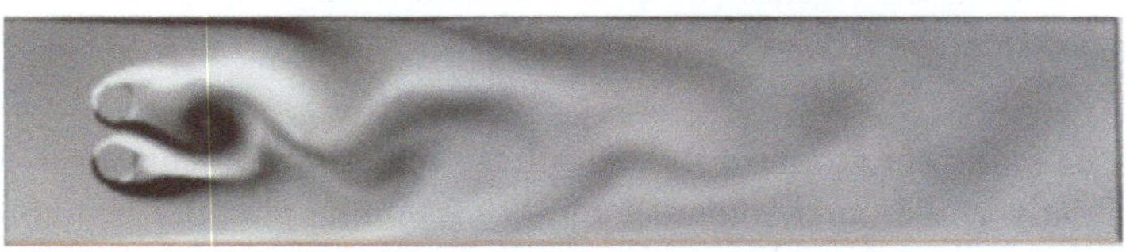

(d) vorticity

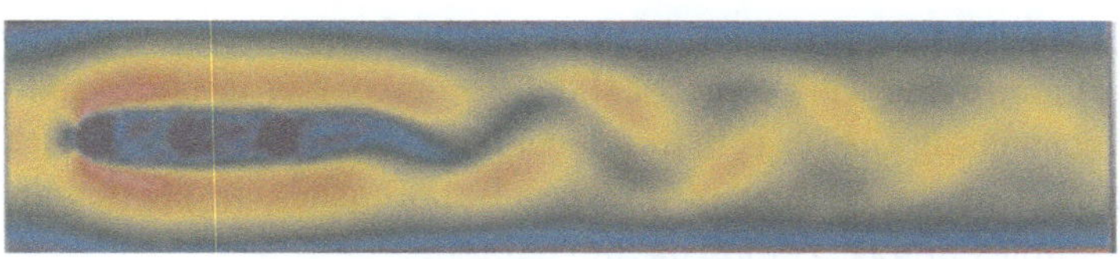

(e) velocity

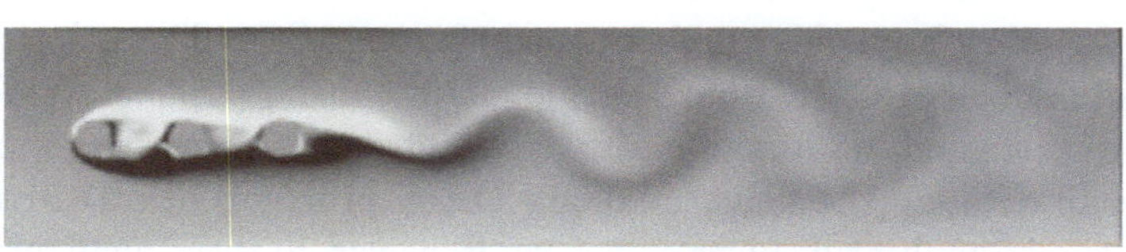

(f) vorticity

Fig. 10. Flow around one, two and three circular cylinders ($Re = 100$)

spond to 'flow around cylinder' while choosing LEVFILT=2, resp., LEVFILT=3 denote the cases that the three highest, resp., the four highest level in the mesh hierarchy contain the internal object.

LEVMAX	Channel Flow	Circle Flow	LEVFILT=3	LEVFILT=2	LEVFILT=1
3	0.13	0.12	0.12	0.12	0.12
4	0.11	0.11	0.11	0.11	0.11
5	0.11	0.13	0.13	0.20	0.64
6	0.10	0.10	0.11	0.45	0.73

Table 1. Multigrid convergence (CC2D with 4 smoothing steps) for Stokes flow

It seems to be sufficient to perform such filter techniques on the three or four finest levels only while on the coarser meshes a simple channel configuration (without any interior object!) can be applied. In these cases, the resulting multigrid rates do not (significantly) differ from the case of applying this filter on all levels: To be on the safe side, the number of smoothing steps should be increased if the filtering process is applied on the finest mesh levels only. This technique of modifying the geometrical details on coarser meshes - taking less objects or coarsening the objects in a level dependent way - has shown to be a powerful tool in the context of multilevel approaches, in particular in 3D (see the examples in `www.featflow.de/album`).

In the next simulations, we employ the same hierarchy of meshes - which do not capture the interior objects - and analyse the nonlinear and linear solution behaviour for a low Reynolds number configurations ($Re \approx 20$) such that the direct steady solver 'CC2D' from FeatFlow [13] can be applied. While the nonlinearity is treated via a standard fixed point-defect correction approach (see [8] for all details), the linear auxiliary Oseen problems are treated with the same coupled multigrid solver as before. However, in these configurations we vary the number of interior objects between 3 and 6. The configuration '6' in Table 2 corresponds to 6 interior circles on <u>all</u> mesh levels, while 'LEVEL' denotes the case that the number of inner circles is related to the (highest) mesh level LEVMAX. The first number in each column denotes the number of necessary nonlinear iteration steps while the second entry corresponds to the total number of linear multigrid sweeps involved.

LEVMAX	LEVEL	6
3	6/11	6/11
4	6/11	6/11
5	6/11	6/11
6	6/11	6/11

Table 2. Snapshots for the resulting pressure for the '6 circle' configurations on the grid level LEVMAX=6 (top) in comparison to '3 circles' on grid level LEVMAX=3; the table shows the corresponding nonlinear and linear solution behaviour.

These numerical tests demonstrate the predicted result that the nonlinear solution behaviour - for this low Reynolds number - is more or less independent of the number of interior objects and the mesh level. Moreover, the linear multigrid solver for the auxiliary Oseen problems is robust against variations of the filtering process and produces the typical iteration numbers as known for standard approaches which directly integrate the interior objects into the computational mesh.

4.2 Analysis of approximation properties: Stationary case

The aim of the following tests is a systematic examination of the approximation properties of the fictitious boundary method, particularly in combination with the iterative filtering approach. We perform the same stationary simulations (flow around cylinder, $Re = 20$) as before, but on several meshes which incorporate the circle into the (coarse) mesh (CIRC1 - CIRC3) and on various meshes which are full channel meshes (CHAN1 - CHAN4). The following Fig. 11 shows different coarse meshes with various degrees of a priori semi-adapted refinements near the location of the cylinder which have been successively refined via connecting opposite midpoints.

CIRC1

CIRC2

CIRC3

CHAN1

CHAN2

CHAN3

CHAN4

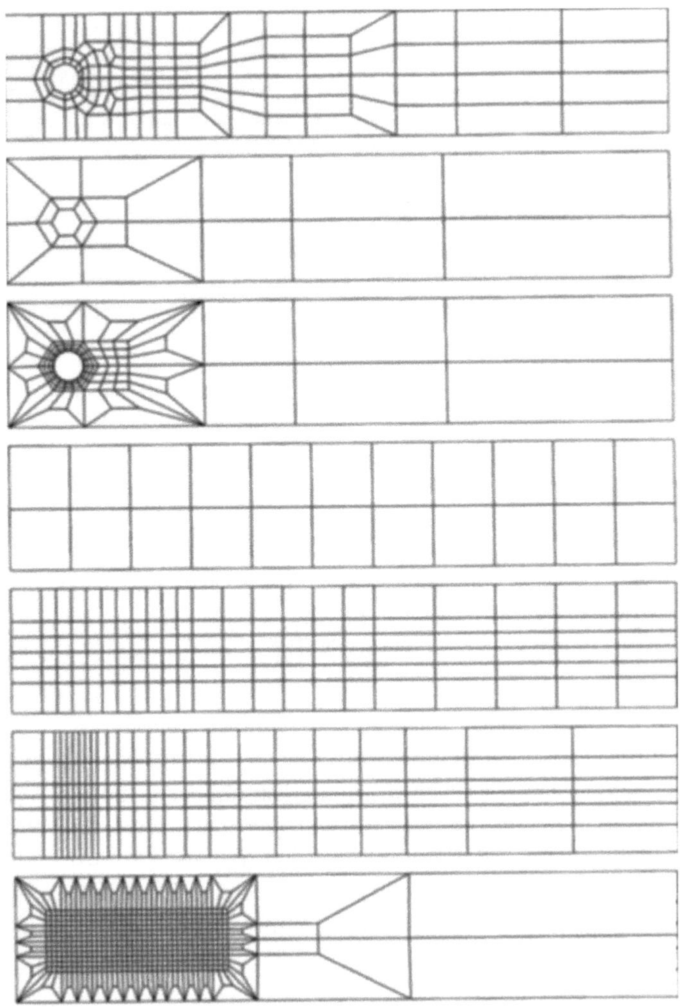

Fig. 11. Different coarse meshes

The following Table 3 shows the results for the velocity (U1,U2) and the pressure (P) in an interior grid point which is part of all types of meshes. Additionally, we measure the 'pressure difference' ΔP which involves the pressure in the point before and behind the circle. In particular, the pressure difference ΔP is a very delicate quantity since it includes the result directly on the interface which is captured by the mesh (CIRC1 - CIRC3) or by the filtering process (CHAN1 - CHAN4).

First of all, we calculated a reference solution (obtained on the mesh CIRC1 with more than 1 million grid points; reference values are also well-known since this is exactly the stationary 2D configuration of the special 'flow around cylinder' benchmark [7]). Based on the reference values, we figure out which level of refinement, that means how many mesh cells, is necessary to obtain an approximation with less than 1%, resp., 5% error. This number can be used as an criterion for rating the approximation properties of the different meshes and boundary implementation techniques since it directly addresses the required mesh width and hence the accuracy of the examined configurations.

5% Accuracy

	CIRC1	CIRC2	CIRC3	CHAN1	CHAN2	CHAN3	CHAN5
U1	2080	5632	32768	5632	8448	9216	9088
U2	2080	352	2048	352	2112	2304	9088
P	2080	1408	8192	352	2112	2304	36352
ΔP	2080	1408	2048	5632	8448	2304	9088

1% Accuracy

	CIRC1	CIRC2	CIRC3	CHAN1	CHAN2	CHAN3	CHAN5
U1	8320	90112	131072	360448	33280	36864	145408
U2	2080	1408	8192	1408	33280	9216	9088
P	8320	22528	131072	1408	2112	9216	581632
ΔP	33280	22528	32768	360448	135168	9216	145408

Table 3. Required number of elements to satisfy the criterion '5% error', resp., '1% error' for the different meshes and control quantities

The results for such low Reynolds number simulations show that an appropriate global grid refinement (for interior control values, in particular in the right part of the channel which is the case for the evaluation of 'P' and 'U1') as well as adequate local mesh adaptation is necessary (in particular for the 'boundary' quantity ΔP). Anyway, the fictitious boundary method with the proposed filtering techniques proves to be competitive with the standard approaches - in terms of necessary grid points, multigrid sweeps and total CPU timing - for such typical CFD applications. In particular, the configurations CIRC1, resp., CIRC2, and the meshes CHAN2, resp., CHAN3 in the context of the fictitious boundary conditions give comparable results, particularly for the weaker, but usually satisfying criterion '5% error'.

4.3 Analysis of approximation properties: Nonstationary case

In the next test cases, we examine the resulting accuracy for the same control quantities, now for a medium range Reynolds number which leads to periodical time-dependent vortex shedding behind the cylinder. Again, this flow configuration follows the well-known 'flow around cylinder' benchmark [7] which results in $Re = 100$. Since we are mainly interested in the spatial accuracy of the fictitious boundary method, in particular w.r.t. capturing the important effects near the interior objects, we try to eliminate the temporal discretization error by choosing very small time steps. Then, we proceed the nonstationary simulations until a fully periodical flow behaviour of all quantities has been observed. Finally, we compare the results for one period (see Fig. 12). To be more precise, we figure out how many levels of global grid refinement, that means the required total number of mesh cells, are necessary to produce a flow behaviour - for one period - such that the agreement with a previously calculated reference solution is 'OK', 'GOOD', resp., 'PERFECT' (see the corresponding pictures in Fig. 12).

The following Table 4 shows the resulting number of mesh cells to satisfy the indicated approximation quality for the different meshes, techniques and quantities: CIRC1 - CIRC3 represent the standard approaches which capture the interior object directly via the mesh and appropriate boundary parametrizations. CHAN1 - CHAN4 lead to analogous results, but for a full channel geometry without capturing the interior objects via grid points; the circle is described via the fictitious boundary conditions together with the proposed filtering techniques.

'OK'

	CIRC1	CIRC2	CIRC3	CHAN1	CHAN2	CHAN3	CHAN4
U2	8320	22528	32768	22528	8448	9216	9088
ΔP	8320	22528	32768	22528	8448	9216	9088

'GOOD'

	CIRC1	CIRC2	CIRC3	CHAN1	CHAN2	CHAN3	CHAN4
U2	33280	90112	131072	90112	33792	36864	36352
ΔP	33280	22528	131072	90112	33792	36864	36352

'PERFECT'

	CIRC1	CIRC2	CIRC3	CHAN1	CHAN2	CHAN3	CHAN4
U2	133120	360448	524887	360448	135168	147456	145408
ΔP	133120	90112	131072	360448	135168	147456	145408

Table 4. Required number of elements to satisfy the different quality criterions for the proposed meshes/filtering techniques

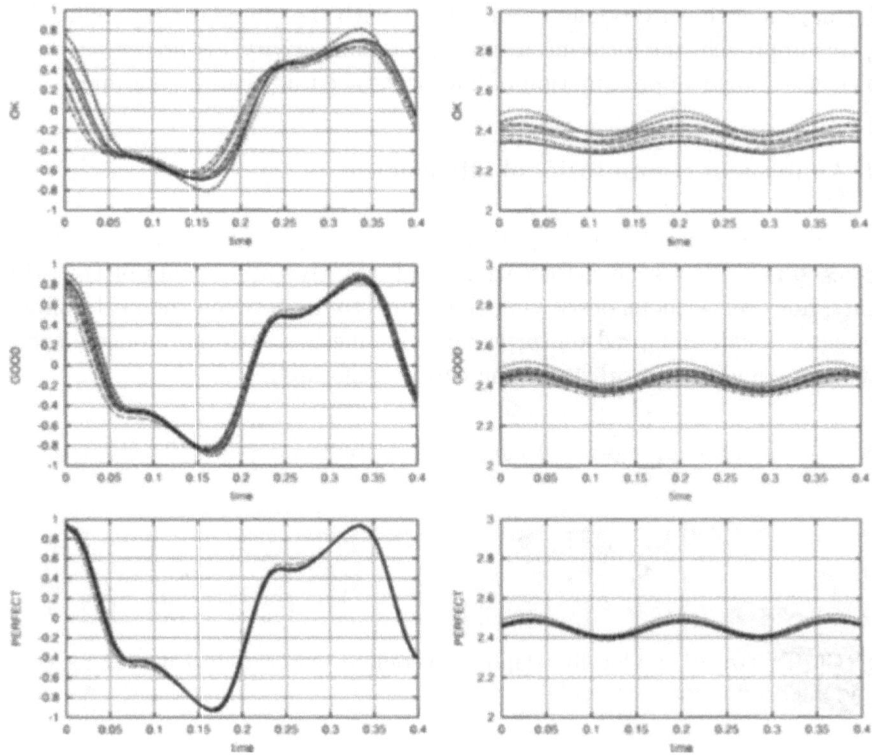

Fig. 12. Fitting of the results w.r.t. the control quantities 'U2' (left) and 'ΔP' (right) for the different categories

The typical results from such nonstationary simulations for medium Re numbers show that a combination of local as well as global grid adaptation is necessary if all relevant flow scales have to be resolved. Moreover, the results also show that for such 'simple' (geometrical) configurations, the fictitious boundary method leads to comparative results as the standard approaches with 'body-fitted' meshes, at least if we compare the number of grid refinements, resp., the required total number of mesh cells/grid points in our computational mesh. Since both implementations are almost identical - the only difference is that we have to apply the filtering process after each matrix-vector application which however costs $O(N)$, N the number of unknowns, arithmetic operations only - the resulting CPU times are more or less identical since the number of nonlinear iterations and linear multigrid sweeps is almost the same (see Table 2). Moreover, it is obvious that the right way to get higher (local) accuracy is via local mesh adaptivity: Since the proposed 'fixed mesh' approach leads to a piecewise constant approximation of the boundaries only, the local mesh size has to be decreased in an appropriate way. Another possibility is a local deformation of the grid points to align them with the boundary segments. Using this approach, the local mesh topology ('connectivity') is preserved while a piecewise linear approximation of 2nd order accuracy can be obtained.

4.4 Analysis of computational efficiency

The following examples in Fig. 13 show (local) generalized tensorproduct meshes which allow such deformations and adaptive movements of the grid points, however without destroying the connectivity between the grid points. Therefore, locally adapted tensorproduct-like meshes with more or less arbitrary non-equidistant grid stretching can be generated.

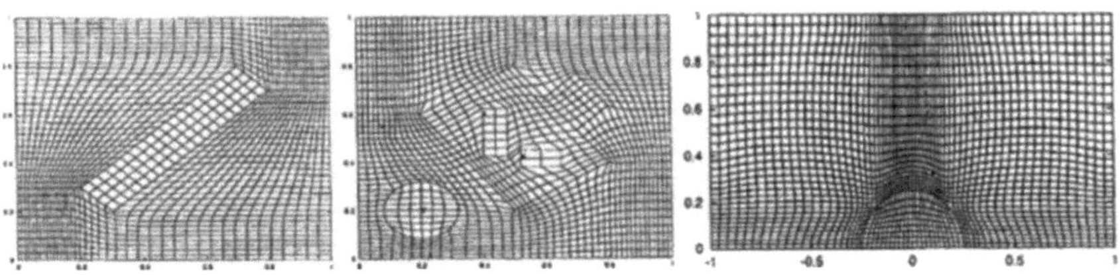

Fig. 13. 'Deformed' and locally adapted generalized tensorproduct meshes for capturing interior interfaces (from [4])

Since our complete computational mesh is built up - on a macro level - by a numerous collection of such generalized tensorproduct meshes, the described mesh adaptation can be performed locally - on certain macros only - such that only a small part of (local) stiffness matrices has to be re-assembled. Moreover, we preserve the full flexibility of standard FEM approaches since the macro structure is allowed to be fully unstructured while the 'substructures' (= macros) consist of many structured parts only which however can be adapted via mesh deformation, mesh alignment or local mesh refinement with hanging node techniques (see [9]). These meshing techniques have been currently implemented into the FEM package FEAST (see [1]) which is a basic tool box for FEM applications on high performance computers.

The underlying main philosophy in FEAST is to recursively split the global problem into a sequence of local problems, preferably on such *macros* which allow the robust and efficient solution of the auxiliary problems not only w.r.t. high numerical efficiency, but also in terms of very high MFLOP/s rates. The number of arithmetic operations tends to be larger than for the standard approach, but these local operations are cheap [9]: *Data moving, not data processing is costly!* However, these ideas must be combined with corresponding hierarchical data and matrix structures, which can exploit the described tensorproduct-like meshes on each *macro* to achieve the high performance rates for the necessary Numerical Linear Algebra components in the local (multigrid) solvers. And, the introduced fictitious boundary/iterative filtering techniques are important ingredients since this combination allows the use of such highly structured, generalized tensorproduct-like meshes, in particular for time-dependent constellations!

To demonstrate such performance considerations, we give an examples with our FeatFlow code which is applied to a prototypical "2D flow around a car" configuration, and we measure the resulting MFLOP/s rates for the MV multiplication inside of the multigrid solver for the momentum equation (see [8] for mathematical and algorithmic details). We use the typical (for FEM approaches) 'two level' (TL) numbering (old vertices preserve their numbers if the mesh is refined), a version of the bandwidth-minimizing Cuthill-McKee (CM) algorithm and an arbitrary 'stochastic' numbering. Hereby, we apply *sparse MV* concepts which are the standard techniques in FEM codes (and others), also well known as 'compact storage' technique: Depending on the programming language, the matrix entries plus index arrays/lists/pointers are stored as long arrays or heaps, containing the 'nonzero elements' only (for instance, see [18] and the literature cited therein). While this *sparse* approach can be applied for general unstructured meshes and arbitrary numberings of the unknowns, no explicit advantage of (possible) highly structured parts can be exploited. Consequently, a massive loss of performance with respect to the possible peak rates may be expected since - at least for large problems with more than 100,000 unknowns - no 'caching in' and 'pipelining' can be exploited such that the higher cost of memory access will dominate the resulting MFLOP/s rates.

The following Table 5 shows that highly structured MV techniques ('sparse banded BLAS' SBB [9]) can be performed much faster since we can exploit vectorization facilities and data locality. Additionally, we can further differ between the case of *variable matrix* entries ('xxx-V') and *constant bands* ('xxx-C') as typical for Poisson-like PDE's.

Summing up, restringing the grid geometry to a collection of generalized tensorproduct meshes, complete (local) multigrid solvers with very sophisticated smoothers can realized which can actually work with several hundreds of MFLOP/s on recent hardware platforms. Moreover, we can incorporate these 'local structures' into very complex FEM simulation tools on arbitrary domains such that modern numerical components can be applied, as for instance *adaptive meshing* and *a posteriori error control*, or generalized *multigrid/domain decomposition* solvers of SCARC type (see [9]). Since our introduced fictitious boundary methods together with the iterative filtering techniques allow the use of fixed, time-independent meshes - up to local adaptation via mesh deformation and alignement - the demonstrated high numerical efficiency of this approach can be linked with the very high computational efficiency of corresponding data structures, such that the combination of such numerical and algorithmic tools promises a significant performance gain for future simulation codes.

Computer	#Unknowns	CM	TL	STO	ILU-CM	ILU-TL	ILU-STO
DEC 21264 (667 MHz) 'ES40'	8,320	147	136	116	90	76	72
	33,280	125	105	100	86	73	63
	133,120	81	71	58	81	52	55
	532,480	60	51	21	40	35	22
	2,129,920	58	47	13	38	30	14
	8,519,680	58	45	10	36	30	11

Computer	#Unknowns	STO (CM)	SBB-V	SBB-C	MGTRI-V	MGTRI-C
DEC 21264	65^2	178 (205)	538	795	370	452
(667 MHz)	257^2	110 (224)	358	1010	314	487
'ES40'	1025^2	11 (78)	158	813	185	401

Table 5. MFLOP/s rates for sparse MV multiplication and ILU smoothing in FeatFlow for different numberings and grid levels; the second table shows corresponding performance results on a generalized tensorproduct grid for the Poisson problem, discretized by conforming bilinear FEM. We compare the standard sparse approach (STO/CM) with the 'sparse banded BLAS' techniques (SBB-V/C) and corresponding multigrid solvers MGTRI with very robust 'linewise Gauß-Seidel smoothers'.

4.5 'Iterative filtering' vs. 'Density-Viscosity blocking'

Finally, we shortly analyse the alternatively proposed fictitious boundary method, namely the addressing of internal objects via prescribing 'huge' values for viscosity, resp., density. It is obvious that this approach seems to be most straightforward since no modification of the source code is required: We only have to prescribe 'appropriate' parameters in the input files. However, there arise very severe difficulties:

1) Typically, the involved multigrid solvers have big problems if such extreme discontinuities arise, in the momentum equation for the velocity (viscosity and density) as well as in the corresponding Pressure-Poisson problem (see [8]) which includes the density. In fact, if the jumps, the means the size of the discontinuities, is larger than $10^4 - 10^6$, we observe severe convergence problems in our numerical simulations which are based on standard geometrical multigrid components.

2) Another severe problem arises since - due to the finite parameters for the viscosity and density - the true boundary condition $\mathbf{u} \cdot \mathbf{n} \neq 0$ between solid and fluid cannot be guaranteed. In practical applications, this defect results in penetration into the interior objects and corresponding (very small) flow velocities even inside of the 'solids'. However, as the following example in Fig. 14 shows, it may happen that the time scale due to such unphysical penetration effects may be much longer than the time scale for the actual flow simulation. Moreover, this time scale is related to the actual size of the viscosity, resp., density changes such that jumps of size $10^4 - 10^6$ and more

may lead to useful simulations. However, one has to keep in mind that larger discontinuities, which lead to a more physical behaviour, lead to more severe convergence problems for the involved multigrid solvers, too.

Moreover, one has to state that the evolution time for such penetration effects is finite as soon as the interior solids are described by huge, but finite values for density and viscosity. While for nonstationary configurations the simulations might lead to useful results (see Fig. 14) due to the differences of the time scales, we observe severe problems for stationary configurations: Since standard Poiseuille flow is a corresponding solution, independent of the values for ρ and ν (Remark: $\mathbf{u} \cdot \nabla \mathbf{u} = 0$, $\nu(x,y)\Delta \mathbf{u} = c\nu(x,y)$ such that $C\nu(x,y)x$ is the corresponding pressure, independent of ρ), the direct stationary approach results in this solution, that means pure channel flow.

3) While the results seem to be quite useful for the velocity and pressure values in interior (grid) points, it is not clear what happens in those vertices/midpoints of the mesh which are directly on the interface between liquid and 'solid' (= interior object). The examples in Fig. 15 show a comparison - on level 6 for CIRC1 and CHAN2 which lead to comparable grid resolution and numbers of elements - for the values in an interior point ('U2', 'P') as well as for the pressure difference ('ΔP') directly on the 'cylinder'. While the results for 'U2' and 'P' lead to very similar results, the pressure difference for the 'density-viscosity blocking' approach is completely wrong and depends massively on the values for density and viscosity. It might be possible to get better values by a better postprocessing, however this study will be subject of future research.

Altogether, it appears that the use of this 'density-viscosity blocking' method may lead to a good agreement of the results with the exact solution, however (up to now) only for interior grid points and the 'right' time scale. Nevertheless, this approach is quite dangerous since the underlying model will lead to wrong results - on the corresponding time scales - which however is very hard to predict by a priori analysis. So, although this approach seems to be very easy to implement and to integrate into existing codes, the problems w.r.t. controlling the effects and solving the resulting systems of equations seem to be very severe.

5 Conclusions

We do not claim that these 'fictitious boundary' methods together with 'iterative filtering' techniques are superior to the typical approach of resolving accurately small-scale structures by boundary parametrizations and corresponding grid adaption. But there are situations which might require the combination of both techniques. Our numerical experience can be concluded in the following "thumb rules":

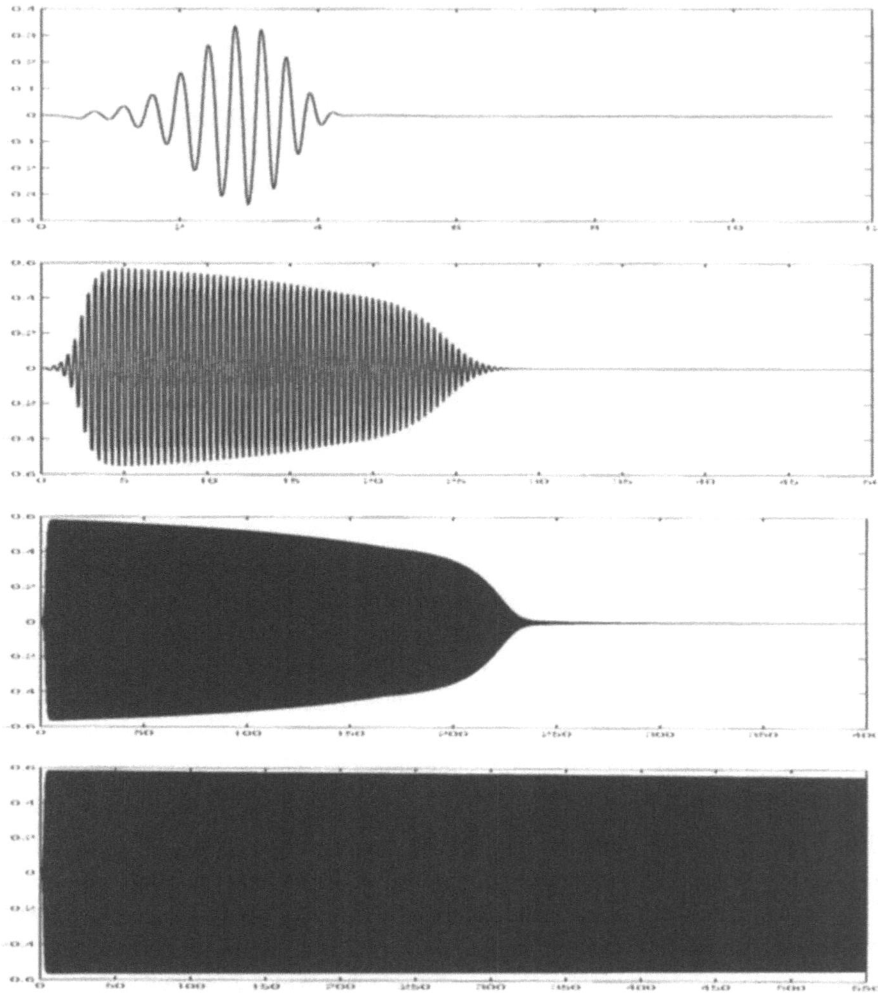

Fig. 14. Nonstationary behaviour of the velocity value 'U2' (mesh CHAN2 and grid level 4) for the density-viscosity jumps of order 10^3 (top), 10^4, 10^5 up to 10^6 (bottom)

If the geometry is not too complex, approximate the boundary by corresponding parametrizations (as piecewisely parametrized functions) and via adapted meshes (for instance, objects which are not too numerous and which have a piecewise smooth surface as circles, squares, etc.).

Try to approximate the boundary parts by a "rough" boundary parametrization which contains already most of the important structure. Additionally, apply the iterative filtering *techniques to resolve the fine structures depending on the granularity of the mesh. This approach may be useful if large-scale objects*

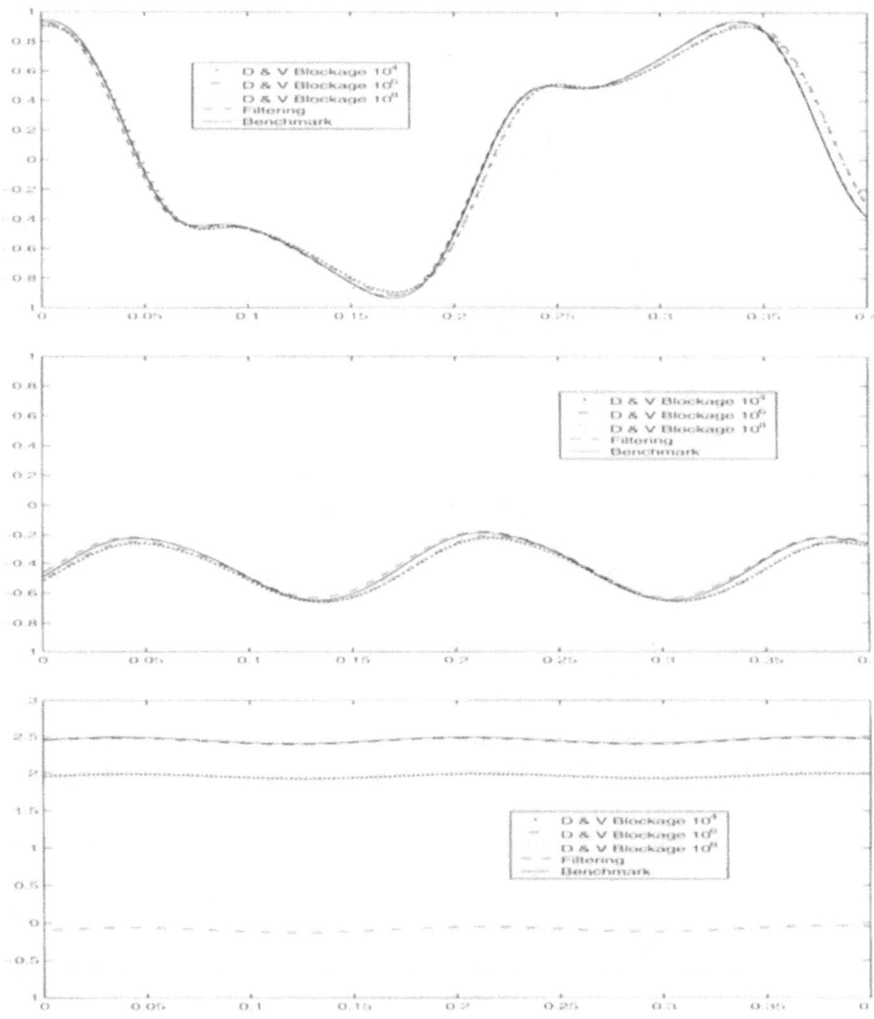

Fig. 15. Comparative study of one period for the velocity value 'U2' (top), for the pressure value 'P' (middle) and for the pressure difference 'ΔP' (bottom)

with complicated surfaces are approximated, for instance mountains, buildings or car shapes with special features.

If the geometrical details are very small and numerous, perform the fictitious boundary *techniques as an alternative. This approach is advantageous if many (> 100) boundary parts are required (see [12] for the example of a 3D heating device with about 1000 small holes at the outflow and 250 apertures in the interior) or if small-scale objects are involved (for instance, cars with antennae).*

In comparison to the other fictitious boundary condition method, based on the 'density-viscosity blockage', the advantages seem to be more clear. While such discontinuities in parameter choices always lead to trouble with approximation properties ($\mathbf{u} \cdot \mathbf{n} \neq 0$ on the interface) and robustness problems of the involved multigrid solvers - depending on the real size of these parameters - the proposed 'iterative filtering' procedure exhibits a much more robust behaviour, is easily implementable and is very flexible w.r.t. various kinds of boundary conditions. In particular for time-dependent boundary parts, for instance moving objects in the interior, this approach shows a very advantageous behaviour w.r.t. robustness, efficiency, flexibility and also accuracy if additionally ALE techniques are included.

Summing up, the combination of **local mesh adaption** and **iterative filtering techniques** together with **fictitious boundary conditions** may lead to the following advantageous numerical behaviour if these techniques are realized together with appropriate implementation strategies:

1. If these techniques are applied with respect to different levels of refinement, the coarse mesh is allowed to consist of few elements only. If the accuracy of approximating the complex boundary parts is simultanously increased with refining the mesh, the full multigrid efficiency can be obtained. Moreover, it is quite straightforward to perform the corresponding modifications in standard (CFD) codes. Especially for complex 3D geometries (look at `www.featflow.de/album`) and for 'moving obstacles', these techniques provide an elegant and 'quick' way to obtain qualitatively accurate results without paying too much for additional implementation or extended run-time behaviour.
2. The combination of semi–adapted meshes (near boundary parts) with these *fictitious boundary* techniques leads to quite accurate results, in particular due to improved approximation results for locally orthogonal meshes. However, it is clear that the accuracy near (curved) boundary parts is locally of first order only due to a piecewise constant approximation of the interface: Much better results - in a quantitative meaning - will be obtained if additionally local mesh deformation or mesh adaptation concepts are applied to guarantee (at least) a linear approximation of the boundaries. However, one has to take care that the remeshing can be performed in a very local way since time-dependent global remeshing can lead to disastrous CPU timings.
3. The resulting number of unknowns is not 'optimal' since components corresponding to some parts of the boundaries with prescribed values are explicitly involved before they are set again to the prescribed boundary values. However, the employed data structures can be locally regular

such that much higher performance in terms of MFLOP/s rates on fast computer platforms can be obtained. (See [9] for a discussion of such approaches which are the basis of our 'hardware-oriented techniques for PDEs'.) The combination of appropriate hierarchical data, solver and matrix structures which can combine the efficiency and accuracy of modern adaptivity concepts and fast multigrid-domain decomposition principles with the huge GFLOP/s rates of modern processors will be the optimal playground for such fictitious boundary and iterative filtering techniques in the future.

Up to now, we have applied these techniques in the case of **explicitely** prescribed boundary parts, resp., movement of solids only. The even more interesting case are free interfaces, multiphase flow, solidification and fluid-structure interaction when the deformation and shape of internal interfaces and boundaries is described implicitely. However, it is obvious that the presented **fictitious boundary methods** together with the **iterative filtering techniques** can be analogously applied. The main difference is that the set of 'active unknowns' for the filtering process has to be calculated in each time step, in most applications via corresponding transport problems involving the fluid velocity $\mathbf{u}$. However, as soon as the position of the boundary parts is given which also means the index numbers in the corresponding coefficient vector, the filtering process can be applied in the same way. Recently, we examine these extensions more carefully and we are implementing such techniques in the context of fluid-structure interaction and particularly solidification and free boundaries.

References

1. Becker, Chr., Kilian, S., Turek, S.: Consequences of modern hardware design for numerical simulations and their realization in FEAST. Proceedings "Euro-Par'99 Parallel Processing" (P. Amesto, P. Berger, M. Dayde, I. Duff, V. Fraysse, L. Giraud, D. Ruiz eds.). Toulouse, France, August/September (1999)
2. Fadlum, E.A., Verzicco, R., Orlandi, P. and Mohd-Yusof, J.: Combined immersed-boundary finite difference methods for three dimensional complex flow simulations. J. Comput. Phys. **161** (2000) 35
3. J. Heywood, R. Rannacher, S. Turek: Artificial boundaries and flux and pressure conditions for the incompressible Navier–Stokes equations. Int. J. Numer. Meth. Fluids **22** (1996)
4. Hyman, J.M., Li, S., Knupp, P., Shaskov, M.: An algorithm for aligning a quadrilateral grid with internal boundaries. J. Comput. Phys. **163** (2000)
5. Koopmann, G. H.: The vortex wakes of vibrating cylinders at low Reynolds numbers. J. Fluid Mech. **28** (1967) 501

6. Rannacher, R. and Turek, S.: A Simple nonconforming quadrilateral Stokes element. Numer. Methods for Partial Differential Equations. **8** (1992) 97
7. Schäfer, M., Turek, S.: *Benchmark computations of laminar flow around cylinder.* in E.H. Hirschel (editor) Flow Simulation with High-Performance Computers II. Volume 52 of Notes on Numerical Fluid Mechanics, Vieweg (1996)
8. Turek, S.: Effecient Solvers for Incompressible Flow Problems. Springer Verlag, Berlin-Heidelberg-New York (1999)
9. Turek, S., Becker, Chr., Kilian, S.: Hardware-oriented Numerics and concepts for PDE software. to appear in: Special Journal Issue for *PDE Software.* International Conference on Computational Science ICCS2002, Amsterdam (2002)
10. Turek, S.: Tools for simulating nonstationary incompressible flow via discretely divergence–free finite element models. Int. J. Numer. Meth. Fluids. **18** (1994) 71
11. Turek,S.: Multigrid techniques for a divergence-free finite element discretization. East-West J. Numer. Math. **2(3)** (1994) 3229
12. Turek, S., Kilian, S: The Virtual Album of Fluid Motion. Springer (Multimedia DVD) ISBN 3-540-14900-7 (2002) (`http://www.featflow.de/album`)
13. Turek, S. et al.: **FEATFLOW** . Finite element software for the incompressible Navier–Stokes equations: User Manual, Release 1.2 (1999)
14. Udaykumar, H.S., Mittal, R., Rampunggoon, P. and Khanna, A.: A sharp interface Cartesian grid method for simulating flows with complex moving boundaries. J. Comput. Phys. **174** (2001) 345
15. Udaykumar, H.S., Kan, H.C., Shyy, W. and Tran-Son-Tay, R.: Multiphase dynamics in arbitrary geometries on fixed Cartesian grids. J. Comput. Phys. **137** (1997) 366
16. Wesseling, P.: Introduction to Multigrid methods. Wiley, New York (1992).
17. Ye, T, Mittal, R., Udaykumar, H.S. and Shyy, W.: An accurate Cartesian grid for viscous incompressible flows with complex immersed boundaries. J. Comput. Phys **156** (1999) 209
18. SPARSKIT, `http://www.cs.umn.edu/Research/arpa/SPARSKIT/sparskit.html`

Numerical Simulation of Three Dimensional Free Surface Flows with Bubbles

Alexandre Caboussat[1], Vincent Maronnier[2], Marco Picasso[1], and Jacques Rappaz[1]

[1] Institut de Mathématiques, Ecole Polytechnique Fédérale de Lausanne, 1015 Lausanne, Switzerland.
[2] Calcom SA, Parc Scientifique, 1015 Lausanne, Switzerland.

Summary. A numerical model is presented for the simulation of free surface flows. The unknowns are the volume fraction of liquid, the velocity and pressure in the liquid and the pressure in the bubbles of gas which can appear in the liquid flow.

The volume fraction of liquid satisfies an advection equation, the pressure in each bubble is uniformly constant in space but depends on time and is computed using the ideal gas law, the velocity and pressure in the liquid satisfy the incompressible Navier-Stokes equations, the gas pressure being imposed as a normal force on the liquid-gas interface.

The numerical method is similar to the one described in [12, 13], advection and diffusion phenomena being solved on two different grids. Numerical results in the frame of mould filling show that the effect of pressure in the bubbles of gas surrounded by the liquid cannot be neglected.

1 Introduction

Industrial processes such as casting, injection or extrusion involve complex free surface phenomena that can nowadays be solved numerically using commercial codes. In three dimensional situations, the motion of the free surface is generally too complex to be handled by front-tracking [10] or Lagrangian methods [7, 8, 18]. The Arbitrary Lagrangian-Eulerian method is also difficult to implement since the selection of the mesh velocity is nontrivial for complex flows.

An alternative is to consider the Eulerian approach, which consists in using a fixed mesh but adding an unknown function φ which is, in our case, the characteristic function of the liquid domain. In the level set approach [1, 26] or in the pseudo-concentration method [3, 4, 11, 17, 28, 29], the function φ is smooth. The velocity and pressure in the liquid and the gas generally satisfy the incompressible Navier-Stokes equations. Therefore the compressiblity effect of gas is neglected.

In this paper, the function φ is the volume fraction of liquid (in fact the liquid characteristic function). It has value one in the liquid region, zero in the surrounding gas and jumps across the interface. The most famous numerical implementation of this model is the so-called Volume Of Fluid (VOF) method, which was originally devised for finite volumes [9] and recently extended to finite elements [16]. The major difficulty in solving this problem is then due to the fact that the volume fraction of liquid φ is discontinuous across the interface. Since the interface moves with the fluid, φ must satisfy (in some weak sense) the following advection equation

$$\frac{\partial \varphi}{\partial t} + \mathbf{v} \cdot \nabla \varphi = 0 \ ,$$

where $\mathbf{v}$ is the velocity field of the fluid governed by the incompressible Navier-Stokes equations. In our model, the velocity field $\mathbf{v}$ is computed only in the liquid region of the cavity. The influence of the gas surrounding the liquid is taken into account by computing its pressure, which can then be applied as an external normal force onto the liquid. The pressure in the gas is computed using law of ideal gases and is constant in space in each connected component of gas. Surface tension effects are neglected as they are not important in mould filling applications.

The numerical method is the one presented in [12, 13]. An implicit splitting algorithm is used to decouple advection and diffusion phenomena in Navier-Stokes equations. Then, these two phenomena are solved using two different grids. Advection phenomena (including the motion of the volume fraction of liquid and the prediction of the fluid velocity) are solved using a fixed, structured grid of cubic cells and a forward characteristic method. On the other hand, diffusion phenomena (the computation of the gas pressure and a generalized Stokes problem in the liquid) are solved using continuous, piecewise linear finite element techniques on a fixed, unstructured mesh of tetrahedrons.

Numerical results in two and three space dimensions have already been presented in [12, 13]. The pressure in the bubbles of gas that may appear in the flow was neglected. The goal of this paper is to present some results when taking into account the influence of the gas pressure on the fluid.

The structure of the paper is the following. In the next section, the governing equations are presented. In Sects. three and four, time and spatial discretizations are proposed. In Sect. five, numerical simulations are presented.

2 The Mathematical Model

2.1 Governing Equations in the Liquid

The model presented in this section is described in [12, 13] except the influence of the surrounding gas on the liquid. Let Λ be a cavity of $\mathbb{R}^3$ in which the fluid must be confined, and let $T > 0$ be the final time of the simulation. For any given time t, let $\Omega(t)$ denote the region occupied by the liquid. Finally, let Q_T be the space-time domain containing the liquid and let Σ_T be the space-time free surface between the liquid and the surrounding gas. The notations are reported in Fig. 1 in the frame of a two dimensional situation, namely the broken dam problem in a confined domain.

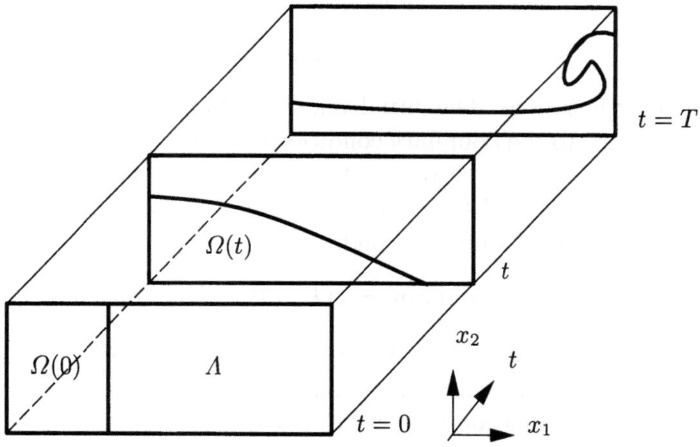

Fig. 1. Calculation domain for the broken dam problem in a confined domain. At initial time, the fluid is at rest on the left part of the cavity. It is then free to move and hits the boundary.

The velocity field $\mathbf{v} : Q_T \to \mathbb{R}^3$ and the pressure field $p : Q_T \to \mathbb{R}$ in the liquid are assumed to satisfy the time-dependent, incompressible Navier-Stokes equations in conservative form and in the presence of a gravity field $\mathbf{g}$, that is

$$\rho \frac{\partial \mathbf{v}}{\partial t} + \rho(\mathbf{v} \cdot \nabla)\mathbf{v} - 2\mu \operatorname{div} \mathbf{D}(\mathbf{v}) + \nabla p = \rho \mathbf{g} \qquad \text{in } Q_T \ , \tag{1}$$

$$\operatorname{div} \mathbf{v} = 0 \qquad \text{in } Q_T \ , \tag{2}$$

where $\mathbf{D}(\mathbf{v}) = \frac{1}{2}(\nabla \mathbf{v} + \nabla \mathbf{v}^T)$ is the rate of deformation tensor.

Let $\varphi : \Lambda \times [0, T] \to \mathbb{R}$ be the volume fraction of liquid. The function φ equals one if liquid is present and zero if it is not and so φ is the characteristic function of Q^T. Since the interface moves with the liquid, the function φ must satisfy (in a weak sense)

$$\frac{\partial \varphi}{\partial t} + \mathbf{v} \cdot \nabla \varphi = 0 \qquad \text{in } Q_T \ . \tag{3}$$

From a Lagrangian point of view, the function φ is constant along the trajectories of the fluid particles. More precisely, $\varphi(X(t), t) = \varphi(X(0), 0)$, where $X(t)$ is the trajectory of a fluid particle, thus $X'(t) = \mathbf{v}(X(t), t)$.

2.2 Initial and Boundary Conditions

The initial conditions are the following. At initial time, the volume fraction of liquid $\varphi(\cdot, 0)$ is given, which defines the liquid region,

$$\Omega(0) = \{x \in \Lambda; \varphi(x, 0) = 1\} \ ;$$

see Fig. 1 for notations. The initial velocity field $\mathbf{v}$ is then prescribed in $\Omega(0)$. Let us now turn to the boundary conditions for the velocity field. It is assumed that the only forces acting on the free surface are the normal forces due to the external pressure of the surrounding gas (we neglect capillary forces or surface tension); the boundary condition is,

$$-p\mathbf{n} + 2\mu \mathbf{D}(\mathbf{v})\mathbf{n} = -(P - P_{\text{atmo}})\mathbf{n} \qquad \text{on } \Sigma_T \ , \tag{4}$$

where $\mathbf{n}$ is the unit normal of the liquid-gas free surface oriented toward the gas, P the pressure in the gas and P_{atmo} the atmospheric pressure. For example, consider the situation of Fig. 2, namely the filling of a two dimensional S-shaped cavity (the numerical experiment is described in Sect. 5). When filling the cavity with liquid, the gas between the valve and the liquid can escape, thus $P = P_{\text{atmo}}$ on the upper part of the liquid-gas interface. However, a fraction of gas is trapped by the liquid and cannot escape, a resulting force acts on the lower part of the liquid-gas interface which implies that the bubble persists during experiment.

On the boundary of the liquid region being in contact with the walls (that is to say the boundary of Λ, see Fig. 1), two cases can be considered. The first case corresponds to noslip or inflow conditions, the three components of the velocity being imposed. In the second case, slip or zero force boundary conditions apply, according to the flow. More precisely, if the fluid pushes against the wall, then slip boundary conditions are imposed; that is, zero normal velocity and zero tangent stress. That means,

$$\begin{aligned} &\text{if} \quad (-p\mathbf{n} + 2\mu \mathbf{D}(\mathbf{v})\mathbf{n}) \cdot \mathbf{n} < 0 \quad \text{then} \\ &\quad \mathbf{v} \cdot \mathbf{n} = 0 \quad \text{and} \quad (-p\mathbf{n} + 2\mu \mathbf{D}(\mathbf{v})\mathbf{n}) \cdot \mathbf{t}_i = 0, \quad i = 1, 2 \ , \end{aligned} \tag{5}$$

where $\mathbf{t}_i$, $i = 1, 2$, are two unit vectors orthogonal to $\mathbf{n}$. On the other side, if the fluid does not push against the wall, then the force is zero. That means,

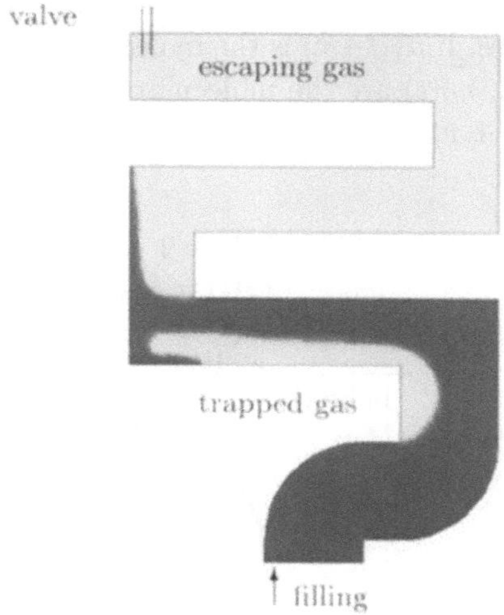

Fig. 2. Filling of an S-shaped cavity. The gas in the upper part of the cavity is free to escape from the valve. The gas trapped by the liquid may exert a force on the liquid.

$$\text{if} \quad (-p\mathbf{n} + 2\mu\mathbf{D}(\mathbf{v})\mathbf{n}) \cdot \mathbf{n} \geq 0 \quad \text{then} \quad -p\mathbf{n} + 2\mu\mathbf{D}(\mathbf{v})\mathbf{n} = 0 \ . \tag{6}$$

We have considered this kind of boundary condition in order to prevent the fluid sticking onto the walls.

2.3 The Model in the Surrounding Gas

Consider again the case of Fig. 2. Some gas is trapped by the fluid and is compressed. In our model, the velocity in the gas is disregarded (indeed, computing the velocity field in the gas would require solving the Euler compressible equations, this being too expensive). The pressure in the gas is computed using the *ideal gas law* at constant temperature :

$$PV = nRT \ , \tag{7}$$

where P is the pressure in the gas, V its volume, n is the molecular fraction, R is the gas constant and T is the (constant) temperature. Moreover the pressure inside each connected component of gas is constant in space. In the following we set $P_i(t)$ and $V_i(t)$ the pressure and volume of bubble i at time t and consider the situations illustrated in Fig. 3 and 4. In Fig. 3 the pressure of the trapped gas at time $t + \tau$ is computed from the relation

$$P_2(t+\tau)V_2(t+\tau) = P_2(t)V_2(t) \ .$$

The situation of Fig. 4 is more complex since two bubbles merge. The pressure at time $t + \tau$ is computed by taking into account the conservation of number of molecules in the bubbles. If $n_1(t+\tau)$ is the molecular fraction of the bubble 1 at time $t+\tau$ and $n_1(t)$, $n_2(t)$ are the molecular fractions of bubbles 1 and 2 at time t, we have $n_1(t+\tau) = n_1(t) + n_2(t)$.

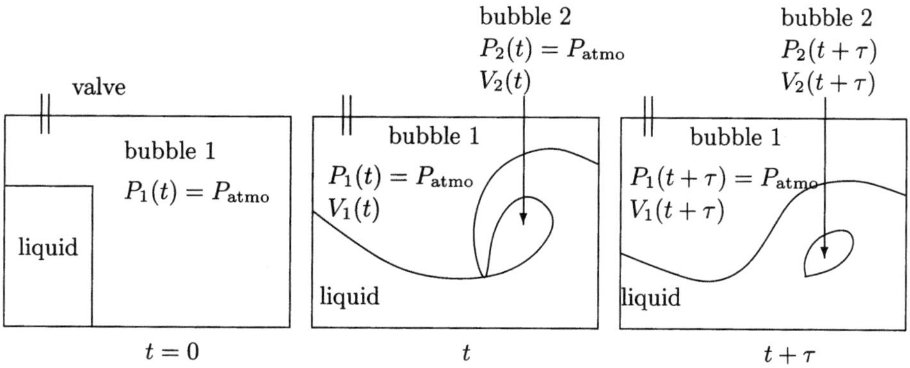

Fig. 3. At time t, air is trapped by the liquid and the pressure equals to atmospheric pressure. ($P_1(t) = P_{atmo}$, $P_2(t) = P_{atmo}$). At time $t+\tau$, the pressure in bubble 2 is computed from the relation $P_2(t+\tau)V_2(t+\tau) = P_2(t)V_2(t)$.

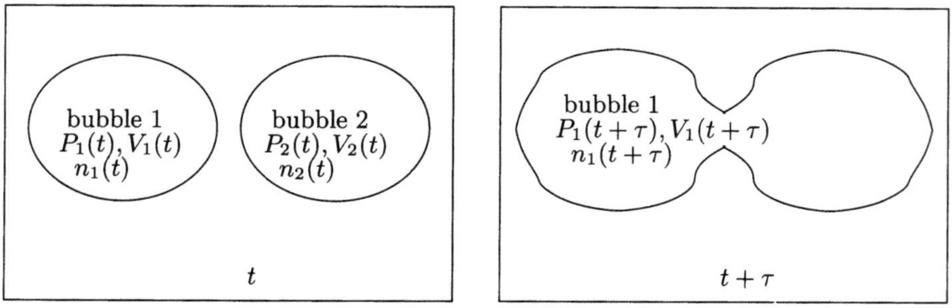

Fig. 4. Merging of two bubbles between time t and time $t+\tau$. The pressure in bubble 1 at time $t+\tau$ is computed from the relation $P_1(t+\tau)V_1(t+\tau) = P_1(t)V_1(t) + P_2(t)V_2(t)$.

3 Time Discretization: an Implicit Splitting Algorithm

As described in [12, 13], a splitting algorithm is used to solve problem (1)-(3), allowing advection and diffusion phenomena to be decoupled.

Let $0 = t^0 < t^1 < t^2 < \ldots < t^N = T$ be a subdivision of the time interval $[0, T]$, define $\tau^n = t^n - t^{n-1}$ the n-th time step, $n = 1, 2, \ldots, N$, τ the largest time step. Given an integer n, assume that approximations φ^{n-1}, $\mathbf{v}^{n-1}$, Ω^{n-1}, P^{n-1} of φ, $\mathbf{v}$, Ω, P at time t^{n-1} respectively, are available. Then φ^n, $\mathbf{v}^n$, Ω^n, P^n are computed by means of a splitting algorithm as illustrated in Fig. 5. Firstly, two advection problems are solved, leading to a prediction of the new velocity $\mathbf{v}^{n-1/2}$ together with the new volume fraction of liquid φ^n, which allows the new computational domain Ω^n to be defined. At this step, the new volumes of gas in bubbles are known and it is possible to compute the new pressure P^n of this gas. Finally, a generalized Stokes problem is solved on Ω^n with boundary conditions (4) (5) (6) and the velocity $\mathbf{v}^n$ and pressure p^n are obtained.

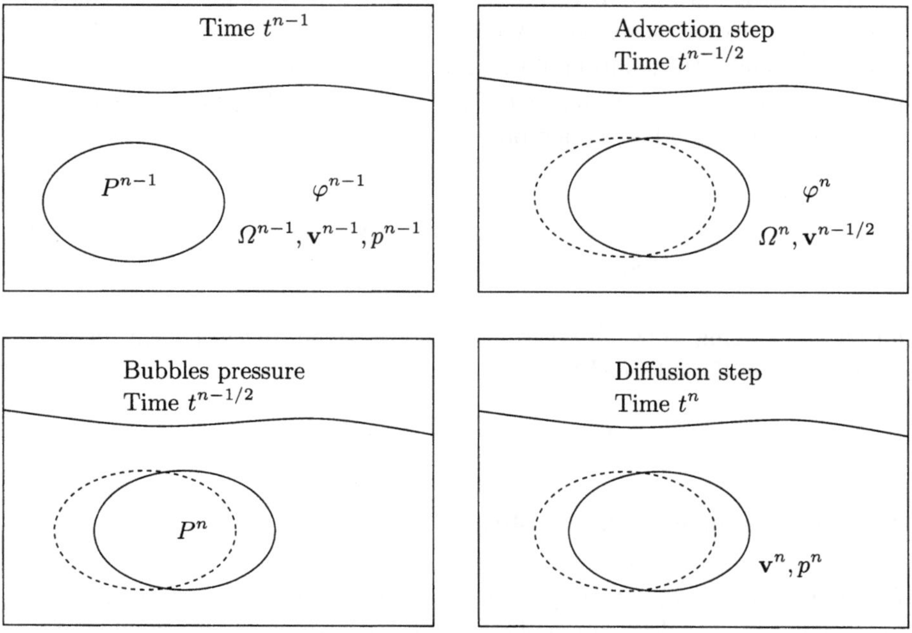

Fig. 5. The splitting algorithm. Two advection problems are solved to determine the new volume fraction of liquid φ^n, the new domain Ω^n and the predicted velocity $\mathbf{v}^{n-1/2}$. Then a constant pressure P^n is computed in each connected component of gas. Finally, a Stokes problem is solved to obtain the velocity $\mathbf{v}^n$ and the pressure p^n in the fluid.

3.1 Advection Step

Solve between time t^{n-1} and t^n the two advection problems

$$\frac{\partial \mathbf{w}}{\partial t} + (\mathbf{w} \cdot \nabla)\mathbf{w} = 0 \ , \tag{8}$$

$$\frac{\partial \psi}{\partial t} + \mathbf{w} \cdot \nabla \psi = 0 \ , \tag{9}$$

with initial conditions

$$\mathbf{w}(t^{n-1}) = \mathbf{v}^{n-1} \ ,$$
$$\psi(t^{n-1}) = \varphi^{n-1} \ .$$

If the effect of the boundary of the cavity Λ is not considered, these two problems can be solved exactly, using the method of Characteristics [21, 22, 23, 30, 32], the trajectories of the velocity field $\mathbf{w}$ being straight lines. Indeed, the trajectories are given by $X'(t) = \mathbf{w}(X(t), t)$, but since $\mathbf{w}$ is constant along the trajectories, we have $X'(t) = \mathbf{w}(X(t^{n-1}), t^{n-1}) = \mathbf{v}^{n-1}(X(t^{n-1}))$. Let $\mathbf{v}^{n-\frac{1}{2}}$ denote the solution of the first advection problem at time t^n ($\mathbf{v}^{n-\frac{1}{2}}$ is a prediction of the velocity field), $\mathbf{v}^{n-\frac{1}{2}}(x) = \mathbf{w}(x, t^n)$. Let φ^n denote the solution of the second advection problem at time t^n, i.e. $\varphi^n(x) = \psi(x, t^n)$. We thus have

$$\mathbf{v}^{n-\frac{1}{2}}(x + \tau^n \mathbf{v}^{n-1}(x)) = \mathbf{v}^{n-1}(x) \ , \tag{10}$$

$$\varphi^n(x + \tau^n \mathbf{v}^{n-1}(x)) = \varphi^{n-1}(x) \ , \tag{11}$$

for all x belonging to Ω^{n-1}. Once φ^n is known in the cavity Λ, then the liquid region at time t^n is defined by :

$$\Omega^n = \{y \in \Lambda; \varphi^n(y) = 1\} \ .$$

3.2 The Pressure in the Bubbles

An approximation P_i^n of the pressure in bubble i at time t^n, $P_i(t^n)$, is computed following the description of Sect. 2.3. Each connected component of gas is obtained from a *coloration algorithm* described hereafter (see Fig. 6).

Let $k(t)$ be the number of connected components of gas at time t, we set $B_i(t)$ the i-th connected component (i.e. bubble number i). Let ξ be a function defined on $\Lambda \times (0, T)$ negative in $\Omega(t)$ and equal to i in bubble $B_i(t)$. At each time step we compute B_i^n, ξ^n and k^n, the approximations of $B_i(t^n)$, $\xi(t^n)$ and $k(t^n)$, as follows.

The function ξ^n is initialized with color 0 on $\Lambda \backslash \Omega^n$, $\xi^n = -1$ on Ω^n, and $k^n = 0$. The gas domain is determined by $\Xi^n = \{x \in \Lambda \backslash \Omega^n\}$ or $\Xi^n = \{x \in \Lambda : \xi^n(x) = 0\}$. The goal is to assign to each point x in the gas an integer value $\xi^n(x) \neq 0$, the so-called *bubble color*. The algorithm is illustrated in Fig. 6 and is the following :

Repeat $\Big\{$

1. Choose a point P in the interior of Ξ^n
2. Solve the following problem

$$\begin{cases} -\Delta u = \delta_P, & \text{in } \Xi^n \ , \\ u = 0, & \text{on } \partial\Xi^n \ , \end{cases} \tag{12}$$

where δ_P is Dirac delta function at point P;

3. Increase the number of colors at time t^n, $k^n := k^n + 1$;
4. Define the new connected component $B^n_{k^n} = \{x; u(x) > 0\}$;
5. Set the new color $\xi^n(x) = k^n$, $\forall x \in B^n_{k^n}$;
6. Define the domain Ξ^n for the next iteration $\Xi^n = \{x \in \Lambda \backslash \Omega^n : \xi^n(x) = 0\}$.

$\Big\}$ until $\Xi^n = \emptyset$.

The bubble coloration algorithm thus requires to solve as many Poisson problems as the number of connected gas components. However, this is much less CPU time consuming than solving a Stokes problem in the liquid. In our numerical experiments, the CPU time required to compute the bubbles pressure was always less than 10% of the total CPU time.

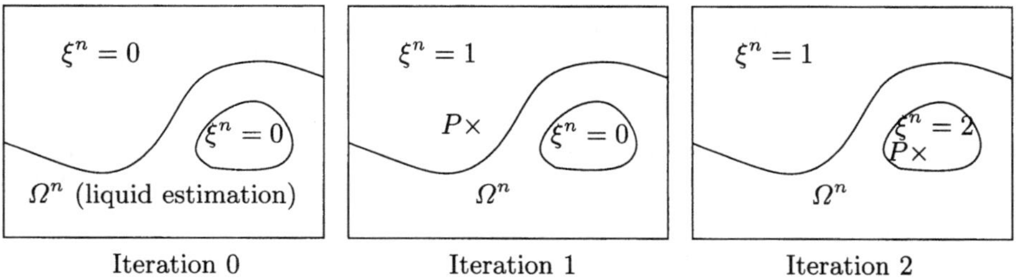

Fig. 6. Coloration algorithm of the bubbles. Initially the function ξ^n equals zero everywhere in the gas. At each iteration of the algorithm a point P is chosen in the domain where $\xi^n = 0$. Then problem (12) is solved and a connected component is found. The algorithm stops when all points in the gas have a color $\xi^n \neq 0$.

3.3 Diffusion Step

Once the connected components of gas are obtained, a constant pressure P^n is computed in each connected component. Then the following generalized Stokes problem is solved :

$$\rho\frac{\mathbf{v}^n - \mathbf{v}^{n-\frac{1}{2}}}{\tau^n} - 2\mu \operatorname{div} \mathbf{D}(\mathbf{v}^n) + \nabla p^n = \rho\mathbf{g} \qquad \text{in } \Omega^n \ , \tag{13}$$

$$\operatorname{div} \mathbf{v}^n = 0 \qquad \text{in } \Omega^n \ , \tag{14}$$

with the boundary conditions (4) (5) (6) described in Sect. 2.2.

4 Space Discretization and Implementation

Advection and diffusion phenomena being now decoupled, two distinct grids can be used. Since the shape of the cavity Λ can be very complex (this is the case in industrial mould filling situations), finite element techniques are well suited for solving (13) (14) on an unstructured mesh. Furthermore, the pressure in the bubbles is also computed on the same unstructured mesh. On the other hand, a structured grid of smaller cubic cells is used to implement (10) (11) in order to have an accurate approximation of volume fraction of liquid φ.

4.1 Advection Step

Assume that the grid is made out of cubic cells of size h, each cell being labeled by indices (ijk). Let φ_{ijk}^{n-1} and $\mathbf{v}_{ijk}^{n-1}$ be the approximate value of φ and $\mathbf{v}$ at the center of cell number (ijk) at time t^{n-1}. According to (10) and (11), the advection step on cell number (ijk) consists in advecting φ_{ijk}^{n-1} and $\mathbf{v}_{ijk}^{n-1}$ by $\tau^n \mathbf{v}_{ijk}^{n-1}$ and then projecting the values on the structured grid. An example of cell advection and projection is presented in Fig. 7 in two space dimensions.

This method can be interpreted as a forward characteristics method, therefore is unconditionally stable with respect to the Courant-Friedrichs-Lewy (CFL) condition, and $O(\tau + h^2/\tau)$ convergent according to the theoretical results available for the Characteristics-Galerkin method [21, 22, 23]. Since the volume fraction of liquid φ is a step function, this method, as any other method, will smooth the discontinuity. Indeed, numerical diffusion is introduced when the values of the advected cells are projected on the grid. Moreover, if the time step is too large, two cells may arrive at the same place, producing numerical (artificial) compression (in fact (8) develops a shock).

In order to enhance the quality of the volume fraction of liquid, two post-processing procedures have been implemented. We refer to [12, 13, 14] for a detailed description in two and three space dimensions. The first procedure reduces numerical diffusion and is a simplified implementation of the SLIC (Simple Linear Interface Calculation) algorithm [2, 19]. The second procedure aims to remove artificial compression, which may result when the time step is too large.

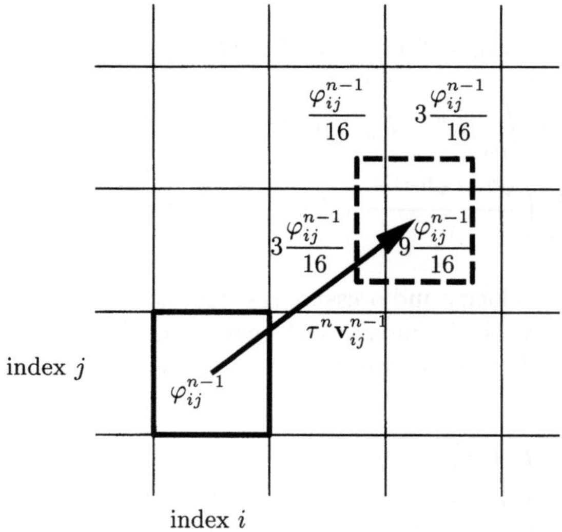

Fig. 7. An example of two dimensional advection of φ_{ij}^{n-1} by $\tau^n \mathbf{v}_{ij}^{n-1}$, and projection on the grid. The advected cell is represented by the dashed lines. The four cells containing the advected cell receive a fraction of φ_{ij}^{n-1}, according to the position of the advected cell.

Once values φ_{ijk}^n and $\mathbf{v}_{ijk}^n$ have been computed on the cells, values of the fraction of liquid φ_P^n and of the velocity field $\mathbf{v}_P^{n-\frac{1}{2}}$ are computed at the nodes P of the finite element mesh. When these values are available at the vertices of the finite element mesh, the liquid region is defined as follows. An element of the mesh is said to be liquid if (at least) one of its vertices P has a value $\varphi_P^n > 0.5$. The computational domain Ω^n used for solving (13) (14) is then defined to be the union of all liquid elements.

4.2 Bubbles

Once the computational liquid domain Ω^n is defined, the gas domain is given by $\Lambda \backslash \Omega^n$. The connected components of gas are recognized using the algorithm explained in Sect. 3.2 on the finite element mesh. The pressure P^n is piecewise constant on each connected component and is computed on the gas domain using the method of Sect. 2.3.

4.3 Diffusion Step

Let us now turn to the finite element techniques used for solving (13) (14), the boundary conditions (4) (8) (9) being those described in Sect. 2.2. The velocity and pressure are continuous, piecewise linear and a stabilized weak formulation is used [5, 6, 27], namely for all test functions $\mathbf{w}$ and q :

$$\int_{\Omega^n} \frac{\mathbf{v}^n - \mathbf{v}^{n-1/2}}{\tau^n} \mathbf{w} dx + 2\mu \int_{\Omega^n} \mathbf{D}(\mathbf{v}^n) : \mathbf{D}(\mathbf{w}) dx - \int_{\Omega^n} p^n \text{ div } \mathbf{w} dx$$
$$-\rho \mathbf{g} \int_{\Omega^n} \mathbf{w} dx + \int_{\Gamma^n} (P^n - P_{\text{atmo}}) \mathbf{n} \mathbf{w} dS - \int_{\Omega^n} \text{div } \mathbf{u}^n q dx$$
$$- \sum_{K \subset \Omega^n} \alpha_K \int_K \left(\frac{\mathbf{v}^n - \mathbf{v}^{n-1/2}}{\tau^n} + \nabla p^n - \rho \mathbf{g} \right) \cdot \nabla q dx = 0 \ .$$

Here $\mathbf{w}$ and q are the velocity and pressure test functions, compatible with the boundary conditions on the boundary of the cavity Λ. The stability coefficient α_K is defined on each tetrahedron K as a function of the local Reynolds number as follows :

$$\alpha_K = \begin{cases} \dfrac{1}{12} \dfrac{h_K^2}{\mu} & \text{if } \dfrac{\rho |\mathbf{v}^{n-1/2}|_\infty h_K}{2\mu} \leq 3 \ , \\ \dfrac{1}{4Re_K} \dfrac{h_K^2}{\mu} & \text{if } \dfrac{\rho |\mathbf{v}^{n-1/2}|_\infty h_K}{2\mu} > 3 \ . \end{cases}$$

Although the above variational formulation corresponds to solving a generalized Stokes problem, numerical experiments reported in [14] have shown that the occurence of Reynolds number in the definition of the stabilization parameter was necessary. A theoretical justification in a simplified framework (for instance Navier-Stokes equations solved with the Characteristics-Galerkin Least Squares method in a given domain) would be useful but is still missing.

The degrees of freedom are the three velocity components and pressure at each vertex of the finite element mesh. At the moment, all the degrees of freedom are stored in a big matrix. The linear system is solved using a BICGSTAB algorithm and a classical incomplete LU preconditioner. Therefore, the method is memory consuming and decoupling/splitting algorithms such as SIMPLE (see for instance [31] for a recent improvement), Shur complement (pressure-matrix method), projection methods (pressure Poisson solvers) or inexact LU decompositions should be investigated. However, the two following constraints should be kept in mind before selecting one of these algorithms : i) the decoupling scheme must remain implicit in order to avoid CFL or Fourier-like stability conditions between the time step and the space step, ii) since natural conditions apply on the free surface, the Stokes problem has to be written in conservative form - $2\mu \text{div } \mathbf{D}(\mathbf{v})$ instead of $\mu \Delta \mathbf{v}$ - so that the velocity components remain coupled when using splitting algorithms. Some of these decoupling algorithms have been studied in [20].

Once the new velocity field $\mathbf{v}^n$ is computed at the vertices of the finite element mesh, values are interpolated at the center of the cells (ijk). The CPU time spent to interpolate the velocity and volume fraction of liquid between the two grids is negligible compared to the CPU time required to solve Stokes problem. From the implementation point of view, the use of two grids requires an extra data structure. For each cell, the tetraedron number containing the cell is required. Moreover, for each node, the number of the nearest cell has to be stored. The corresponding data structure is set once for all at the beginning of the computation. The size of the cubic cells should be as small as possible, in order to avoid numerical diffusion when advecting the volume fraction of liquid, whereas the size of the finite element mesh can be larger. Numerical experiments reported in [12, 13, 14] have shown that choosing the size of the cubic cells three to five times smaller than the size of the finite element mesh is a good compromise between precision and memory requirements.

5 Numerical Results

Two and three dimensional results have been proposed in [12] and [13], the influence of gas not being considered. In the sequel, an experiment demonstrating the influence of gas is proposed, namely the S-shape channel. All the computations were performed on a PC with Intel Pentium III 1 Ghz CPU and 512 Mb Memory. The results were post-processed with the CalcoSoft^{TM} software.

An S-shape channel lying between two horizontal plates is filled. The experiment is reported in [25]. A valve is located at the top of the channel, as in Fig. 2, allowing gas to escape. The dimensions of the channel are 0.17×0.24 m. Water is injected with velocity 8.7 m/s. In three dimensions, the thickness of the geometry is 0.008 m. The 2D finite element mesh has 3483 vertices and 6418 triangles, while the cells grid contains 49896 cells. The 3D finite element mesh is built using five layers of the 2D mesh (see Fig. 8) and contains 20898 nodes and 96270 elements, while the 3D grid contains $1'544'000$ cells. The simulation required 760 time steps of size $\tau = 0.00007$ s to reach the final time 0.00532 s.

The CPU time for the simulations in two space dimensions is approximatively 14 minutes without the bubbles computing and 15 minutes with the bubbles computing. In three space dimensions, these CPU times become 319 minutes without taking into account the gas effect and 344 minutes with the bubbles computing. Most of the CPU time is spent to solve Stokes problem, bubbles computations requiring less than 10% of the total CPU time.

In Fig. 9 the experiment is compared to 2D and 3D computations when the influence of the surrounding gas is not taken into account. Indeed, during the computation, the pressure in the gas remains equal to P_{atmo}. Therefore, the bubbles of gas disappear faster in the computation than in the experiment (see for instance the second row of Fig. 9).

The influence of gas is taken into account in Fig. 10. Clearly, the bubbles of gas remain longer in the computations, thus showing that the influence of gas cannot be neglected.

Fig. 8. The 3D mesh used for the computations of the S-shape channel.

6 Conclusion

A two- and three-dimensional free surface flow solver has been presented. The volume fraction of liquid is used to describe the interface. The equations of incompressible flows are solved only in the fluid and the influence of gas is taken into account. A splitting algorithm allows diffusion and convection phenomena to be decoupled. Diffusion is solved with finite elements using an unstructured mesh of tetrahedrons, whereas convection is solved on a grid made out of small cubic cells. Numerical results show the efficiency of this approach, particularly in mould filling situations. Moreover, numerical results show that the pressure in the bubbles of gas has to be taken into account.

We are now investigating the introduction of the surface tension effects in the model. Since the volume fraction of liquid is a step function, the VOF method is not very well suited (compared to level set techniques for instance) to compute the curvature of the free surface. Nevertheless, a number of techniques are available and are constantly being improved [24]. Preliminary results have been obtained but are not included in this paper for the following two reasons: i) due to poor convergence rate, the mesh size has to be very small which prevents accurate computations in 3D, ii) our main interest is to reproduce injection casting processes and the surface tension effects can be neglected for these processes, as the Reynolds number is usually much larger than the Capillary number.

References

1. Y. C. Chang, T. Y. Hou, B. Merriman, S. Osher : A level set formulation of Eulerian interface capturing methods for incompressible fluid flows. J. Comput. Phys. **124 (2)** (1996) 449–464
2. A. J. Chorin : Flame Advection and Propagation Algorithms. J. Comput. Phys. **35** (1980) 1–11
3. R. Codina, U. Schäfer, E. Oñate : Mould Filling Simulation using Finite Elements. Int. J. Numer. Meth. Heat Fluid Flow. **4** (1994) 291–310
4. G. Dhatt, D. M. Gao, A. Ben Cheikh : A Finite Element Simulation of Metal Flow in Moulds. Int. J. Numer. Meth. Eng. **30** (1990) 821–831
5. L. P. Franca, S. L. Frey : Stabilized Finite Element Methods : II. The Incompressible Navier-Stokes Equations. Comp. Meth. Appl. Mech. Eng. **99** (1992) 209–233
6. L. P. Franca and T. J. R. Hughes : Convergence Analyses of Galerkin Least Squares Methods for Symmetric Advective-Diffusive Forms of the Stokes and Incompressible Navier-Stokes Equations. Comp. Meth. Appl. Mech. Eng. **105** (1993) 285–298
7. P. Hansbo : The Characteristic Streamline Diffusion Method for the Time-Dependent Incompressible Navier-Stokes Equations. Comp. Meth. Appl. Mech. Eng. **99** (1992) 171–186
8. P. Hansbo : Lagrangian Incompressible Flow Computations in Three Dimensions by Use of Space-Time Finite Elements. Int. J. Numer. Meth. Fluids. **20** (1995) 989–1001
9. C. W. Hirt, B. D. Nichols : Volume of Fluid (VOF) Method for the Dynamics of Free Boundaries. J. Comput. Phys. **39** (1981) 201–225
10. R. J. LeVeque, K-M. Shyue : Two Dimensional Front Tracking Based on High Resolution Wave Propagation Methods. J. Comput. Phys. **123** (1996) 354–368
11. N. Lock, M. Jaeger, M. Medale, R. Occelli : Local mesh adaptation technique for front tracking problems. Int. J. Numer. Meth. Fluids. **28** (1998) 719–736
12. V. Maronnier, M. Picasso, J. Rappaz : Numerical simulation of free surface flows. J. Comput. Phys. **155 (2)** (1999) 439–455
13. V. Maronnier, M. Picasso, J. Rappaz : Numerical Simulation of Three Dimensional Free Surface Flows. accepted in Int. J. Numer. Meth. Fluids.
14. V. Maronnier : Simulation numérique d'écoulements de fluides incompressibles avec surface libre. PHD Thesis **2248**, Département de Mathématiques, Ecole Polytechnique Fédérale de Lausanne. (2000)
15. J. C. Martin, W. J. Moyce : An Experimental Study of the Collapse of Liquid Collumns on a Rigid Horizontal Plate. Philos. Trans. Roy. Soc. London Ser. **A 244** (1952) 312–324
16. F. Mashayek, N. Ashgriz : A Hybrid Finite-Element-Volume-Of-Fluid Method for Simulating Free Surface Flows and Interfaces. Int. J. Numer. Meth. Fluids. **20** (1995) 1367–1380

17. M. Medale, M. Jaeger : Numerical simulations of incompressible flows with moving interfaces. Int. J. Numer. Meth. Fluids **24** (1997) 615-638
18. F. Muttin, T. Coupez, M. Bellet, J. L. Chenot : Lagrangian Finite-Element Analysis of Time-Dependent Viscous Free-Surface Flow using an Automatic Remeshing Technique: Application to Metal Casting Flow. Int. J. Numer. Meth. Eng. **36** (1993) 2001–2015
19. W. F. Noh, P. Woodward : SLIC (Simple Line Interface Calculation). Springer-Verlag, Lectures Notes in Physics. **59** (1976) 330–340
20. M. Picasso and J. Rappaz : Stability of time-splitting schemes for the Stokes problem with stabilized finite elements. Numerical Methods for Partial Differential Equations, **17-6** (2001) 632-656
21. O. Pironneau : Finite Element Methods for Fluids, Wiley, Chichester (1989)
22. O. Pironneau, J. Liou, T. Tezduyar : Characteristic-Galerkin and Galerkin/Least-Squares Space-Time Formulations for the Advection-Diffusion Equation with Time-Dependent Domains. Comp. Meth. Appl. Mech. Eng. **100** (1992) 117–141
23. A. Quarteroni, A. Valli : Numerical Approximation of Partial Differential Equations. Springer Series in Computational Mathematics. **23** (1991)
24. Renardy, Yuriko and Renardy, Michael : PROST: a parabolic reconstruction of surface tension for the volume-of-fluid method. J. Comput. Phys. **183 (2)** (2002) 400–421
25. M. Schmid, F. Klein : Einflußder Wandreibung auf das Füllverhalten Dünner Platten. Preprint, Steinbeis Transferzentrum, Fachhochschule Aachen. (1996)
26. M. Sussman, A.S. Almgren, J.B. Bell, P. Colella, L.H. Howell, M. L. Welcome : An adaptive level set approach for incompressible two-phase flows . J. Comput. Phys. **148 (1)** (1999) 81–124
27. T. E. Tezduyar, S. Mittal, S. E. Ray, R. Shih : Incompressible Flow Computations with Stabilized Bilinear and Linear Equal-Order-Interpolation Velocity-Pressure Elements. Comp. Meth. Appl. Mech. Eng. **95** (1992) 221–242
28. E. Thompson : Use of Pseudo-Concentrations to Follow Creeping Viscous Flows during Transient Analysis. Int. J. Numer. Meth. Fluids. **6** (1986) 749–761
29. S. O. Unverdi, G. Tryggvason : Computations of multi-fluid flows Physica D. **60** (1992) 70-83
30. H. Wang, H.K. Dahle, R.E. Ewing, M.S. Espedal, R.C. Sharpley, S. Man : An ELLAM scheme for advection-diffusion equations in two dimensions. SIAM J. Sci. Comput. **20 (6)** (1999) 2160–2194 (electronic)
31. C. Vuik, A. Saghir, G.P. Boerstoel : The Krylov accelerated SIMPLE(R) method for flow problems in industrial furnaces. Int. J. Num. Meth. Fluids. **33** (2000) 1027–1040
32. T. Yabe, T. Ishikawa, P.Y. Wang, I. Aoki, Y. Kadota, F. Ikeda : A universal solver for hyperbolic equations by cubic-polynomial interpolation. II. Two- and three-dimensional solvers. Comput. Phys. Comm. **66 (2-3)** (1991) 233–242

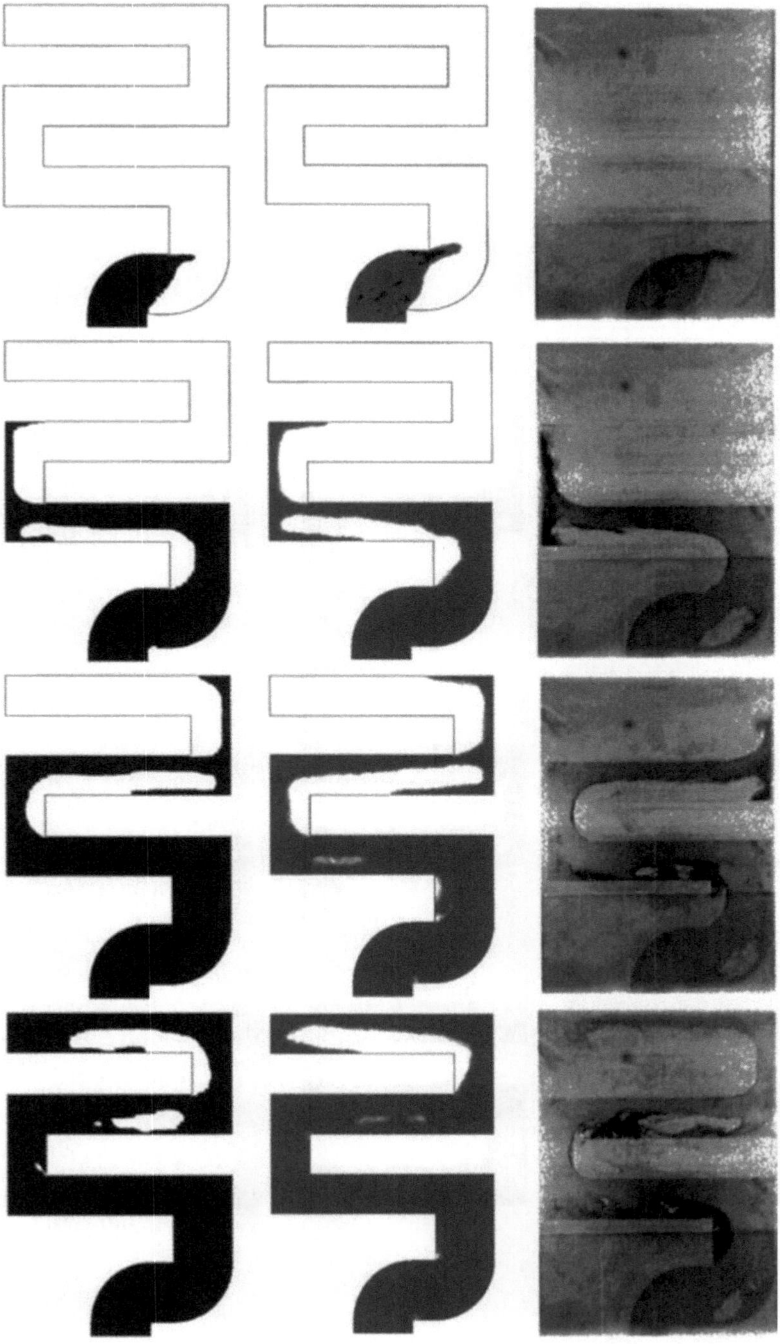

Fig. 9. S-shaped channel without influence of gas : left : 2D results, middle : 3D results in a cut, right : experimental results [25]. First row : time equals 7.15 ms, second row : 25.3 ms, third row : 39.3 ms and fourth row : 53.6 ms.

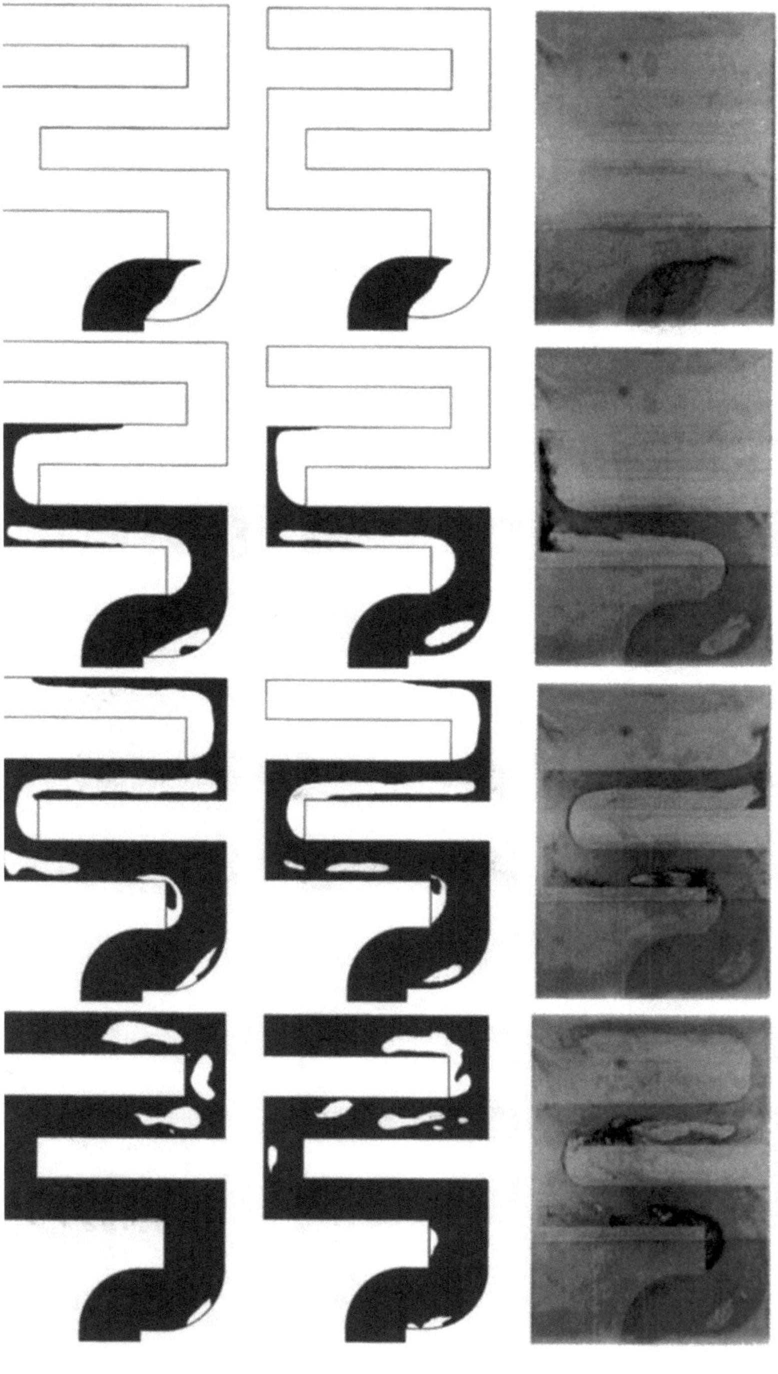

Fig. 10. S-shaped channel with influence of gas : left : 2D results, middle : 3D results in a cut, right : experimental results [25]. First row : time equals 7.15 ms, second row : 25.3 ms, third row : 39.3 ms and fourth row : 53.6 ms.

A Computational Comparison of Two FEM Solvers for Nonlinear Incompressible Flow

Jaroslav Hron, Abderrahim Ouazzi, and Stefan Turek

Institute of Applied Mathematics, University of Dortmund,
Vogelpothsweg 87, 44227 Dortmund, Germany

Summary. In this comparative study, we examine the influence of two different FEM discretization techniques (conforming Q_2/P_1, nonconforming $\tilde{Q}_1/Q_0$ Stokes element) and solution procedures (nonlinear Newton variants and multigrid vs. Krylov-space solvers for the linear subproblems) onto the approximation properties and particularly the total efficiency of corresponding CFD simulation tools. We discuss algorithmic details and give numerical results for laminar incompressible flow examples including non-Newtonian behavior of *power law* type.

1 Introduction

The finite element pair Q_2/P_1, being potentially of 3rd order accuracy due to the biquadratic polynomials for the velocity (and the linear pressure approximation), is one of the most popular discretization techniques in the CFD community. In most applications, it is employed in a fully coupled approach which applies an outer Newton-like iteration for the discretized Navier-Stokes equations while the resulting linear subproblems of Oseen-type (as a non-symmetric generalization of Stokes problems) are treated with (single-grid) Krylov-space methods: Due to the indefinite and particularly non-symmetric character of the equations, preconditioned BiCGSTAB [17] and GMRES [10] variants are the most popular solvers for the corresponding linear systems.

As an alternative, see for instance the realization in the FEATFLOW software package [15], special nonconforming FEM elements like the rotated multilinear/piecewise constant $\tilde{Q}_1/Q_0$ Stokes element have been developed which provide an optimal approximation error of one order less. However, for such finite elements, special hierarchical multigrid techniques have been successfully realized.

AS a consequence, such 'linear' (low order) finite elements seem to require much more degrees of freedom to satisfy a prescribed accuracy than compared with the quadratic Q_2/P_1 ansatz, while they show at the same time a much superior behavior w.r.t. the efficiency of the involved solvers. Therefore, the aim of this paper is to analyze the numerical characteristics of both popular approaches and to give an absolute comparison for non-trivial flow configurations in 2D which are based on the (nonlinear) incompressible Navier-Stokes equations:

'What is the total efficiency in terms of CPU timing, if we take into account the different numbers of unknowns and iteration steps which are required for a comparatively accurate simulation?'

The paper is organized as follow: In section §2 we present the problem formulations, while in §3 the two Stokes elements, the conforming Q_2/P_1 element pair and the stabilized version of the rotated bilinear $\tilde{Q}_1/Q_0$ Stokes element for problems involving deformation tensor formulation, are introduced. Two Newton linearization techniques, based on a continuous approach which first calculates the continuous Jacobi matrix and then performs the FEM discretization, in contrast to a discrete variant based on 'divided differences' for the discretized nonlinear stiffness matrices, are developed in §4. A brief description of the employed linear solvers, namely preconditioned GMRES and a local 'multilevel pressure Schur complement' smoother of so-called Vanka-type inside of a direct multigrid approach are presented in §5. Finally, section §6 is devoted to a discussion of algorithmic details and to show numerical results for different flow configurations which all are related to the standard 'Flow around cylinder' CFD benchmark [12].

2 Problem Formulation

In the underlying Navier-Stokes problem, the Cauchy stress tensor is given by $\boldsymbol{\sigma} = 2\nu(D_{\mathrm{II}}(\boldsymbol{u}))\mathbf{D}(\boldsymbol{u}) - p\mathbf{I}$, where p is the pressure; $\mathbf{D}(\boldsymbol{u}) = \frac{1}{2}(\nabla\boldsymbol{u} + \nabla^T\boldsymbol{u})$ is the rate of deformation tensor, $\boldsymbol{u}$ being the velocity; $\nu(\cdot)$ is the viscosity which can depend on the second invariant of the rate deformation tensor $D_{\mathrm{II}}(\boldsymbol{u}) = \frac{1}{2}\,\mathrm{tr}(\mathbf{D}^2(\boldsymbol{u}))$. Then, depending on the chosen viscosity function, the following prototypical models are used in certain applications (with appropriate parameters $\nu_0, \nu_\infty, \lambda, \alpha$):

- 'power law' defined for $\nu(z) = \nu_0 z^{\frac{\alpha}{2}}$
- 'Carreau law' defined for $\nu(z) = \nu_\infty + (\nu_0 - \nu_\infty)(1 + \lambda z)^{\frac{\alpha}{2}}$
- 'Bingham law' defined for $\nu(z) = z^{-\frac{1}{2}}$

In the following, we will mainly concentrate on the 'power law' model with varying parameter α. Moreover, we recently performed similar tests for models with a pressure-dependent viscosity [7, 5], as for instance the 'Schaeffer law' [11] with $\nu(z) = pz^{-\frac{1}{2}}$ which is a special model for granular flow.

Altogether, the velocity $\boldsymbol{u}$ and the pressure p satisfy the following generalized incompressible Navier-Stokes equations:

$$\frac{\partial \boldsymbol{u}}{\partial t} + \boldsymbol{u} \cdot \nabla \boldsymbol{u} - \operatorname{div}\left(2\nu(D_{\mathrm{II}}(\boldsymbol{u}))\mathbf{D}(\boldsymbol{u})\right) + \nabla p = \boldsymbol{f}, \quad \operatorname{div} \boldsymbol{u} = 0 \tag{1}$$

Let us consider a stationary flow governed by the generalized Navier-Stokes problem (1) in a bounded domain $\Omega \subset \mathbb{R}^2$. If we restrict the set V of test functions to be divergence-free and if we take the constitutive laws into account, the above equations (1) lead to:

$$\int_\Omega 2\nu(D_{\mathrm{II}}(\boldsymbol{u}))\mathbf{D}(\boldsymbol{u}) : \mathbf{D}(\boldsymbol{v})\, dx + \int_\Omega (\boldsymbol{u} \cdot \nabla \boldsymbol{u})\boldsymbol{v}\, dx = \int_\Omega \boldsymbol{f}\boldsymbol{v}\, dx, \ \forall \boldsymbol{v} \in V \tag{2}$$

Some existence and uniqueness statements for (2) are given in [1], [8]. It is straightforward to penalize the constraint $\operatorname{div} \boldsymbol{v} = 0$ to derive the equivalent mixed formulations of (2):

Find $(\boldsymbol{u}, p) \in X \times M$ such that

$$\begin{aligned} \int_\Omega 2\nu(D_{\mathrm{II}}(\boldsymbol{u}))\mathbf{D}(\boldsymbol{u}) : \mathbf{D}(\boldsymbol{v})\, dx + \int_\Omega (\boldsymbol{u} \cdot \nabla \boldsymbol{u})\boldsymbol{v}\, dx \quad + \int_\Omega p \operatorname{div} \boldsymbol{v}\, dx & \\ = \int_\Omega \boldsymbol{f}\boldsymbol{v}\, dx, \qquad & \forall \boldsymbol{v} \in X, \\ \int_\Omega q \operatorname{div} \boldsymbol{u}\, dx = 0, \qquad & \forall q \in M, \end{aligned} \tag{3}$$

with the spaces $X = \mathbf{H}_0^1(\Omega)$ and $M = L_0^2(\Omega)$. For the following considerations, we introduce the bilinear forms:

$$\langle \mathbf{A}(\boldsymbol{w})\boldsymbol{u}, \boldsymbol{v} \rangle = \int_\Omega 2\nu(D_{\mathrm{II}}(\boldsymbol{w}))\mathbf{D}(\boldsymbol{u}) : \mathbf{D}(\boldsymbol{v})\, dx \tag{4}$$

$$\langle \mathbf{N}(\boldsymbol{w})\boldsymbol{u}, \boldsymbol{v} \rangle = \int_\Omega (\boldsymbol{w} \cdot \nabla \boldsymbol{u})\boldsymbol{v}\, dx \tag{5}$$

$$\langle \mathbf{B}q, \boldsymbol{v} \rangle = \int_\Omega q \operatorname{div} \boldsymbol{v}\, dx \tag{6}$$

Then, we can rewrite our generalized flow problems in the following more compact form:

Find $(\boldsymbol{u}, p) \in X \times M$ such that

$$\begin{aligned} \langle \mathbf{A}(\boldsymbol{u})\boldsymbol{u}, \boldsymbol{v} \rangle + \langle \mathbf{N}(\boldsymbol{u})\boldsymbol{u}, \boldsymbol{v} \rangle + \langle \mathbf{B}p, \boldsymbol{v} \rangle = \int_\Omega \boldsymbol{f}\boldsymbol{v}\, dx, \quad & \forall \boldsymbol{v} \in X, \\ \langle \mathbf{B}q, \boldsymbol{u} \rangle = 0, \quad & \forall q \in M. \end{aligned} \tag{7}$$

3 Discretization Aspects

3.1 The conforming Stokes Element Q_2/P_1

We discretize the continuous problem (3) by the standard Galerkin finite element method, hereby approximating the Ω by a domain Ω_h with piecewise linear boundary which is equipped with a quadrilateral mesh $\mathcal{T}_h$. On this mesh we define the finite dimensional spaces V_h and P_h for the velocity and the pressure approximations as

$$V_h = \{\boldsymbol{v}_h \in \mathbf{H}_0^1(\Omega_h)^2, \quad \boldsymbol{v}_{h|T} \in Q_2(T)^2 \;\; \forall T \in \mathcal{T}_h, \quad \boldsymbol{v}_h = \mathbf{0} \text{ on } \partial\Omega_h\}, \quad (8)$$

$$P_h = \{p_h \in L^2(\Omega_h), \quad p_{h|T} \in P_1(T) \;\; \forall T \in \mathcal{T}_h\}. \quad (9)$$

By $Q_2(T)$ we denote the standard biquadratic space on the quadrilateral T which, when transformed by the bilinear transformation to the reference quadrilateral $T_{\text{ref}} = (-1,1)^2$, is defined by

$$Q_2(T_{\text{ref}}) = \text{span}\{1, \xi, \eta, \xi\eta, \xi^2, \eta^2, \xi^2\eta, \xi\eta^2, \xi^2\eta^2\}, \quad (10)$$

with the 9 local degrees of freedom located at the vertices, midpoints of the edges and in the center of the quadrilateral. The space $P_1(T)$ consists of linear functions defined on the reference element by

$$P_1(T_{\text{ref}}) = \text{span}\{1, \xi, \eta\}, \quad (11)$$

with the function value and both partial derivatives in the center of the quadrilateral as its 3 local degrees of freedom. Recently, we modified the ansatz via using a non-parametric version for $P_1(T)$ working without transformation to the reference element.

For smooth solutions, the approximation error for the velocity in the L^2-norm is of order $O(h^3)$ which can be easily demonstrated for prescribed polynomials or for smooth data on appropriate domains. However, the actual approximation properties for local quantities as drag and lift coefficient, for instance on the surface of objects in the flow field, is not so clear. And, it is by far not obvious what to do in the case of moderate or even high Reynolds numbers: Due to the dominance of the convective term, certain stabilization techniques like FEM upwinding or streamline-diffusion are non-trivial to construct, and the resulting approximation properties, particularly for realistic flow configurations, is not clear. Therefore, we will follow the 'standard approach' for this Stokes element, namely performing pure central differencing - which is the shown Galerkin discretization without any additional stabilization. Since the discussed flow configurations are in the low Reynolds number regime, this approach is justified. However, the reader should keep in mind that the extension of our results to higher Reynolds numbers is not clear at all.

3.2 The nonconforming Stokes Element $\tilde{Q}_1/Q_0$

Again, we consider a subdivision $T \in \mathcal{T}_h$ consisting of quadrilaterals in the domain $\Omega_h \in \mathbb{R}^2$, and we employ the rotated bilinear *Rannacher-Turek* element [9]. For any quadrilateral T, let (ξ, η) denote a local coordinate system obtained by joining the midpoints of the opposing faces of T. Then, in the *non-parametric* case, we set on each element T

$$\tilde{Q}_1(T) := \text{span}\left\{1, \xi, \eta, \xi^2 - \eta^2\right\}. \tag{12}$$

The degrees of freedom are determined by the following nodal functionals $\{F_\Gamma^{(a,b)}(\cdot), \Gamma \subset \partial\mathcal{T}_h\}$, with

$$F_\Gamma^a := |\Gamma|^{-1} \int_\Gamma v d\gamma \quad \text{or} \quad F_\Gamma^b := v(m_\Gamma) \quad (m_\Gamma \text{ midpoint of edge } \Gamma) \tag{13}$$

such that the finite element spaces can be written as

$$\begin{aligned} W_h^{a,b} := \{&v \in L^2(\Omega_h), v \in \tilde{Q}_1(T), \forall T \in \mathcal{T}_h, v \text{ continuous w.r.t. all} \\ &\text{nodal functionals } F_{\Gamma_{i,j}}^{a,b}(\cdot), \text{ and } F_{\Gamma_{i0}}^{a,b}(v) = 0, \forall \Gamma_{i0}\}. \end{aligned} \tag{14}$$

Here, $\Gamma_{i,j}$ means all inner edges sharing the two elements i and j, while Γ_{i0} denotes the boundary edges of $\partial\Omega_h$. In this paper, we always employ version 'a' with the integral mean values as degrees of freedom. Then, the corresponding discrete functions will be approximated in the spaces

$$V_h := W_h^a \times W_h^a\,, \; L_h := \left\{q_h \in L_0^2(\Omega), q_{h|T} = \text{const.}\,, \forall T \in \mathcal{T}_h\right\}. \tag{15}$$

Due to the nonconformity of the discrete velocities, the classical discrete Korn's inequality is not satisfied which is important for problems involving the symmetric part of the gradient. Therefore, appropriate edge-oriented stabilization techniques (see [3], [4], [7]) have to be included which directly treat the jump across the inter-elementary boundaries via adding the following bilinear form

$$\sum_{edgesE} \frac{1}{|E|} \int_E [\varphi_i][\varphi_j] d\sigma \tag{16}$$

for all basis functions φ_i and φ_j of $W_h^{a,b}$. Taking into account an additional relaxation parameter $s = s(\nu)$, the corresponding stiffness matrices are defined via the edge-oriented stabilization term:

$$\langle \mathbf{S} \boldsymbol{u}, \boldsymbol{v} \rangle = s \sum_{E \in E_I \cup E_D} \frac{1}{|E|} \int_E [\boldsymbol{u}][\boldsymbol{v}] d\sigma \tag{17}$$

Here, the jump of a function $\boldsymbol{u}$ on an edge E is given by

$$[\boldsymbol{u}] = \begin{cases} \boldsymbol{u}^+ \cdot \boldsymbol{n}^+ + \boldsymbol{u}^- \cdot \boldsymbol{n}^- & \text{on internal edges } E_I \,, \\ \boldsymbol{u} \cdot \boldsymbol{n} & \text{on Dirichlet boundary edges } E_D, \\ 0 & \text{on Neumann boundary edges } E_N, \end{cases} \tag{18}$$

where $\boldsymbol{n}$ is the outward normal to the edge and $(\cdot)^+$ and $(\cdot)^-$ indicate the value of the generic quantity $(\cdot)$ on the two elements sharing the same edge. For a detailed description of the analysis for this stabilization technique, we refer to [3] and [4], while a description of related numerical and algorithmic approaches can be found in [7].

In contrast to the described conforming Stokes Element Q_2/P_1, the approximation error for the velocity in the L^2-norm is of order $O(h^2)$ for the 'linear' nonconforming element pair $\tilde{Q}_1/Q_0$. However, the realization and analysis of appropriate stabilization techniques for the case of higher Reynolds numbers is possible and is reported in several papers (see for instance [6] and [14]). Moreover, these special upwinding and streamline-diffusion techniques have been realized in the FEATFLOW software which has been successfully applied to a wide range of realistic flow configurations (see [16]). On the other hand, such stabilization techniques are - in contrast to a pure Galerkin discretization resp. to central differencing for the convective terms - very closely related to special additional parameters which have to be chosen by the user in a more or less clever way: We will shortly discuss the influence of such 'free' parameters onto the global approximation behavior and hence the total efficiency.

4 Nonlinear Solvers

4.1 Discrete Newton solver

After applying the discretization method to the system (1) where the approximations belong to the finite dimensional spaces (8) and (9), we obtain a system of nonlinear algebraic equations which can be represented as

$$\mathcal{R}(\boldsymbol{x}) = \mathbf{0}, \tag{19}$$

where $\boldsymbol{x}$ represents the vector of the coefficients corresponding to the unknowns $(\boldsymbol{v}_h, p_h)$. To solve this system we apply a Newton iteration with damping which results in iterations of the form

$$\boldsymbol{x}^{n+1} = \boldsymbol{x}^n + \omega^n \left[\frac{\partial \mathcal{R}(\boldsymbol{x}^n)}{\partial \boldsymbol{x}}\right]^{-1} \mathcal{R}(\boldsymbol{x}^n). \tag{20}$$

This iteration is repeated until a prescribed convergence criterion is satisfied, i.e., a certain norm of the residual term $\|\mathcal{R}(\boldsymbol{x}^n)\|$ is small enough. The damping parameter $\omega^n \in [-1, 0)$ is chosen such that

$$\mathcal{R}(\boldsymbol{x}^{n+1}) \cdot \boldsymbol{x}^{n+1} \leq \mathcal{R}(\boldsymbol{x}^n) \cdot \boldsymbol{x}^n. \tag{21}$$

This damping significantly improves the robustness of the Newton iteration in the case when the current approximation $\boldsymbol{x}^n$ is not close enough to the final solution and the Newton method without damping is not guaranteed to converge.

The Jacobian matrix $\left[\frac{\partial \mathcal{R}(\boldsymbol{x}^n)}{\partial \boldsymbol{x}}\right]$ is not computed exactly in our code, instead its approximation is computed using divided differences as

$$\left[\frac{\partial \mathcal{R}(\boldsymbol{x}^n)}{\partial \boldsymbol{x}}\right]_{ij} \approx \frac{\mathcal{R}_i(\boldsymbol{x}^n + \varepsilon \boldsymbol{e}_j) - \mathcal{R}_i(\boldsymbol{x}^n - \varepsilon \boldsymbol{e}_j)}{2\varepsilon}, \tag{22}$$

where $\boldsymbol{e}_j$ is the vector with unit j-th component and zero otherwise. The parameter ε can be fixed or can be modified according to some norm of the solution $\|\boldsymbol{x}^n\|$ or the norm of the update in the previous step, i.e., $\|\boldsymbol{x}^n - \boldsymbol{x}^{n-1}\|$. In the subsequent numerical tests, we will compare this discrete approach with the following continuous method, not only w.r.t. the nonlinear convergence behavior, but also checking the sensitivity of the overall nonlinear behavior in relation to this 'free' parameter ε.

4.2 Continuous Newton solver

Diffusive term

In this approach, the nonlinearity is handled on the continuous variational level. Let $\boldsymbol{u}^l$ being the initial state, the (continuous) Newton method consists of finding $\boldsymbol{u} \in V$ such that

$$\begin{aligned}
&\int_\Omega 2\nu(D_{\mathrm{II}}(\boldsymbol{u}^l))\mathbf{D}(\boldsymbol{u}) : \mathbf{D}(\boldsymbol{v})dx \\
&\qquad + \int_\Omega 2\nu'(D_{\mathrm{II}}(\boldsymbol{u}^l))[\mathbf{D}(\boldsymbol{u}^l) : \mathbf{D}(\boldsymbol{u})][\mathbf{D}(\boldsymbol{u}^l) : \mathbf{D}(\boldsymbol{v})]dx \\
&\qquad = \int_\Omega \boldsymbol{f}\boldsymbol{v}dx - \int_\Omega 2\nu(D_{\mathrm{II}}(\boldsymbol{u}^l))\mathbf{D}(\boldsymbol{u}^l) : \mathbf{D}(\boldsymbol{v})dx, \quad \forall \boldsymbol{v} \in V\,, \quad (23)
\end{aligned}$$

where $\nu'(\cdot)$ is the derivative of $\nu(\cdot)$. To see this, set $\boldsymbol{y} = \mathbf{D}(\boldsymbol{u}^l), \boldsymbol{z} = \mathbf{D}(\boldsymbol{u}), F(\boldsymbol{y}) = \nu(\frac{1}{2}|\boldsymbol{y}|^2)\boldsymbol{y}$ and $f(t) = F(\boldsymbol{y} + t\boldsymbol{z})$, so that

$$\partial_j F_i(\boldsymbol{y}) = \nu'(\frac{1}{2}|\boldsymbol{y}|^2)y_j y_i + \nu(\frac{1}{2}|\boldsymbol{y}|^2)\delta_{ij}\,, \quad (24)$$

where δ_{ij} stands for the standard Kronecker symbol. Having

$$\begin{aligned}
f_i'(t) &= \textstyle\sum_j \partial_j F_i(\boldsymbol{y} + t\boldsymbol{z})z_j \\
&= \nu'(\tfrac{1}{2}|\boldsymbol{y} + t\boldsymbol{z}|^2)\langle \boldsymbol{y}, \boldsymbol{z}\rangle y_i + \nu(\tfrac{1}{2}|\boldsymbol{y} + t\boldsymbol{z}|^2)z_i\,,
\end{aligned} \quad (25)$$

we decrease t towards zero, such that we obtain the Frechet derivative:

$$\nabla \cdot [2\nu(D_{\mathrm{II}}(\boldsymbol{u}^l))\mathbf{D}(\boldsymbol{u}) + 2\nu'(D_{\mathrm{II}}(\boldsymbol{u}^l))(\mathbf{D}(\boldsymbol{u}^l) : \mathbf{D}(\boldsymbol{u}))\mathbf{D}(\boldsymbol{u}^l)] \quad (26)$$

Finally, the weak formulation of (26) leads to the relation given in (23). Therefore, the algorithm consists of finding $(\boldsymbol{u}, p)$ as solution of the nonlinear discrete system

$$\begin{cases} \mathbf{A}(\boldsymbol{u}^l)\boldsymbol{u} + \delta_d \mathbf{A}^*(\boldsymbol{u}^l)\boldsymbol{u} + \mathbf{B}p = R_u(\boldsymbol{u}^l, p^l), \\ \mathbf{B}^T\boldsymbol{u} \qquad\qquad\qquad\qquad\quad = R_p(\boldsymbol{u}^l, p^l), \end{cases} \quad (27)$$

where $R_u(\cdot,\cdot)$ and $R_p(\cdot,\cdot)$ denote the corresponding nonlinear residual terms for the momentum and continuity equations, and the matrix $\mathbf{A}^*(\boldsymbol{u}^l)$ is related to the bilinear form

$$\langle \mathbf{A}^*(\boldsymbol{u}^l)\boldsymbol{u}, \boldsymbol{v}\rangle = \int_\Omega 2\nu'(D_{\mathrm{II}}(\boldsymbol{u}^l))[\mathbf{D}(\boldsymbol{u}^l) : \mathbf{D}(\boldsymbol{u})][\mathbf{D}(\boldsymbol{u}^l) : \mathbf{D}(\boldsymbol{v})]dx. \quad (28)$$

Convective term

The corresponding Newton linearization applied to the convective term $\langle \mathbf{N}(\boldsymbol{u})\boldsymbol{u}, \boldsymbol{v}\rangle$ leads to the additional bilinear form

$$\langle \mathbf{N}(\boldsymbol{u}^l)\boldsymbol{u}, \boldsymbol{v}\rangle + \delta_c \langle \mathbf{N}^*(\boldsymbol{u}^l)\boldsymbol{u}, \boldsymbol{v}\rangle, \; \forall \boldsymbol{v} \in X\,, \tag{29}$$

with $\langle \mathbf{N}^*(\boldsymbol{u}^l)\boldsymbol{u}, \boldsymbol{v}\rangle = \int_\Omega (\boldsymbol{u} \cdot \nabla \boldsymbol{u}^l)\boldsymbol{v}dx$ for all $\boldsymbol{v} \in X$. Then, the complete algorithm consists of finding $(\boldsymbol{u}, p)$ as solution of the following linear matrix-vector system

$$\begin{cases} \mathbf{A}(\boldsymbol{u}^l)\boldsymbol{u} + \delta_d \mathbf{A}^*(\boldsymbol{u}^l)\boldsymbol{u} + \mathbf{N}(\boldsymbol{u}^l)\boldsymbol{u} + \delta_c \mathbf{N}^*(\boldsymbol{u}^l)\boldsymbol{u} + \mathbf{B}p = R_u(\boldsymbol{u}^l, p^l), \\ \mathbf{B}^T \boldsymbol{u} \qquad\qquad\qquad\qquad\qquad\qquad\qquad\qquad = R_p(\boldsymbol{u}^l, p^l). \end{cases} \tag{30}$$

Once the auxiliary solution $(\boldsymbol{u}, p)$ is determined, the current approximation is updated, with ω^l as appropriate relaxation parameter:

$$(\boldsymbol{u}^{l+1}, p^{l+1}) = (\boldsymbol{u}^l, p^l) + \omega^l(\boldsymbol{u}, p) \tag{31}$$

The full Newton method is supplied for $\delta_d = 1$ and $\delta_c = 1$, while the fixpoint method corresponds to $\delta_d = 0$ and $\delta_c = 0$.

It is well-known that the 'standard' discretization of the convective part, i.e. with central discretization, may lead to numerical problems, namely to:

- stiffness matrices not of positive type ('M-matrix properties'), so the iterative schemes such as multigrid will not lead to efficient convergence results.
- oscillations and deteriorations of the discrete solutions which have purely numerical character.

To stabilize the convective term in the FEM context, there are two approaches which are widely used in the CFD community: *upwind* schemes and *streamline-diffusion* techniques. However, these two stabilization techniques may veil the (potentially) quadratic convergence of the Newton methods, unless if we compute the Jacobian matrix for the stabilization terms, too. For streamline-diffusion, the additional stabilization of the Newton linearization may be supplied in the same manner as it was done previously for the nonlinear viscosity, while upwind stabilization techniques require special treatment of $\langle \mathbf{N}^*(\boldsymbol{u}^l)\boldsymbol{u}, \boldsymbol{v}\rangle$. We will show the resulting nonlinear and linear convergence behavior with Newton and fixpoint methods for both upwind and streamline-diffusion stabilization techniques.

5 Linear Solvers

This section is devoted to give a brief description of the involved solution techniques for the resulting linear systems. For the nonconforming Stokes element $\tilde{Q}_1/Q_0$, a 'local pressure Schur complement' preconditioner (see [13] and [14]) as generalization of so-called 'Vanka smoothers' (see [18]) is constructed on patches Ω_i which are ensembles of one or several mesh cells, and this local preconditioner is embedded as global smoother into an outer block Jacobi/Gauss-Seidel iteration which acts directly on the coupled systems of Stokes, resp. of Oseen type. If we denote by $\tilde{R}_u$ and $\tilde{R}_p$ the discrete residuals for the complete momentum and continuity equation which include the complete stabilization term due to the bilinear form $\mathbf{S}$ as defined in (17), one smoothing step in defect-correction notation can be described as

$$\begin{bmatrix} u^l \\ p^l \end{bmatrix} = \begin{bmatrix} u^{l-1} \\ p^{l-1} \end{bmatrix} + \omega^l \sum_i \begin{pmatrix} \mathbf{F} + \mathbf{S}^*_{|\Omega_i} & \tilde{\mathbf{B}}_{|\Omega_i} \\ \tilde{\mathbf{B}}^T_{|\Omega_i} & 0 \end{pmatrix}^{-1} \begin{bmatrix} \tilde{R}_u(u^{l-1}, p^{l-1}) \\ \tilde{R}_p(u^{l-1}, p^{l-1}) \end{bmatrix} \tag{32}$$

with matrix $\mathbf{F} = \tilde{\mathbf{A}} + \delta_d \tilde{\mathbf{A}}^* + \tilde{\mathbf{N}} + \delta_c \tilde{\mathbf{N}}^*$. For the preconditioning step, only a part of the full matrix, i.e. $\mathbf{F} + \mathbf{S}^*$ instead of $\mathbf{F} + \mathbf{S}$, is involved (see [7] for the details). All other components in the multigrid approach, that means intergrid transfer, coarse grid correction and coarse grid solver, are quite standard and are based on the underlying hierarchical mesh hierarchy and the properties of the nonconforming finite elements [14].

For the biquadratic conforming Stokes Element Q_2/P_1, a standard GMRES method [10] as typical iterative single grid solver with ILU preconditioning is applied. The preconditioner is taken from the library developed in [2]. It is clear that multigrid might be preferable, too, but up to now, corresponding Vanka-like smoothers for this Q_2/P_1 ansatz are not yet available; in fact, they are the target of intensive research studies in our projects. Anyway, this GMRES solver is probably the most common approach in the CFD community, besides using BiCGSTAB [17], such that it is quite important to perform these comparisons with the multigrid solver.

6 Numerical Comparisons

In this section, we present the results from our numerical experiments with various parameters settings and also different physical models for the both proposed discretization and solution frameworks. The computational domain is always a channel geometry with an interior circle as shown via the coarsest mesh in the following figure; this configuration is following the standard CFD benchmark 'Flow around cylinder' [12]. The reader should know the the position of the circle is slightly non-symmetric w.r.t. the height such that the lift coefficient is non-homogeneous.

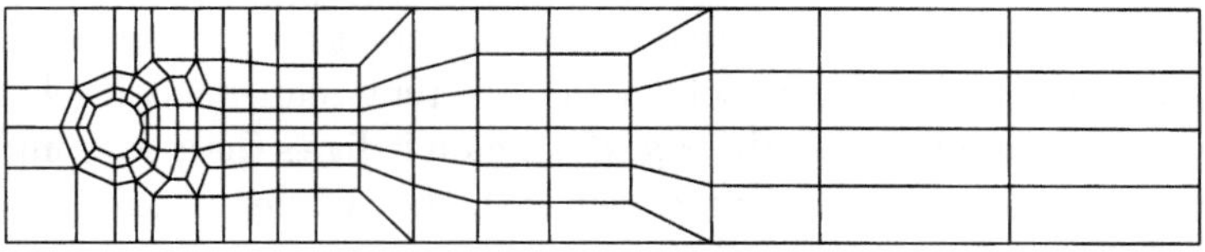

Mesh information			$\tilde{Q}_1/Q_0$	Q_2/P_1
Level	Elements	Vertices	Total unknowns	Total unknowns
1	156	130	702	1533
2	572	520	2686	5927
3	2184	2080	10608	23295
4	8528	8320	42016	92351
5	33696	33280	167232	367743
6	133952	133120	667264	–

Table 1. Coarse mesh and geometrical details for the 'Flow around cylinder' configuration for the both Stokes elements $\tilde{Q}_1/Q_0$ and Q_2/P_1 on the different levels of mesh refinement.

6.1 Constant viscosity test case

The first computational tests are performed for the Newtonian Navier-Stokes equations, i.e. with constant viscosity, so the only source of nonlinearity is the convective term. To be in the limit of steady flows, the Reynolds number - by variations of the viscosity parameter only - is kept to be less than $Re = 50$ which is near the critical point to turn into (periodically) oscillating flow behavior. Although the cases $Re = 20$ (the standard test in [12]) and also $Re = 50$ appear to be far away from transport-dominated flows, nevertheless the involved convective term requires special treatment: While the simulations show that the approximation results for such 'medium' Reynolds numbers seem to be the best for no stabilization at all, that means applying central differencing or the pure Galerkin formulation, the involved multigrid routines are much more sensitive. Therefore, in the case of the rotated bilinear nonconforming Stokes element $\tilde{Q}_1/Q_0$, we apply the two standard techniques of streamline-diffusion (SD) and upwind (UPW) stabilization (see [14] for details of the discretization). However, such methods require additional 'free' parameters as the local mesh width, the relation to the local Re number and an additional relaxation parameter, which all of them lead to the non-trivial task of parameter tuning, particularly in the case of nonlinear flow models and/or on non-uniform meshes. In contrast, the conforming Q_2/P_1 element is employed - as usually - without any stabilization which seems to be sufficient in our case of low up to medium *Re* numbers. However, we also have to state that, up to now, it is by far not clear of how to develop corresponding stabilization techniques which can be adapted to the potentially available higher order accuracy of this Stokes element in the case of much larger *Re* numbers.

The following tables show the resulting approximation properties w.r.t. the measured drag and lift coefficient on the different refinement levels, the required nonlinear iteration numbers #NN to gain 8 digits (here we compare the continuous resp. discrete Newton solver with the simpler fixpoint approach) and the corresponding averaged number #NL of linear iterations per nonlinear cycle to gain 2 digits. We apply the described multigrid solver for the $\tilde{Q}_1/Q_0$ element, while for the conforming Q_2/P_1 element, the preconditoned GMRES solver has been used. These approximation and solution properties are shown with the corresponding CPU times.

The results from these tables are quite prototypical for similar configurations with low up to medium *Re* numbers and can be concluded as follows:

1. Both discretizations lead to the same results if we apply sufficiently many grid refinement steps; as expected, the quadratic Q_2/P_1 approach is more accurate which can be summarized by the following thumb rules:

   'The linear $\tilde{Q}_1/Q_0$ approach requires - depending on the specific choice of the stabilization technique for the convective term - about 1 or 2 further levels of grid refinement to produce a comparable accuracy as the quadratic Q_2/P_1 ansatz. That means, the required number of elements is about 5 - 20 times larger which can result in approximately up to 10 times more unknowns (see Table 1). However, keep in mind that the corresponding differences regarding the amount of RAM storage are much smaller since the matrix stencil for the higher-order FEM is significantly bigger!'
2. We clearly see the typical effect of 'optimal' multigrid: the linear convergence rates are in the range of $\rho_{MG} \approx 0.1$ (only 2 - 4 steps to gain two digits are required) and quite independent of the mesh size and also the underlying *Re* numbers. It is remarkable that the solution of the nonsymmetric Oseen equations is often 'easier' than solving symmetric Stokes problems!
   In contrast, the GMRES solver shows the expected mesh-dependent convergence behavior which is quantitatively much worse than the corresponding multigrid approach. Refining the mesh does not only lead to bigger problem sizes, but also to significantly more linear iteration sweeps due to the increasing condition number of the discrete matrices.
3. The advantage of the more complex Newton solvers in comparison to the 'simple' fixpoint approach is not so obvious: Even in the $Re = 50$ case, the required number of nonlinear fixpoint steps is only 2 - 3 times larger

$\tilde{Q}_1/Q_0$ - SD			
Level	Stokes	$Re = 20$	$Re = 50$
3	3.0560×10^3	5.7365×10^0	3.9976×10^0
4	3.1191×10^3	5.6679×10^0	3.8516×10^0
5	3.1364×10^3	5.6161×10^0	3.7572×10^0
6	3.1409×10^3	5.5925×10^0	3.7167×10^0
$\tilde{Q}_1/Q_0$ - UPW			
Level	Stokes	$Re = 20$	$Re = 50$
3	3.0560×10^3	5.5833×10^0	3.9580×10^0
4	3.1191×10^3	5.5737×10^0	3.7662×10^0
5	3.1364×10^3	5.5769×10^0	3.7085×10^0
6	3.1409×10^3	5.5788×10^0	3.6972×10^0
Q_2/P_1 - central			
Level	Stokes	$Re = 20$	$Re = 50$
2	3.1237×10^3	5.5569×10^0	3.6677×10^0
3	3.1374×10^3	5.5728×10^0	3.6867×10^0
4	3.1411×10^3	5.5777×10^0	3.6925×10^0
5	3.1421×10^3	5.5791×10^0	3.6940×10^0

$\tilde{Q}_1/Q_0$ - SD			
Level	Stokes	$Re = 20$	$Re = 50$
3	2.8240×10^1	0.6431×10^{-2}	-1.0423×10^{-2}
4	3.0158×10^1	0.8341×10^{-2}	-1.1131×10^{-2}
5	3.0684×10^1	0.9691×10^{-2}	-1.0977×10^{-2}
6	3.0820×10^1	1.0281×10^{-2}	-1.0850×10^{-2}
$\tilde{Q}_1/Q_0$ - UPW			
Level	Stokes	$Re = 20$	$Re = 50$
3	2.8238×10^1	0.4754×10^{-2}	-1.3578×10^{-2}
4	3.0157×10^1	0.9536×10^{-2}	-9.9338×10^{-3}
5	3.0684×10^1	1.0455×10^{-2}	-1.0010×10^{-2}
6	3.0820×10^1	1.0593×10^{-2}	-1.0479×10^{-2}
Q_2/P_1 - central			
Level	Stokes	$Re = 20$	$Re = 50$
2	2.9854×10^1	0.9512×10^{-2}	-1.1764×10^{-2}
3	3.0578×10^1	1.0609×10^{-2}	-1.0894×10^{-2}
4	3.0788×10^1	1.0616×10^{-2}	-1.0739×10^{-2}
5	3.0846×10^1	1.0617×10^{-2}	-1.0730×10^{-2}

Table 2. Drag coefficient (top) and lift coefficient (bottom) for different *Re* numbers and for various levels of mesh refinement for the $\tilde{Q}_1/Q_0$ (with streamline-diffusion and upwinding for the stabilization of the convective term) and the Q_2/P_1 approach (with no stabilization at all).

than compared with the Newton variants. However, in contrast to the linear solvers, the nonlinear convergence behavior is only related to the

$\tilde{Q}_1/Q_0$ - SD							
		Stokes		$Re = 20$		$Re = 50$	
Level	Nonlinear scheme	#NN/#NL	CPU	#NN/#NL	CPU	#NN/#NL	CPU
3	Fixpoint	4/2	9	11/3	19	17/4	41
	Newton	4/3	9	8/3	18	9/3	19
4	Fixpoint	4/3	26	10/3	59	16/3	110
	Newton	4/3	27	6/3	41	7/3	53
5	Fixpoint	4/3	113	9/3	206	14/3	346
	Newton	4/3	109	5/2	114	6/3	137
6	Fixpoint	4/4	551	8/3	671	12/2	989
	Newton	4/4	522	5/3	562	6/3	601

$\tilde{Q}_1/Q_0$ - UPW							
		Stokes		$Re = 20$		$Re = 50$	
Level	Nonlinear scheme	#NN/#NL	CPU	#NN/#NL	CPU	#NN/#NL	CPU
3	Fixpoint	4/3	9	10/4	35	16/4	53
	Newton	4/3	9	8/4	26	11/5	46
4	Fixpoint	4/3	25	10/3	78	15/3	143
	Newton	4/3	25	6/3	47	7/3	71
5	Fixpoint	4/3	106	10/3	246	14/3	386
	Newton	4/3	107	5/3	116	6/3	174
6	Fixpoint	4/4	505	8/2	633	12/3	1194
	Newton	4/4	516	5/3	525	6/3	535

Q_2/P_1 - central							
		Stokes		$Re = 20$		$Re = 50$	
Level	Nonlinear scheme	#NN/#NL	CPU	#NN/#NL	CPU	#NN/#NL	CPU
2	Fixpoint	4/24	8	12/35	26	18/38	40
	Newton	4/24	7	6/32	11	7/44	14
3	Fixpoint	4/52	49	12/67	172	17/78	271
	Newton	4/51	42	6/69	76	7/81	102
4	Fixpoint	4/134	400	11/116	1029	16/149	1878
	Newton	4/135	393	6/122	534	7/163	885
5	Fixpoint	5/586	9726	11/348	12457	16/315	16193
	Newton	4/541	7184	6/302	6326	6/330	6250

Table 3. Corresponding results for the total number of nonlinear iterations (#NN) and the averaged number of linear sweeps (#NL) per nonlinear cycle together with the elapsed CPU time.

properties of the underlying continuous (!) problem, that means the *Re* number in our case. This is in contrast to the linear problems which are mainly depending on the condition numbers of the matrices and hence on the mesh width resp. the mesh topology.

Scheme-Level	Drag	Lift	#NN/#NL	CPU
Q_2/P_1 - central - LEV2	0.9512×10^{-2}	5.5569×10^{0}	6/32	11
Q_2/P_1 - central - LEV3	1.0609×10^{-2}	5.5728×10^{0}	6/69	76
Q_2/P_1 - central - LEV4	1.0616×10^{-2}	5.5777×10^{0}	6/122	534
Q_2/P_1 - central - LEV5	1.0617×10^{-2}	5.5791×10^{0}	6/302	6326
$\tilde{Q}_1/Q_0$ - UPW - LEV3	0.4754×10^{-2}	5.5833×10^{0}	8/4	26
$\tilde{Q}_1/Q_0$ - UPW - LEV4	0.9536×10^{-2}	5.5737×10^{0}	6/3	47
$\tilde{Q}_1/Q_0$ - UPW - LEV5	1.0455×10^{-2}	5.5769×10^{0}	5/3	116
$\tilde{Q}_1/Q_0$ - UPW - LEV6	1.0593×10^{-2}	5.5788×10^{0}	5/3	525

Table 4. Direct comparison of the Q_2/P_1 approach vs. the $\tilde{Q}_1/Q_0$ Stokes element (with upwinding) for $Re = 20$ (standard stationary CFD benchmark configuration in [12]).

As a first summary, we can state that - as expected - the quadratic Q_2/P_1 is more accurate for the considered low up to medium Re numbers which lead to steady state solutions. That means, the 'simpler' $\tilde{Q}_1/Q_0$ approach may require approximately 5 - 20 times more grid points. On the other hand, the corresponding direct multigrid solver for the $\tilde{Q}_1/Q_0$ element pair shows the typical 'optimal' convergence behavior which is in contrast to the preconditioned GMRES solver: The single-grid approach shows convergence rates which clearly depend on the mesh width and which are quantitatively much worse.

Therefore, concerning the total efficiency, the '$\tilde{Q}_1/Q_0$ + multigrid' combination is (still) preferable. It is obvious that a higher order discretization has a high potential for the future, but the comparisons demonstrate that discretization and solver aspects have to be treated as a unit! Hence, the analogous development of robust and efficient multigrid techniques for the Q_2/P_1 and similar high-order Stokes elements is a must for future research and will have high priority in our own research activities.

6.2 Non-constant viscosity test case

Since the examined nonlinearity related to the convective term in the case of a steady state solution is quite moderate, our next tests are concerned with the discussed nonlinear viscosity functions. As a first consequence, the equivalence between the gradient and the deformation tensor formulations for the diffusive part are not valid any more, so the stabilized version of the rotated bilinear element pair $\tilde{Q}_1/Q_0$ is required.

The following tables show the results for the regularized *power paw* model, that means $\nu(\boldsymbol{u}) = (\varepsilon + D_{\mathrm{II}}(\boldsymbol{u}))^{-\alpha/2}$, $\varepsilon = 10^{-2}$, which has been integrated into both described Navier-Stokes solvers; the geometrical configuration corresponds to the previous 'Flow around cylinder' configuration, too. Again, we calculate the lift and drag coefficients, and additionally the mean pressure difference $\Delta\, p$ on the cylinder (see [12]). Due to the fact that the considered flow is very slow, the computation was done without the convective term such that - for numerical test reasons - the nonlinear viscosity is the only source of nonlinearity. However, the calculations show that the strength of the nonlinear behavior for the chosen parameters $\alpha = 0.5$ and particularly $\alpha = 0.9$ is quite significant.

The results for these non-Newtonian test cases are quite similar as those for the Newtonian flow model: Again, the quadratic Q_2/P_1 approach is more accurate while the corresponding GMRES solver for the linear subproblems in each Newton step shows the well-known grid dependent convergence behavior. In contrast, the performed multigrid solver in the $\tilde{Q}_1/Q_0$ context is robust against variations in the mesh width and also w.r.t. the problem-specific parameter α which mainly determines the corresponding nonlinear behavior.

$\nu(\boldsymbol{u}) = (\varepsilon + D_{\mathrm{II}}(u))^{-\alpha/2}$					
Level	Stokes element	Drag	Lift	$\Delta\, p$	#NN/#NL
$\alpha = 0.5$					
2	$\tilde{Q}_1/Q_0$	1409.60	11.86	22.41	6/2
	Q_2/P_1	1607.10	14.48	25.38	6/27
3	$\tilde{Q}_1/Q_0$	1554.10	13.70	23.99	8/2
	Q_2/P_1	1629.10	14.27	25.16	8/65
4	$\tilde{Q}_1/Q_0$	1594.20	14.25	24.56	9/2
	Q_2/P_1	1635.80	14.39	25.09	8/140
5	$\tilde{Q}_1/Q_0$	1615.60	14.43	24.81	8/2
	Q_2/P_1	1637.60	14.44	25.07	9/723
6	$\tilde{Q}_1/Q_0$	1626.20	14.46	24.94	8/2
$\alpha = 0.9$					
2	$\tilde{Q}_1/Q_0$	819.49	2.5201	14.22	13/2
	Q_2/P_1	920.59	2.1805	16.76	17/37
3	$\tilde{Q}_1/Q_0$	917.89	3.1958	15.54	14/2
	Q_2/P_1	941.51	3.3310	16.02	14/93
4	$\tilde{Q}_1/Q_0$	916.02	3.7381	15.74	12/2
	Q_2/P_1	953.94	3.9217	15.82	19/294
5	$\tilde{Q}_1/Q_0$	935.13	3.9954	15.82	15/3
	Q_2/P_1	957.64	4.0587	15.87	18/1162
6	$\tilde{Q}_1/Q_0$	946.22	4.0592	15.85	13/5

Table 5. Comparison of approximation results (drag, lift pressure difference) and of the total number of nonlinear iterations (#NN) and the averaged number of linear sweeps (#NL) per Newton step.

6.3 Examination of the Newton variants

In the final simulations, we examine the characteristics of the chosen nonlinear solvers of Newton type. To be precise, in the case of the continuous Newton approach we demonstrate the importance of including <u>all</u> terms of the Jacobi matrix in contrast to the simpler fixpoint approach which utilizes the partial differential operator only, hereby evaluated for the results of the last nonlinear iteration step (see in (30)).

	$\alpha = 0.5$				$\alpha = 0.9$			
	Newton		Fixpoint		Newton		Fixpoint	
Level	#NN/#NL	CPU	#NN/#NL	CPU	#NN/#NL	CPU	#NN/#NL	CPU
2	6/2	11	19/2	37	13/2	60	74/2	228
3	8/2	46	18/2	126	14/2	215	114/2	1359
4	9/2	199	16/2	389	12/2	740	125/2	5871
5	8/2	707	14/2	1517	15/3	3957	119/2	21735
6	8/2	2885	12/2	5047	13/5	25926	109/2	80438

Table 6. Direct comparison of continuous Newton vs. fixpoint method for the treatment of the nonlinearity with the $\tilde{Q}_1/Q_0$ Stokes element.

The results in Table 6 show the corresponding comparison for the nonconforming $\tilde{Q}_1/Q_0$ element. In both cases, for the full Newton and the simplified fixpoint approach, the described multigrid solvers for the linear auxiliary problems have been applied to gain 2 digits of relative accuracy. While for the moderate case of $\alpha = 0.5$, the differences are not so significant, the harder test case $\alpha = 0.9$ shows the importance of a fully developed Newton solver: Differences of order 4 - 8 w.r.t. the number of nonlinear iterations occur while the averaged number of linear sweeps remains (more or less) constant such that big differences in the total efficiency are typical for such flow problems.

Finally, we discuss the sensitivity of the discrete Newton approach via divided differences regarding the involved step-length parameter ε, the mesh width h and the problem-dependent parameter α which is responsible for the strength of the nonlinearity. As explained in (22), the discrete Jacobi matrix is calculated via:

$$\left[\frac{\partial \mathcal{R}(\boldsymbol{x}^n)}{\partial \boldsymbol{x}}\right]_{ij} \approx \frac{\mathcal{R}_i(\boldsymbol{x}^n + \varepsilon \boldsymbol{e}_j) - \mathcal{R}_i(\boldsymbol{x}^n - \varepsilon \boldsymbol{e}_j)}{2\varepsilon} \tag{33}$$

Therefore, in the following tables we examine more carefully the resulting discrete Jacobi matrices w.r.t. the discussed parameters ε, h and α, and we show how the 'outer' nonlinear convergence behavior as well as the 'inner' linear solution behavior is influenced by the resulting matrix.

$\varepsilon \setminus$ TOL	10^{-1}	10^{-2}	10^{-3}	10^{-4}
10^{-2}	40*/ 4 [60]	32/ 7 [49]	30/ 13 [48]	21/ 20 [36]
10^{-3}	14/ 12 [22]	10/ 16 [16]	8/ 24 [14]	8/ 29 [15]
10^{-4}	10/ 15 [16]	8/ 19 [13]	6/ 27 [11]	7/ 29 [13]
10^{-5}	10/ 15 [16]	8/ 19 [13]	6/ 27 [11]	7/ 29 [13]
10^{-6}	10/ 15 [16]	8/ 19 [13]	6/ 27 [11]	7/ 29 [13]
10^{-7}	10/ 15 [16]	8/ 19 [13]	6/ 27 [11]	7/ 29 [13]
10^{-8}	10/ 15 [16]	8/ 19 [13]	6/ 27 [11]	7/ 29 [13]

$\varepsilon \setminus$ TOL	10^{-1}	10^{-2}	10^{-3}	10^{-4}
10^{-2}	40*/ 7 [280]	40*/ 14 [305]	40*/ 22 [335]	37/ 36 [358]
10^{-3}	33/ 24 [288]	14/ 35 [134]	10/ 57 [120]	11/ 67 [145]
10^{-4}	11/ 40 [113]	10/ 40 [103]	8/ 65 [105]	8/ 70 [114]
10^{-5}	11/ 38 [113]	9/ 43 [99]	8/ 65 [109]	8/ 70 [110]
10^{-6}	11/ 38 [114]	9/ 43 [97]	8/ 65 [105]	8/ 70 [111]
10^{-7}	21/ 25 [189]	21/ 35 [212]	13/ 58 [167]	14/ 78 [207]
10^{-8}	21/ 23 [185]	15/ 40 [160]	11/ 62 [146]	15/ 82 [234]

$\varepsilon \setminus$ TOL	10^{-1}	10^{-2}	10^{-3}	10^{-4}
10^{-2}	40*/ 11 [1306]	40*/ 23 [1548]	40*/ 51 [2055]	40*/ 84 [2720]
10^{-3}	40*/ 33 [1745]	31/ 48 [1561]	16/ 102 [1314]	18/ 129 [1877]
10^{-4}	15/ 85 [1124]	11/ 104 [983]	9/ 134 [999]	9/ 199 [1399]
10^{-5}	11/ 101 [976]	9/ 100 [807]	8/ 140 [941]	9/ 204 [1423]
10^{-6}	11/ 100 [954]	9/ 100 [798]	8/ 139 [920]	9/ 204 [1426]
10^{-7}	10/ 104 [898]	15/ 97 [1320]	10/ 138 [1148]	13/ 219 [2241]
10^{-8}	40*/ 29 [1814]	40*/ 156 [5174]	40*/ 199 [6244]	40*/ 274 [8108]

$\varepsilon \setminus$ TOL	10^{-1}	10^{-2}	10^{-3}	10^{-4}
10^{-2}	diverg.	40*/ 57 [10153]	40*/ 122 [17075]	40*/ 245 [31638]
10^{-3}	40*/ 35 [8251]	40*/ 82 [13550]	26/ 214 [18679]	22/ 348 [25640]
10^{-4}	20/ 181 [13244]	16/ 264 [14637]	14/ 432 [20002]	9/ 755 [22325]
10^{-5}	11/ 296 [11331]	8/ 492 [12978]	9/ 723 [22105]	8/ 944 [25220]
10^{-6}	25/ 148 [13947]	10/ 374 [12388]	14/ 383 [18271]	11/ 590 [22030]
10^{-7}	40*/ 255 [37031]	27/ 479 [44067]	20/ 757 [50167]	19/ 811 [50788]
10^{-8}	40*/ 252 [36063]	40*/ 245 [28289]	40*/356 [47587]	40*/ 472 [62539]

Table 7. Direct comparison of the results for the discrete Newton approach with the Q_2/P_1 Stokes element for $\alpha = 0.5$; the parameters ε and TOL as step-length parameter, resp., as stopping criterion are varied. The results are shown for grid refinement level 2 (top) up to level 5 (bottom). We always provide the total number of nonlinear iterations (#NN), the averaged number of linear sweeps (#NL) per Newton step and the elapsed CPU time in seconds (in brackets). The maximum number of nonlinear Newton steps is 40.

ε \ TOL	10^{-1}	10^{-2}	10^{-3}	10^{-4}
10^{-2}	40*/ 6 [62]	40*/ 15 [67]	40*/ 27 [74]	40*/ 34 [79]
10^{-3}	26/ 15 [43]	19/ 23 [34]	20/ 35 [39]	19/ 43 [40]
10^{-4}	16/ 20 [28]	15/ 28 [28]	18/ 37 [37]	19/ 43 [40]
10^{-5}	15/ 20 [26]	14/ 27 [26]	17/ 37 [35]	19/ 43 [40]
10^{-6}	16/ 21 [28]	14/ 27 [26]	17/ 37 [34]	19/ 43 [40]
10^{-7}	16/ 21 [28]	14/ 27 [26]	17/ 37 [35]	19/ 43 [40]
10^{-8}	16/ 22 [28]	14/ 27 [26]	17/ 37 [35]	19/ 43 [41]

ε \ TOL	10^{-1}	10^{-2}	10^{-3}	10^{-4}
10^{-2}	diverg.	40*/ 24 [353]	40*/ 43 [435]	40*/ 69 [547]
10^{-3}	40*/ 10 [297]	26/ 44 [286]	21/ 75 [305]	18/ 99 [326]
10^{-4}	40*/ 21 [352]	17/ 70 [243]	16/ 92 [275]	15/ 114 [304]
10^{-5}	35/ 33 [354]	15/ 70 [219]	14/ 93 [243]	14/ 115 [286]
10^{-6}	33/ 37 [348]	15/ 70 [220]	14/ 93 [244]	14/ 115 [296]
10^{-7}	40*/ 34 [411]	29/ 79 [461]	29/ 88 [495]	18/ 104 [342]
10^{-8}	40*/ 32 [403]	40*/ 50 [483]	31/ 89 [525]	32/ 115 [656]

ε \ TOL	10^{-1}	10^{-2}	10^{-3}	10^{-4}
10^{-2}	diverg.	diverg.	40*/ 86 [3047]	40*/ 158 [5078]
10^{-3}	diverg.	40*/ 70 [2726]	40*/ 119 [4015]	37/ 214 [6100]
10^{-4}	31/ 63 [2113]	40*/ 78 [3050]	20/ 263 [4055]	20/ 431 [6486]
10^{-5}	40*/ 82 [3060]	16/ 208 [2721]	19/ 294 [4380]	18/ 454 [6198]
10^{-6}	40*/ 82 [3087]	16/ 214 [2733]	19/ 296 [4435]	18/ 454 [6314]
10^{-7}	21/ 86 [1831]	18/ 205 [3058]	19/ 296 [4532]	18/ 455 [6472]
10^{-8}	diverg.	diverg.	diverg.	diverg.

ε \ TOL	10^{-1}	10^{-2}	10^{-3}	10^{-4}
10^{-2}	40*/ 13 [5997]	40*/ 68 [12117]	40*/ 239 [35923]	40*/678 [119583]
10^{-3}	40*/ 32 [8081]	40*/179 [28095]	40*/ 334 [44798]	40*/ 661 [86097]
10^{-4}	40*/ 71 [13974]	33/ 368 [45204]	24/ 795 [70727]	18/1429 [82560]
10^{-5}	40*/ 90 [14493]	37/ 476 [59433]	18/1162 [77307]	17/1818 [112571]
10^{-6}	40*/902 [133602]	31/1250 [127105]	37/1387 [132800]	35/1798 [169726]
10^{-7}	†	†	†	†
10^{-8}	†	†	†	†

Table 8. Direct comparison of the results for the discrete Newton approach with the Q_2/P_1 Stokes element for $\alpha = 0.9$; the parameters ε and TOL as step-length parameter, resp., as stopping criterion are varied. The results are shown for grid refinement level 2 (top) up to level 5 (bottom). We always provide the total number of nonlinear iterations (#NN), the averaged number of linear sweeps (#NL) per Newton step and the elapsed CPU time in seconds (in brackets). The maximum number of nonlinear Newton steps is 40 († - maximum CPU time exceeded).

The results clearly indicate that the right choice of the step-length parameter ε might be a delicate task: Choosing ε too big, then the advantageous quasi-quadratic convergence behavior may get lost, while very small parameter values for ε can lead to divergence, due to numerical instabilities. Additionally, the range of 'good' parameter values ε may also depend on the mesh width h and much more significantly on the nonlinearity of the problem, which is determined in our case via the parameter α; being outside of the range of admissible values, the increase in necessary CPU time can be extremely big.

	$\alpha = 0.5$		$\alpha = 0.9$	
Level	#NN/#NL	CPU	#NN/#NL	CPU
2	6/27	11	17/37	35
3	8/65	109	14/93	243
4	8/140	941	19/294	4380
5	9/723	22105	18/1162	77307

Table 9. Summarized presentation of the results for the discrete Newton approach with the Q_2/P_1 Stokes element for $\alpha = 0.5$ and $\alpha = 0.9$ on the different mesh levels. The calculations are performed for our 'preferred' configuration $\varepsilon = 10^{-5}$ and $TOL = 10^{-3}$ which seems to give the most robust convergence behavior.

Additionally, we analyze the influence of the prescribed tolerance parameter TOL which determines the relative accuracy for solving the linear subproblems. Here, the consequences are not so dramatic: Choosing 'big' values for TOL might destroy the quadratic nonlinear convergence properties, while very small parameters will not further improve the global nonlinear convergence behavior. However, this 'non-optimal' choice can significantly increase the total CPU time such that the recommendation is to gain between 2 - 4 digits of relative defect improvement in each nonlinear iteration as stopping criterion for the linear problems.

7 Conclusion

In this comparative study, we analyzed the influence of two different FEM discretization techniques (conforming Q_2/P_1, nonconforming $\tilde{Q}_1/Q_0$ elements) and solution procedures (nonlinear Newton variants, multigrid and Krylov-space solvers for the linear subproblems) onto the approximation properties and particularly the total efficiency of corresponding simulation tools for incompressible fluids. We discussed algorithmic details and provided numerical results for laminar steady-state flow examples including Newtonian and non-Newtonian behavior of *power law* type.

The following two approaches have been compared which both are quite popular in the CFD community; additionally, both of them are representative for typical methodologies which are in use for such types of flow simulations:

1. The Q_2/P_1 Stokes element with central differencing for the convective term and with a discrete nonlinear Newton solver based on 'divided differences' together with a preconditioned single-grid GMRES method for the auxiliary linear problems of Oseen type:

This pair of FEM functions leads due to the involved biquadratic polynomials to a higher order of accuracy than compared with linear approaches, at least for the considered low up to medium *Re* numbers; in contrast, for higher *Re* numbers, the question of stabilizing the discretized convective terms in the FEM context while maintaining the higher order accuracy has not been discussed and is one of the current research topics in the mathematical CFD community.

While the discrete Newton method leads in most cases - up to the question of stabilizing the 'divided differences' approach - to quite satisfying results concerning the numerical stability and efficiency, the involved GMRES solver as prototype for a non-multigrid Krylov-space method shows the expected dependence on mesh size which leads to huge numbers of linear iteration steps if the mesh width has to be decreased by accuracy reasons.

2. The nonconforming rotated bilinear $\tilde{Q}_1/Q_0$ Stokes element with upwinding resp. streamline-diffusion stabilization for the convective term and with a continuous nonlinear Newton solver together with an adapted multigrid method for the resulting Oseen-type subproblems:

The resulting accuracy of this 'linear' approach is lower such that up to 5 - 10 times more grid points are needed to produce results of similar quality as with the Q_2/P_1 approach. However, highly sophisticated FEM discretization techniques for the convective terms are available which can be directly applied for convection-dominated flows while maintaining a similar accuracy and robustness as for the considered low *Re* number flow configurations.

The advantage of this approach is - besides the continuous Newton solver for the nonlinearity which however can be performed for the Q_2/P_1 scheme, too - the already realized multigrid solver for the linearized problems. After a careful optimization of all multigrid components in relation to the underlying FEM spaces, the typical robust and very efficient convergence behavior of multigrid approaches can be obtained such that the linear problems are solved with a numerical efficiency which is more or less independent of the problem size, the mesh topology and also the type of problems, as for instance Stokes or Oseen equations for different *Re* numbers, coming from Newtonian and non-Newtonian models.

Concerning the question of the 'total efficiency', that means the amount of CPU seconds to provide a certain accuracy for a given flow problem, the answer due to our massive test calculations should be as follows: the higher order Q_2/P_1 approach requires significantly less grid points, however due to the missing multigrid solvers, the corresponding CPU time is huge. Altogether, the simpler nonconforming $\tilde{Q}_1/Q_0$ Stokes element together with the already realized multigrid components seems to be the better compromise at the moment, if we analyze the available flexibility and efficiency of this combination: 'Multigrid is a must for incompressible flow problems!'

As an outlook, it is clear that the examined Q_2/P_1 approach shows a huge potential for future improvement for complex fluid simulations, but the potentially higher order approximation properties have to be combined with adequate multigrid solvers. This must be one of the main tasks for Mathematics and Numerics in the CFD community and will be one of the major future research topics. Realizing the vision of a 'robust and flexible higher order FEM discretization' with 'accurate FEM treatment of the convective terms' together with 'adaptive local grid refinement' on the basis of 'rigorous a posteriori error control mechanisms for user-specified functionals' in combination with 'efficient nonlinear Newton-like and linear multigrid-like solvers', that should be the aim for future work!

References

1. J. Baranger and K. Najib. Analyse numerique des écoulements quasi-newtoniens dont la viscosity obeit la loi puissance ou la loi de carreau. *Numer. Math.*, 58:35–49, 1990.
2. R. Bramley and X. Wang. *SPLIB: A library of iterative methods for sparse linear systems.* Department of Computer Science, Indiana University, Bloomington, IN, 1997. `http://www.cs.indiana.edu/ftp/bramley/splib.tar.gz`.
3. C. S. Brenner. Korn's inequalities for piecewise H^1 vector fields. *IMI Preprint Series*, 5:1–21, 2002.
4. P. Hansbo and M. G. Larson. A simple nonconforming bilinear element for the elasticity problem. In *CIMNE*, Barcelona, Spain, 2001.
5. J. Hron, J. Málek, J. Nečas, and K. R. Rajagopal. Numerical simulations and global existence of solutions of two dimensional flows of fluids with pressure and shear dependent viscosities. *Math. Comp. Sim.*, 2002. to appear.
6. D. Kuzmin, M. Möller, and S. Turek. Multidimensional fem-fct schemes for arbitrary time-stepping. *J. Comp. Phys.*, 2002. (to appear).
7. A. Ouazzi, R. Schmachtel, and S. Turek. Multigrid methods for stabilized nonconforming finite elements for incompressible flow involving the deformation tensor formulation. *J. Numer. Math.*, 10:235–248, 2002.
8. A. Prohl and M. Růžička. On fully implicit space-time discretization for motions of incompressible fluids with shear-dependent viscosity: the case $p \leq 2$. *SIAM J. Numer. Anal.*, 39:214–249, 2001.

9. R. Rannacher and S. Turek. A simple nonconforming quadrilateral stokes element. *Numer. Meth. Part. Diff. Equ.*, 8:97–111, 1992.
10. Y. Saad and M. H. Schultz. GMRES: A generalized minimal residual algorithm for solving nonsymmetric linear systems. *SIAM J. Sci. Stat. Comput.*, 7:856–869, 1986.
11. D. Schaeffer. Instability in the evolution equations describing incompressible granular flow. *J. Diff. Eq.*, 66:19–50, 1987.
12. M. Schäfer and S. Turek. Benchmark computations of laminar flow around cylinder. In E.H. Hirschel, editor, *Flow Simulation with High-Performance Computers II*, volume 52 of *Notes on Numerical Fluid Mechanics*, pages 547–566. Vieweg, 1996.
13. R. Schmachtel. *Robuste lineare und nichtlineare Lösungsverfahren für die inkompressiblen Navier-Stokes-Gleichungen.* PhD thesis, University of Dortmund, 2002. (to be published).
14. S. Turek. *Efficient solvers for incompressible flow problems: An algorithmic and computational approach*, volume 6 of *LNCSE*. Springer, Berlin, 1999.
15. S. Turek et al. **FEATFLOW** – *Finite element software for the incompressible Navier-Stokes equations: User Manual, Release 1.2*, 1999. `http://www.featflow.de`.
16. S. Turek and S. Kilian. *The Virtual Album of Fluid Motion.* Multimedia DVD. Springer, 2002. `http://www.featflow.de/album`.
17. H. A. van der Vorst. Bi-CGSTAB: A fast and smoothly converging variant of Bi-CG for the solution of nonsymmetric linear systems. *SIAM J. Sci. Stat. Comput.*, 13(2):631–644, 1992.
18. S. P. Vanka. Implicit multigrid solutions of navier-stokes equations in primitive variables. *J. Comp. Phys.*, 65:138–158, 1985.

Algebraic Multigrid for Industrial Semiconductor Device Simulation

Tanja Clees** and Klaus Stüben

Fraunhofer SCAI, Schloss Birlinghoven, 53754 St. Augustin, Germany

Summary. In this paper, strategies for solving systems of partial differential equations by algebraic multigrid are discussed. In particular, a general framework for so-called point-based approaches is outlined. Several applications from industrial semiconductor process and device simulation have been investigated, and detailed results for two industrially relevant devices are presented. It is shown that this framework allows to construct robust and fast algebraic multigrid approaches even for cases, where iterative one-level solvers of the type commonly used in such applications exhibit bad convergence or even fail.

1 Introduction

Classical algebraic multigrid (AMG) [1, 2] is known to provide very efficient and robust solvers or preconditioners for large classes of matrix problems,

$$Au = f \ ,$$

an important one being the class of (sparse) linear systems with matrices A which are "close" to being M-matrices. Problems like this widely occur in connection with discretized *scalar* elliptic partial differential equations (PDEs). In such cases, classical AMG is very mature and can handle millions of variables much more efficiently than any one-level method. Since explicit information on the geometry (such as grid data) is not needed, AMG is especially suited for unstructured grids both in 2D and 3D. In fact, the coarsening process is directly based on the connectivity pattern reflected by the matrix A, and interpolation is constructed based on the matrix entries.

However, extensions of classical AMG are required to efficiently solve *systems* of PDEs involving two or more scalar *functions* (called *unknowns* in the following). This is because classical AMG realizes a *variable-based approach* which

** formerly Tanja Füllenbach

does not distinguish between different unknowns. Unless the coupling between different unknowns is very week, such an approach cannot work efficiently for systems of PDEs where, in general, the corresponding matrix A is far from being an M-matrix.

In the past, several ways to generalize AMG have been investigated, and there is still an ongoing rapid development of new AMG and AMG-like approaches. For a review, we refer to [3]. However, there is no unique and best approach yet. In fact, many problems cannot be tackled at all yet. All approaches seem to have their range of applicability but all of them may fail to be efficient in certain other applications. In this paper, the focus is on extensions of AMG which are direct generalizations of the classical approach.

We first want to recall a rather popular AMG approach to solve systems of PDEs, the so-called *unknown-based approach*, which is very similar to the variable-based approach except that all unknowns are treated separately. To be more specific, let us assume the variables to be ordered by unknowns, that is, $Au = f$ has the form

$$\begin{bmatrix} A_{[1,1]} & \cdots & A_{[1,n_u]} \\ \vdots & \ddots & \vdots \\ A_{[n_u,1]} & \cdots & A_{[n_u,n_u]} \end{bmatrix} \begin{bmatrix} u_{[1]} \\ \vdots \\ u_{[n_u]} \end{bmatrix} = \begin{bmatrix} f_{[1]} \\ \vdots \\ f_{[n_u]} \end{bmatrix} \tag{1}$$

where $n_u > 1$ denotes the number of unknowns of the given system of PDEs, $u_{[n]}$ denotes the vector of variables corresponding to the n-th unknown, and the matrices $A_{[n,m]}$ reflect the couplings between the n-th and the m-th unknown. Using this notation, coarsening the set of variables which correspond to the n-th unknown is strictly based on the connectivity structure reflected by the submatrix $A_{[n,n]}$, and interpolation is based on the corresponding matrix entries. In particular, interpolation to any variable i involves only coarse-level variables corresponding to the same unknown as i. The Galerkin matrices, however, are usually computed w.r.t. all unknowns.

This unknown-based approach has been proposed already in the very early papers on AMG (see [1]). It is certainly the simplest approach for solving PDE systems. By now a lot of experience has been gained with this approach which, in practice, works quite efficiently for many applications. Compared to the variable-based approach, the only additional information required is information about the correspondence between variables and unknowns. The unknown-based approach is mainly used for applications where the diagonal matrix blocks $A_{[n,n]}$ are close to being M-matrices. The essential additional condition for the approach to work is that smoothing the individual equations is sufficient to cause the resulting error to be smooth separately for each unknown. One advantage of this approach is that it can easily cope with anisotropies which are different between the different unknowns. Another advantage is that unknowns can virtually be distributed arbitrarily across mesh points. However, this approach will become inefficient, for instance, if the coupling between different unknowns is too strong.

In this paper, we focus on applications for which classical (variable-based) AMG [1, 2] and also the unknown-based approach do not work. In particular, in Section 2, we outline a flexible framework for constructing new, "point-based" AMG approaches to solve various types of PDE systems. In contrast to the unknown-based approach, a point-based approach operates (i.e. coarsens and interpolates) on the level of points rather than variables. Recent results for industrial applications in semiconductor device simulation are presented in Section 3, showing that suitable point-based AMG approaches yield efficient solution processes. These results are obtained by our code SAMG [4] which is a generalization - based on the framework mentioned before - of the (scalar) code RAMG described in detail in [2].

2 A General Framework for Point-Based Approaches

We talk about a *point-based approach* if, geometrically speaking, coarsening takes place on the level of grid points (rather than variables as before). The resulting hierarchy of grid points is then used for all unknowns, that is, each unknown is associated with the same hierarchy. (Note that this is different from the unknown-based approach where each unknown is associated with its own hierarchy.) Clearly, for any point-based approach, AMG needs additional information on the variable-to-point mapping. We assume corresponding information to be available to AMG.

Remark: Since we have the solution of PDEs in mind, we think of points as being real physical (grid) points in space. However, we want to point out that, from AMG's point of view, it is not important whether "points" really correspond to physical points. Instead, one may think of the nodes of a graph representing the connectivity structure of A. Regarding a point-based approach, it is only relevant for AMG to know whether there are "blocks" of variables (corresponding to different unknowns) which may be treated (coarsened and interpolated) simultaneously.

2.1 Primary Matrix

The framework which we propose allows for many different types of point-based approaches, and it depends on the concrete class of applications which of these possible approaches is meaningful. The idea which is common to all these approaches is that the (point) coarsening (i.e. splitting) process is performed based on some auxiliary (sparse) $(n_p \times n_p)$-matrix $P = (p_{kl})$, called the *primary matrix*, with n_p denoting the number of points. The resulting hierarchy of points is then assigned to all unknowns. For this process to make sense, the primary matrix employed should reflect the physical connectivity (the general structure as well as the strength of connections) of neighboring variables reasonably well, *simultaneously for all unknowns.*

A special point-based approach, sometimes called "block approach", has already been introduced in the very early paper [1] and has been further investigated, for instance, in [6]. To be more specific, we assume the variables to be ordered pointwise, that is, $Au = f$ has the form

$$\begin{bmatrix} A_{(1,1)} & \cdots & A_{(1,n_p)} \\ \vdots & \ddots & \vdots \\ A_{(n_p,1)} & \cdots & A_{(n_p,n_p)} \end{bmatrix} \begin{bmatrix} u_{(1)} \\ \vdots \\ u_{(n_p)} \end{bmatrix} = \begin{bmatrix} f_{(1)} \\ \vdots \\ f_{(n_p)} \end{bmatrix}, \tag{2}$$

where $u_{(k)}$ denotes the "block" of variables located at point k and the $A_{(k,l)}$ represents the "block coupling" between $u_{(k)}$ and $u_{(l)}$. If all unknowns are defined at all points, the $A_{(k,l)}$ are $(n_u \times n_u)$-matrices. Generally, however, they can be smaller and are not necessarily square.

Block coarsening corresponds to defining the primary matrix P by

$$p_{kl} = -||A_{(k,l)}|| \quad (k \neq l) \quad \text{and} \quad p_{kk} = -\sum_{l \neq k} p_{kl} \tag{3}$$

with $||.||$ denoting a suitable norm. Various different norms have been considered in practice, including the maximum, Schur and row sum norm. For the applications considered in Section 3.2, the performance of the employed point-based approach was not sensitively influenced by the concrete choice of the norm. However, in general, it depends on the application (whether block coarsening is meaningful and) which norm is most suitable.

Remark: In (3), p_{kk} has been selected to enforce P to become weakly diagonally dominant. Although this is very natural, it is not really necessary, and there are other choices of p_{kk} which may be advantageous. For instance, the setting $p_{kk} = ||A_{(k,k)}||$ takes also the "size" of the diagonal blocks $A_{(k,k)}$ into account. Sometimes this setting has advantages, for instance, if the matrix P is not only used for the coarsening but also for constructing the interpolation (see next section). However, as far as we have seen for applications of the type discussed in Section 3.2, the performance of the resulting AMG approach does not sensitively depend on the concrete choice of p_{kk}.

Depending on the type of application, there are many other possibilities for defining a primary matrix. Often, this can be done automatically as part of AMG's setup phase. In other cases, it may be better to let the user of AMG provide a reasonable matrix himself, based on his knowledge of the underlying physics of the given problem. In all cases, a primary matrix can usually be interpreted as describing the connectivity structure of some (auxiliary) scalar *primary unknown.* Clearly, the primary unknown should represent the connectivity structure of all "real" unknowns in the given system of PDEs reasonably well.

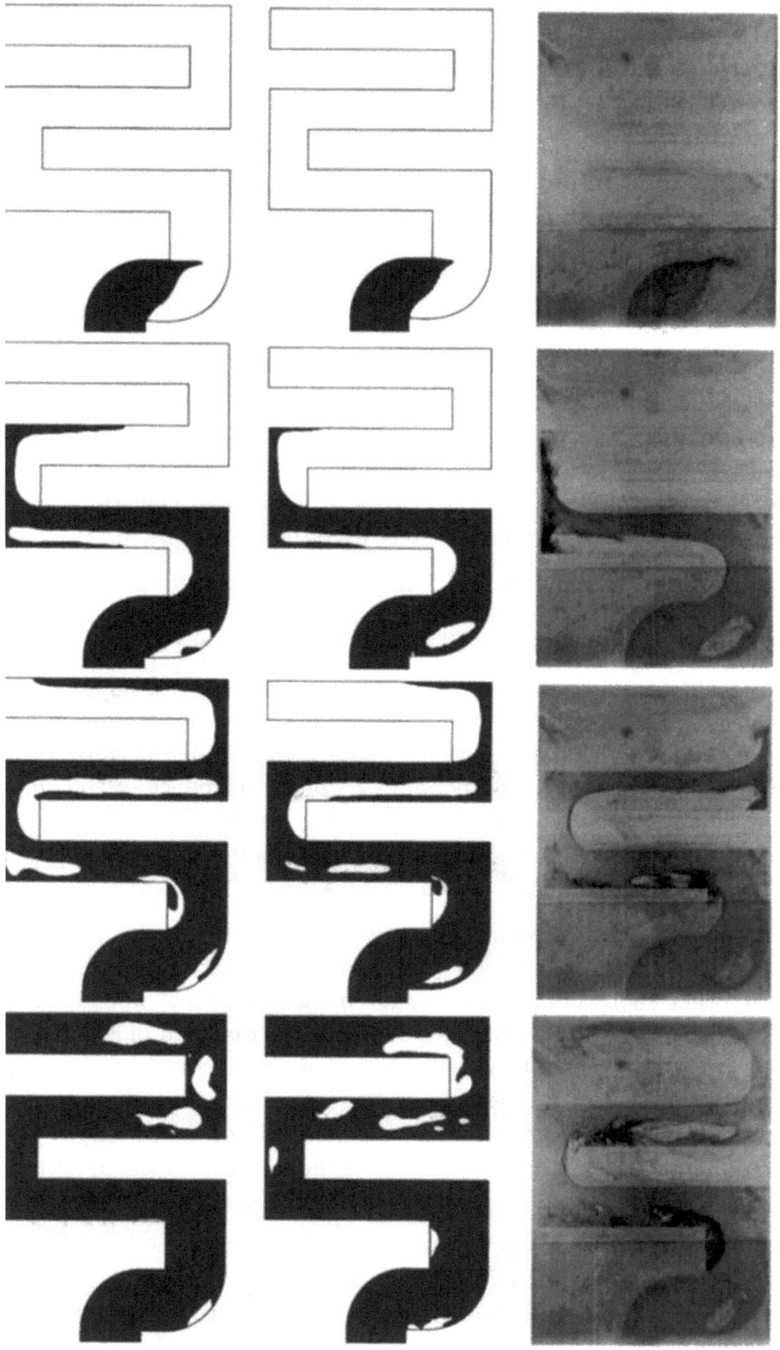

Fig. 10. S-shaped channel with influence of gas : left : 2D results, middle : 3D results in a cut, right : experimental results [25]. First row : time equals 7.15 ms, second row : 25.3 ms, third row : 39.3 ms and fourth row : 53.6 ms.

2.2 Interpolation

The main purpose of a primary matrix is the definition of an AMG hierarchy in terms of a nested sequence of points. In addition to the (recursive) coarsening process, also interpolation operators have to be constructed recursively. In practice, there are various possibilities to generalize the interpolation approaches used in classical AMG.

First, the use of *block interpolation* seems most natural, in particular, if P is defined according to (3). That is, formulas to interpolate the error $e_{(k)}$ at a point k are constructed by approximating the block equations

$$e_{(k)} = -A_{(k,k)}^{-1} \sum_{l \neq k} A_{(k,l)} e_{(l)} \tag{5}$$

in a way which is analogous to the classical, variable-wise approaches to define the interpolation. This idea has already been outlined in [1]. In [6, 7], direct analoga of the classical *direct* and *standard* interpolation formulas (see [2]) are discussed.

Generally, *direct* analoga of the classical interpolation formulas (described in [1, 2]) are obtained by replacing scalar matrix entries by block-coupling matrices $A_{(k,l)}$. All resulting formulas involve in particular inverses of (sums of) block-coupling matrices. Since, in general, it cannot be assumed that these inverses exist, such block interpolation formulas are not robust enough without modification. One possibility to overcome this disadvantage is the replacement of "critical inverses" by inverses of suitable regular diagonal matrices.

In any case, however, the construction of a block interpolation is computationally very expensive. In practice, much simpler types of interpolation often lead to more efficient AMG processes. Thus, besides block interpolations (with modification), we consider (variable-wise defined) interpolation formulas which are either

- separate for each unknown ("u-interpolation") or
- the same for each unknown ("s-interpolation").

Typical ways to define the interpolation *weights* are based on entries in the original matrix A, based on distances and/or positions of points, or based on entries in the primary matrix P. We cannot go into further details here but just want to mention that classical interpolation schemes, as described in [2], such as *direct*, *standard* or *multi-pass interpolation*, can be generalized to this setting in a straightforward way. Also the concept of *aggressive coarsening* carries over. For more details, see [4].

Remark: To derive cheaper formulas, another possibility would be the replacement of *all* matrices $A_{(k,l)}$ by their diagonals $D_{(k,l)}$ in block interpolation schemes, as done in [6, 7]. The emerging interpolations resemble variable-wise defined formulas again.

3 Industrial Semiconductor Device Simulation

The general framework outlined in Section 2 formally allows the definition of various concrete algorithms. It seems clear that there exists no unique AMG procedure which will work satisfactorily for all systems of PDEs. Instead, major work consists in developing concrete algorithms separately for certain classes of industrial applications. In this section, we consider the application of point-based AMG to semiconductor device simulation.

Semiconductor process and device simulation aim at the approximative computation of the final shape and doping profile of a semiconductor device and its electrodynamic behavior, respectively. Due to the complexity of the models and grids used, industrial semiconductor simulation is increasingly recognized as an important and challenging area. Occurring PDE systems include *stress governing* and *reaction-diffusion equations* in process simulation and *drift-diffusion equations* in device simulation, all of which exhibit different numerical difficulties. That straightforward unknown-based AMG is suitable to speed up stress simulations has already been shown in [8]. For reaction-diffusion and drift-diffusion equations, the situation is considerably more complicated. For occurring matrices, classical iterative one-level solvers converge only slowly or even break down, classical (variable-based) AMG usually diverges, and also straightforward unknown-based AMG is not sufficient any more. In contrast to this, suitable point-based AMG approaches, accelerated by BiCGstab or GMRes, can cause remarkable speedups. For exemplary matrices arising from reaction-diffusion systems, this has already been discussed in [5]. Also some promising preliminary results for drift-diffusion systems have been presented there. In the following, we want to discuss the application of point-based AMG approaches to drift-diffusion systems in more detail.

3.1 Drift-Diffusion Systems

The most important results of a device simulation, at least from an engineer's point of view, are *current-voltage characteristics* (*IV-characteristics*) of the device considered. IV-characteristics reflect the dependence of (global) currents on voltages applied to contacts of the device. Currents can be computed from the electrostatic potential ψ and the electron and hole carrier concentrations n and p (which can be regarded as "intermediate" results of a device simulation).

There exists an extended hierarchy of semiconductor models (see [9]), ranging from quasi-hydrodynamic to kinetic and classical to quantum models. The simplest quasi-hydrodynamic model is the standard drift-diffusion system. Compared to more involved models, this simple model might provide less accurate local potentials and concentrations, but often predicts (current densities and) global quantities sufficiently accurately with much less computational effort. Because of this, the standard drift-diffusion system is of highest relevance for industrial simulation and will be considered in the following.

The Stationary Drift-Diffusion System.

In an industrial environment, possible steady states of the devices sufficiently long after switching processes are typically investigated. For this purpose, a stationary drift-diffusion system consisting of so-called basic semiconductor equations has to be solved for the electrostatic potential ψ and the electron and hole carrier concentrations n and p:

$$-\nabla \cdot (\epsilon_\Sigma \nabla \psi) + q(n - p - C) = 0 \qquad \text{Poisson(-type) equation,} \qquad (6)$$
$$-\nabla \cdot J_n + qR(\psi, n, p) = 0 \qquad \text{electron continuity equation,} \qquad (7)$$
$$\nabla \cdot J_p + qR(\psi, n, p) = 0 \qquad \text{hole continuity equation,} \qquad (8)$$

where J_n and J_p are the electron and hole current densities, respectively,

$$J_n = -q(\mu_n n \nabla \psi - D_n \nabla n) \qquad \text{electron current relation,} \qquad (9)$$
$$J_p = -q(\mu_p p \nabla \psi + D_p \nabla p) \qquad \text{hole current relation.} \qquad (10)$$

ϵ_Σ denotes the permittivity, q the elementary charge, $C = C(x)$ the net impurity concentration (i.e. the doping profile), $R = R_p(\psi, n, p, ...)$ the recombination-generation term, $\mu_n = \mu_n(x, \nabla\psi, ...) > 0$ and $\mu_p = \mu_p(x, \nabla\psi, ...) > 0$ the electron and hole mobilities, respectively, $D_n > 0$ and $D_p > 0$ the electron and hole diffusivities, respectively (for which Einstein's relations $D_n = \frac{k_B T}{q}\mu_n$ and $D_p = \frac{k_B T}{q}\mu_p$ are usually assumed to hold), k_B the Boltzmann constant and T the device temperature. All of these quantities are assumed to be given, in general as functions.

By inserting (9) and (10) into equations (7) and (8), respectively, and multiplying the resulting equations by $1/q$, we obtain the following second-order (elliptic) partial differential equations for ψ, n and p:

$$-\nabla \cdot (\epsilon_\Sigma \nabla \psi) + q(n - p - C) = 0 \ , \qquad (11)$$
$$-\nabla \cdot (D_n \nabla n - \mu_n n \nabla \psi) + R = 0 \ , \qquad (12)$$
$$-\nabla \cdot (D_p \nabla p + \mu_p p \nabla \psi) + R = 0 \ . \qquad (13)$$

Generally, the complete simulation domain $\Lambda \subset \mathbb{R}$ ($d = 2$ or 3) consists of two parts, $\Lambda = \Sigma \cup \Omega$. The first part, Σ, represents the union of all material regions where the above described coupled system of stationary semiconductor equations holds. This is typically the case in the semiconductor regions (often doped silicon, the wafer). The second part, Ω, is defined to be the union of regions for which it is assumed that (nearly) no charge carrier currents can occur as, for instance, in insulating regions. In Ω, the above PDE system (11)-(13) degenerates to the Laplace equation,

$$-\nabla \cdot (\epsilon_\Omega \nabla \psi) = 0 \quad \text{in } \Omega \ , \tag{14}$$

where ϵ_Ω represents the permittivity of the corresponding material layers. For example in case of a MOSFET (a metal oxide semiconductor field effect transistor), Ω represents the gate oxide. Here, the interface between Σ and Ω is the semiconductor/oxide-interface. Ω may also be empty, as in case of a diode.

An n-domain (or n-doped region) is defined to be a subdomain of Σ in which $C(x) > 0$ holds. Analogously, a p-domain (or p-doped region) is defined to be a subdomain in which $C(x) < 0$ holds. The interface between an n- and an adjacent p-domain is called pn-junction. Such adjacent pn-domains, forming local diodes, determine the modes of operation of a device to a large extent.

The boundary of Λ can be split into two disjoint parts, $\partial\Lambda = \partial\Lambda_p \dot{\cup} \partial\Lambda_a$. $\partial\Lambda_p$ represents those parts of $\partial\Lambda$ which correspond to real "physical" boundary segments, i.e. interfaces to insulating material and contacts. $\partial\Lambda_a$ consists of artificial boundary segments, which are introduced, for example, to reduce the simulation domain and to obtain a "self-contained" device, i.e. to separate it from adjacent devices, if it is embedded in an integrated circuit. Different types of boundary and interface conditions are prescribed on different parts of $\partial\Lambda$ and the interfaces between Σ and Ω, depending on the physical properties of the neighboring material and the type of contacts (see [10], for instance).

The simplified sketch in Fig. 1 illustrates the simulation domain Λ in case of an n-MOSFET. In particular, Σ, Ω, the n-domains (source and drain), the p-domain, the physical boundaries (i.e. the contacts to source and drain and the gate contact) and interfaces to isolating material (other than Ω), the artificial boundaries and the interface between Σ and Ω are shown.

Layer Behavior and Conditioning.

A detailed description of the basic semiconductor equations and their mathematical analysis can be found in [9, 10, 11], for instance. We just want to summarize some critical aspects.

It is well-known that ψ, n and p are very different in magnitude and show layer behavior. In order to analyze this behavior, one usually scales the system (11) - (13) by so-called "singular perturbation scaling" factors (see [11]). If we denote the scaled quantities by the same symbols as the original ones, equations (12) and (13) keep their form but (11) transforms into

$$\lambda^2 \Delta\psi - (n - p - C) = 0 \tag{15}$$

with λ^2 being a very small constant. As a consequence, this formulation reveals the singular perturbation character of the system. By a singular perturbation analysis, it can be shown that layers occur at pn-junctions, semiconductor/oxide interfaces and so-called Schottky contacts (in reconciliation

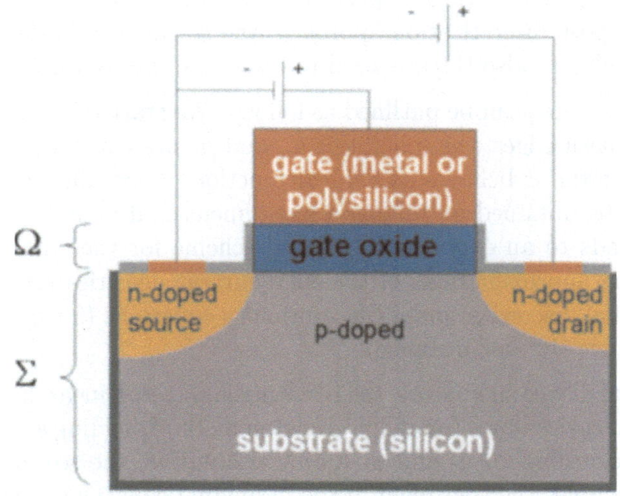

Fig. 1. Simplified sketch of the simulation domain Λ in case of an n-MOSFET. Depicted are Σ, Ω, the p-domain, the two n-domains (source and drain), isolating material (grey) (other than the gate oxide Ω), the contacts (red) and the artificial boundaries (green).

with physics and numerical results). Outside of these regions, ψ, n and p are moderately varying functions.

A further analysis of (11) - (13) yields statements on the conditioning of these three equations. At least for moderate applied bias (e.g. a moderately large voltage applied to one of the contacts), the Poisson equation (11) is well-conditioned with respect to ψ. The electron continuity equation (12) (the hole continuity equation (13)) is well-conditioned with respect to n (p) only if every n- and p-domain has a contact. However, if an n- or p-domain has no contact (a so-called floating region), the continuity equation for the majority carrier concentration of this region is ill-conditioned. The errors might be amplified by a factor of $O(\lambda^{-4})$ and, therefore, floating regions can produce great numerical difficulties in computing carrier concentrations. An industrially very important example for a floating region is the channel domain in a device (e.g. a FinFET) fabricated by silicon-on-insulator (SOI) technology. Since SOI is one of the standard technologies nowadays, such problematic floating regions occur quite often in device simulation.

Discretization and Linearization.

While the discretization of the Poisson(-type) equation (11) is straightforward, the discretization of the continuity equations, which can be characterized as special diffusion-convection-reaction equations, is crucial for an efficient solution of the drift-diffusion system. In practice, the system is discretized by a box method (BM) on (nearly perfect) Delaunay meshes. To gain stability, a special purpose discretization approach, the so-called Scharfetter-Gummel (SG) approach (see also [11]), is used to discretize the continuity equations.

The SG-BM scheme can be outlined as follows. We start with the assumption that, along mesh edges, the mobilities μ_n and μ_p are constant, and the electrostatic potential ψ behaves as a linear function. Approximations of J_n and J_p can then be obtained by solving a one-dimensional boundary value problem. This leads to an exponentially fitted scheme for the current relations. The emerging approximations of the edge current densities are employed to obtain the final discretization of the continuity equations (for more details on the SG-BM, see [12], for instance).

In commercial device simulators, the resulting highly nonlinear discrete system is typically linearized by a (modified) Newton method yielding a "fully coupled approach" regarding ψ, n, and p. As an alternative, the so-called Gummel iteration decouples the treatment of the individual equations of the full drift-diffusion system: one iteration step consists of solving Poisson's equation (11) for ψ, then solving the electron continuity equation (12) for n, and finally the hole continuity equation (13) for p (see [10], for example).

Since the discretized and linearized systems inherit the layer behavior and conditioning of the original equations, we have to expect layer behavior near pn-junctions, Schottky contacts and semiconductor/oxide interfaces and ill-conditioned continuity equations in floating regions. In spite of the fact that the original system (11)-(13) is usually scaled by DeMari factors (see [10]), ψ, n, p and their discrete analoga are still very different in magnitude.

Due to all these reasons, the arising drift-diffusion matrices are ill-conditioned and often nearly singular even if the underlying problem is far away from a possibly existing bifurcation point. As a consequence, this leads to great difficulties in solving the sequence of matrix equations efficiently. Due to this, the linear, iterative one-level solvers usually employed in industry today, namely ILU- (or ILUT-)preconditioned CGS or BiCGstab, exhibit an unsatisfactory performance in general. A promising possibility to obtain more robust and more efficient methods is the investigation of *hierarchical* approaches such as multigrid.

By now, linear multigrid methods have been developed only for solving the three individual partial differential equations arising during a Gummel(-type) iteration. The most difficult part there is the solution of discretized and linearized continuity equations. For example, in [13], a geometric multigrid method for this type of equation was investigated and successfully applied to some examples on structured grids. In contrast to this, in this paper, we are interested in the solution of the matrix equations occurring in the fully coupled approach. This approach is typically used in modern (commercial) device simulators because it is often more favorable than a Gummel(-type) iteration. In summary, the matrices considered here arise from the fully coupled approach for drift-diffusion systems discretized by the SG-BM on unstructured grids.

In the next section it is demonstrated that a robust and rapidly converging point-based AMG method for the drift-diffusion matrices can be obtained using the framework described in Section 2. It should be noted that, due to our experience, classical (variable-based) [1, 2] and unknown-based AMG methods are not robust enough for this kind of matrices. They often diverge and are therefore not discussed in the remainder of this paper.

Remark: In particular until the early nineties, the application of nonlinear geometric multigrid methods (full approximation schemes (FAS)) was investigated as a further approach to solve drift-diffusion systems (see the references given in [13]). These approaches do not carry over in a straightforward way to the more sophisticated models and unstructured grids used in modern commercial device simulators.

3.2 Numerical Results

According to the different modelling situations in the two parts Σ and Ω of Λ, the finally resulting matrices, A, consist of two parts which are very different by nature. On the one hand, the part A_Ω of A which corresponds to Laplace's equation (on Ω) does not pose problems. Since the submatrix describing the ψ-to-ψ couplings is an M-matrix, and there are only a few couplings to n and p (namely across the interface to Σ), A_Ω is especially suited for AMG.

On the other hand, the system of PDEs given on Σ is tightly coupled. In particular, the corresponding part A_Σ of A is dominated by the couplings *to* the potential ψ. To be more specific[3], either the submatrix $A_{[2,1]}$ or $A_{[3,1]}$ (of A_Σ), depending on the majority carrier concentration, contains a significant part of the largest couplings (measured by absolute value).

Because of this tight coupling between the different PDEs in Σ, unknown-based AMG fails for such applications - although all diagonal block matrices $A_{[n,n]}$ $(n = 1, \ldots, 3)$ are essentially M-matrices. Instead, it has turned out that point-based AMG (used as a preconditioner for BiCGstab) leads to a very robust approach if, for example, the following components are chosen:

- ILU(0) as the smoother,
- a primary matrix based on norms[4] (3),
- an s-interpolation with weights being based on the entries of P.

[3] assuming the unknowns ψ, n and p to be numbered 1,2 and 3, respectively.

[4] For all drift-diffusion systems tested by now, the performance of the resulting approach was neither sensitively influenced by the concrete choice of the norm nor the concrete choice of p_{kk} (see Section 2.1).

Generally, to demonstrate the performance of AMG for a given device, it is not sufficient to look at its performance in solving just a few selected linear systems arising as part of a whole simulation. In fact, hundreds of linear systems have to be solved during a full simulation series, most of which have different properties. Consequently, to obtain a clear picture of the performance of AMG, one has to consider full simulation series.

In device simulators such a TAURUS [14], a simulation series for a given device consists of many individual simulations of drift-diffusion systems which differ, for example, in their respective boundary conditions. More precisely, each simulation run starts with the zero bias step, a step in which all voltages are set to zero. Afterwards, several *bias ramps* are applied to the device. For instance, Fig. 7 shows the sequence of bias ramps applied in case of a particular FinFET device: In the first 21 simulation steps (the first ramp), the gate voltage is gradually increased from 0 to 1V, keeping the drain voltage fixed at 0.05V. During the next 10 bias steps (the second ramp), the gate voltage is fixed at 1V, and the drain voltage is increased step by step to a value of 1V. For each individual simulation step (i.e. for each applied bias) of such a series, a Newton process is employed to solve the discretized problem, and, within each Newton step, a few matrix solves are necessary. Therefore, during a whole simulation run, several hundred linear systems have to be solved. In (commercial) simulators such as TAURUS, sophisticated control mechanisms are integrated to detect and "repair" possibly occurring difficulties during the linear and nonlinear iterations. In the worst case, this means a bias step rejection and step size reduction.

In order to demonstrate the performance of AMG, we have replaced the iterative linear TAURUS solver (an ILU-CGS method) by SAMG. To demonstrate the effects of coarse-level corrections, we also compare the performance of AMG with that of the corresponding one-level method, i.e. the smoother ILU (both accelerated by BiCGstab).

Up to now, we have tested examples from several classes of industrially relevant semiconductor devices, including a diode, "conventional" MOSFETs[5], STIs[6], an EEPROM[7], several FinFETs[8] and a power bipolar transistor (for general information about these devices, see [15, 16], for instance).

In all these cases, the outlined AMG approach was able to solve the arising matrix equations efficiently (see also [5]). In the following, we will present detailed results for two exemplary cases, an EEPROM and a FinFET. Table 1 shows details on the concrete test cases and dimensions of the arising matrix problems. The layout and doping profile of the FinFET example is depicted in

[5] metal oxide semiconductor field effect transistors.

[6] shallow trench isolated transistors.

[7] electrically erasable programmable read-only memory cell.

[8] double-gate MOSFET structures in which a thin, fin-shaped body is straddled by the gate forming two self-aligned channels that run along the sides of the fin.

Fig. 2. The bias ramps for the EEPROM are depicted in Fig. 3, the ramps for the FinFET in Fig. 7. Each of the remaining graphs shows results (discussed below) for a simulation series. Note that the matrices arising during a full simulation series are always numbered consecutively in these graphs.

Table 1. Details on the device examples (dim: spatial dimension, n_Σ: number of regions in Σ, n_Ω: number of regions in Ω, n_p: number of points, n_v: number of variables, n_A: number of stored matrix elements ($n_A \geq$ number of non-zeroes)).

Example	dim	n_Σ	n_Ω	n_p	n_v	n_A
EEPROM	3D	1	9	10,493	15,415	310,361
FinFET	3D	2	5	69,092	97,530	1,443,940

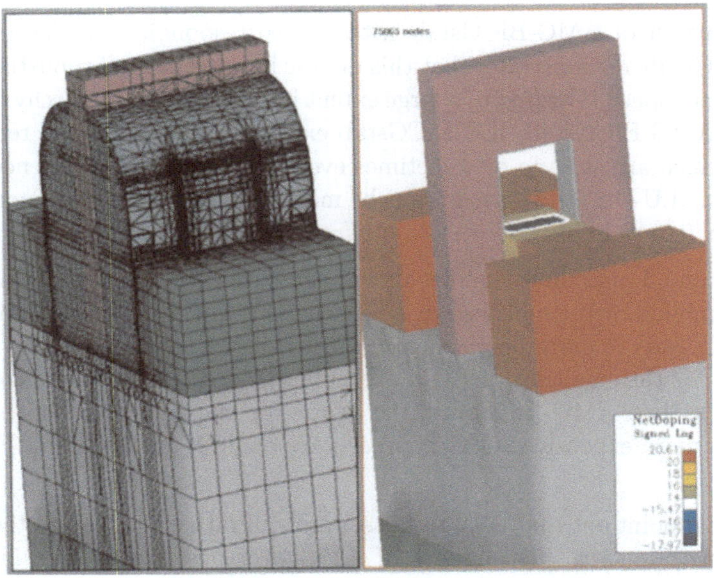

Fig. 2. FinFET example: layout and grid (left), doping profile of the wafer (right). Courtesy of Synopsys Inc.

The FinFET case represents the simulation of a modern device on a moderately large grid. Since a FinFET is fabricated on an SOI wafer, numerical difficulties arise due to the occurring floating region. The EEPROM example, which is rather small in terms of variables, was chosen because it exhibits an additional difficulty: for bias ramp (6), the system is extended by one equation (an algebraic condition) leading to a row with a zero diagonal. Such exceptional rows have to be treated separately during the AMG setup phase. Here, the corresponding row was simply excluded from the coarsening process.

For the examples considered, it can be observed that the linear TAURUS solver (ILU-CGS) does often not fulfill the prescribed convergence criterion (i.e. a residual reduction of at least 10^{-3} within a maximum number of iterations). This is depicted in Figs. 4 and 8 (a value above the yellow line means a violation of the criterion, a value above the red line means divergence of the linear solver for the current matrix). Especially in the EEPROM case, the L2-norm of the last residual, denoted by $||r_e||_2$, is often more than 10^5 times larger than the first residual, $||r_0||_2$.

In contrast to this, the AMG approach shows a stable and fast convergence behavior for both examples. The convergence criterion is fulfilled in all cases, and hence, instead of $||r_e||_2/||r_0||_2$, average residual reduction factors, ρ, are depicted in Figs. 6 and 10. The average residual reduction factors are usually lower than 0.5 and often much better, and less matrix solves were necessary during the Newton steps (see Tables 2 and 3), especially in case of the larger example, i.e. the FinFET.

A comparison of AMG-BiCGstab with the corresponding one-level solver, ILU-BiCGstab, demonstrates that this drastic improvement of robustness and convergence speed is caused to a large extent by employing a hierarchy: In contrast to AMG-BiCGstab, ILU-BiCGstab exhibits average residual reduction factors which are close to or sometimes even larger than 1 (divergence). Additionally, ILU-BiCGstab needs (much) more matrix solves, even more than ILU-CGS (see Tables 2 and 3).

Tables 2 and 3 also show timings for full simulation runs, including meshing and assembling of the matrices (the test character of the TAURUS interface to SAMG leads to extra overhead for transfer of the matrix data to SAMG). Whereas for the smaller EEPROM example the TAURUS solver is considerably faster than AMG-BiCGstab, the efforts for employing the more robust but a bit more expensive AMG approach are paid off for the larger FinFET example.

One should point out that both test cases are still rather small - and too small to demonstrate "real" advantages of AMG over one-level solvers in terms of computational speed. However, since AMG clearly shows a very robust behavior and fast convergence, it can be expected that for increasingly larger problem sizes AMG will be increasingly more efficient than one-level solvers.

Table 2. EEPROM example. Timings and number of necessary matrix solves. "total" is the total wall-clock time in hours needed for the whole simulation run. "SAMG" is the part of "total" which SAMG needed to solve the matrices.

approach	♯ matrices	total	SAMG
TAURUS solver	538	2.38	
AMG-BiCGstab	520	3.81	2.19
ILU-BiCGstab	560	4.31	2.56

Table 3. FinFET example. Timings and number of necessary matrix solves. Description as for Table 2.

approach	♯ matrices	total	SAMG
TAURUS solver	157	4.46	
AMG-BiCGstab	100	4.27	3.25
ILU-BiCGstab	216	11.49	9.15

4 Conclusions

AMG approaches for solving systems of PDEs were presented and discussed. Especially a general framework for point-based approaches was described, which employs a primary matrix to construct a point-based coarsening. Several possibilities for selecting a primary matrix and for the computation of the final interpolation weights were outlined. Recent results for applications in semiconductor device simulation were presented, which demonstrate that robust and rapidly converging point-based AMG methods can be obtained using this framework. The tested point-based methods have the potential to replace one-level preconditioners of the type commonly used in industrial device simulators today.

Acknowledgement.

The authors would like to thank Synopsys Inc. for providing a test license of TAURUS with an interface to SAMG and the examples mentioned.

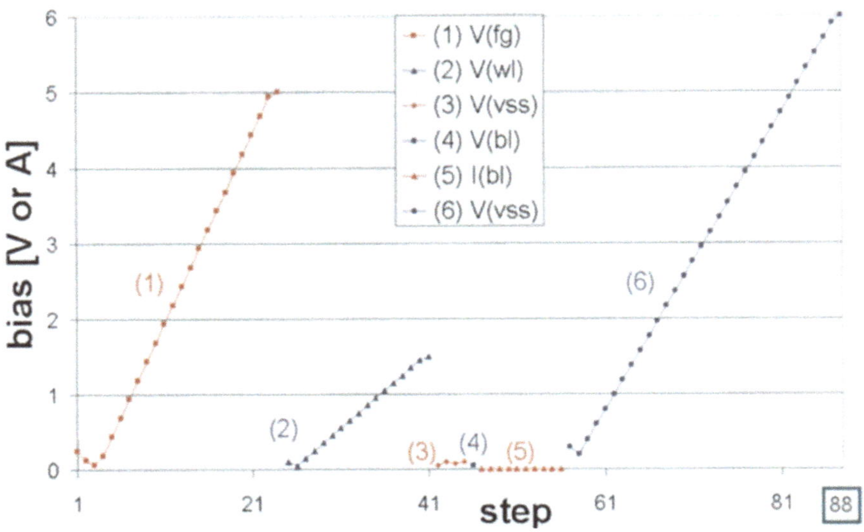

Fig. 3. EEPROM example. Bias steps.

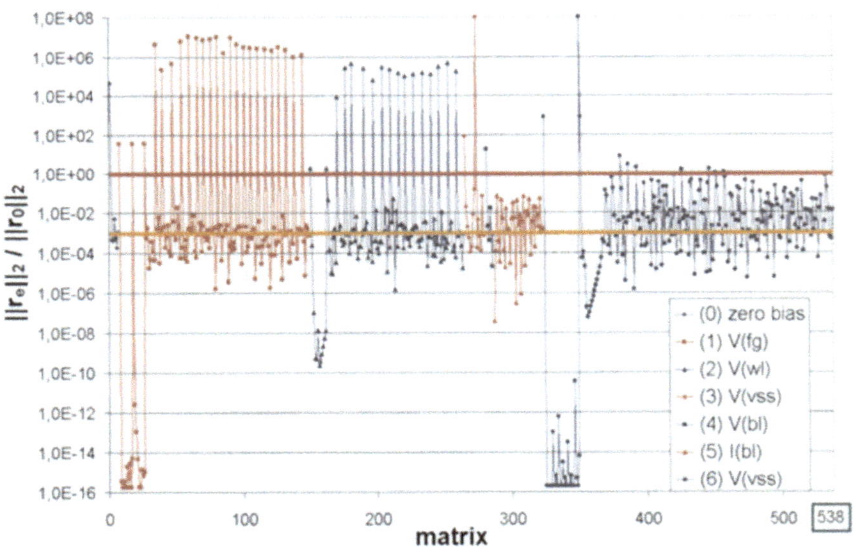

Fig. 4. EEPROM example. Convergence results for TAURUS solver: Last residual, $||r_e||_2 = ||Au_e - f||_2$, divided by first residual, $||r_0||_2 = ||Au_0 - f||_2$.

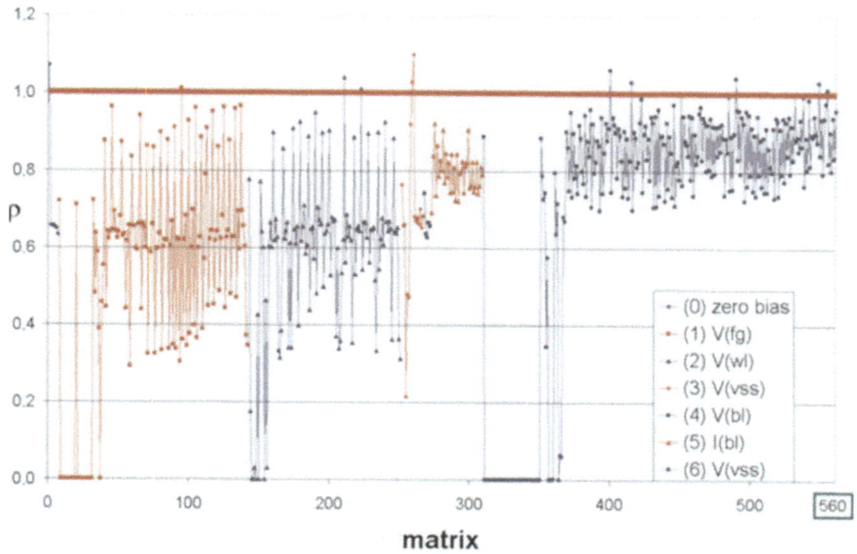

Fig. 5. EEPROM example. Results for ILU-BiCGstab: Average residual reduction factors, ρ.

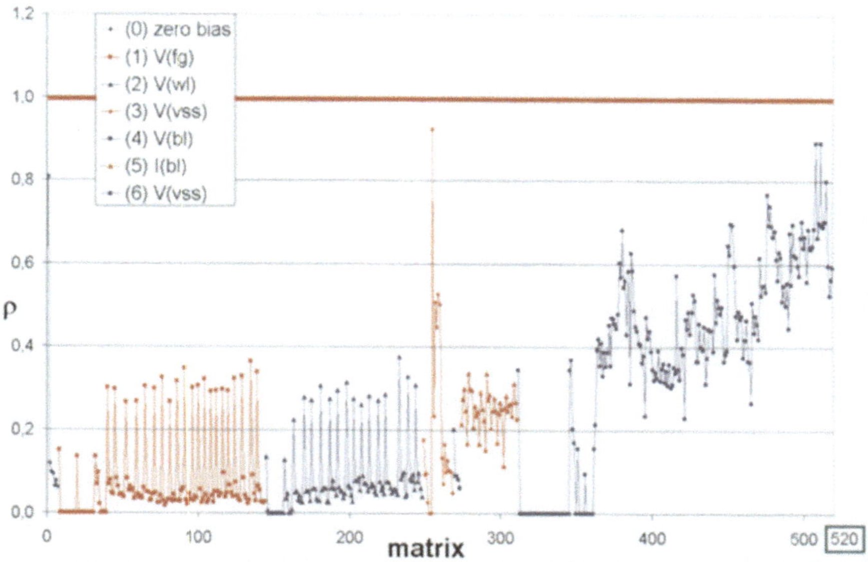

Fig. 6. EEPROM example. Results for AMG-BiCGstab: Average residual reduction factors, ρ.

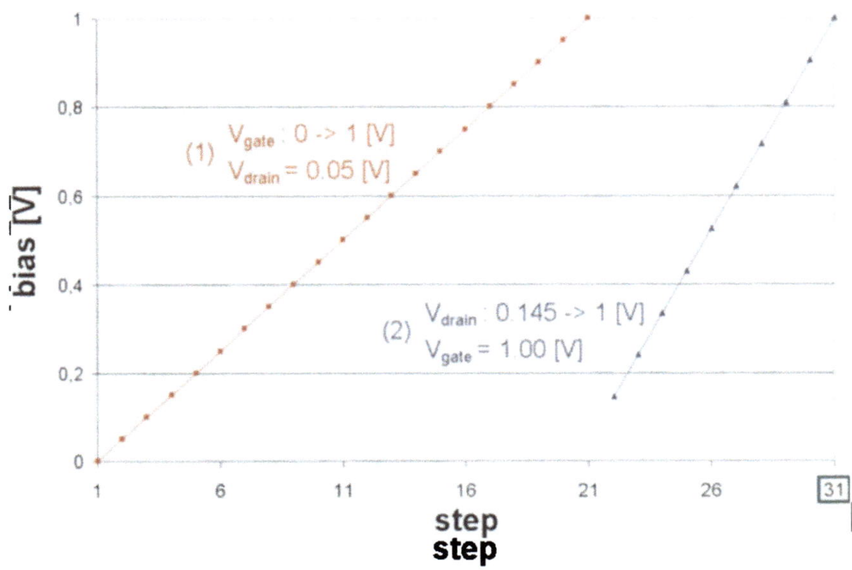

Fig. 7. FinFET example. Bias steps.

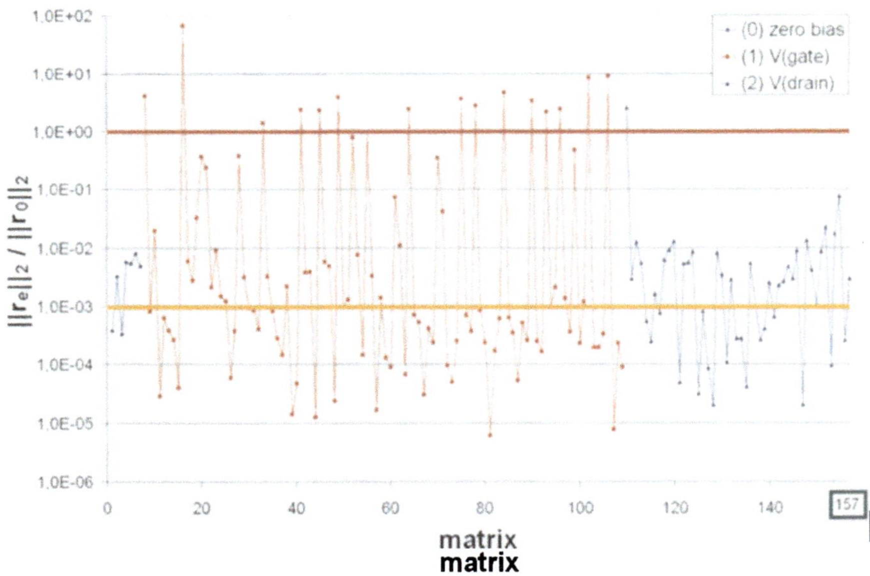

Fig. 8. FinFET example. Convergence results for TAURUS solver: Last residual, $||r_e||_2 = ||Au_e - f||_2$, divided by first residual, $||r_0||_2 = ||Au_0 - f||_2$.

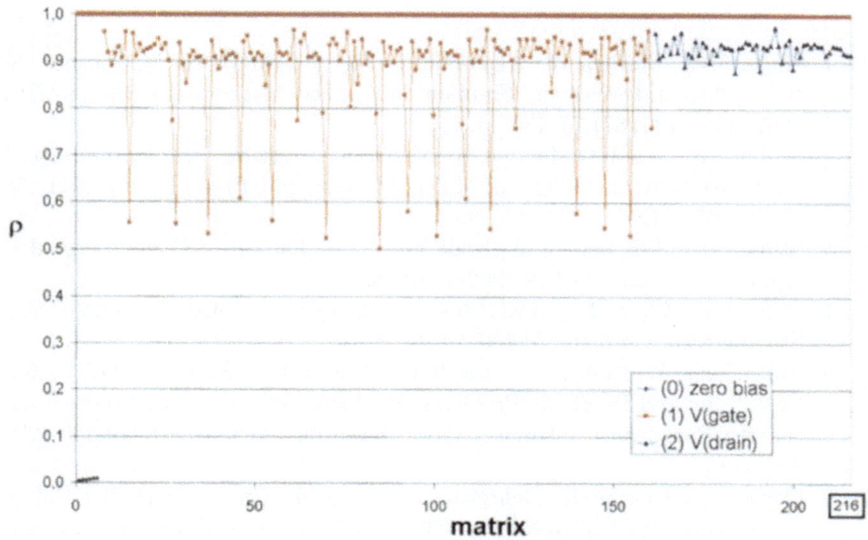

Fig. 9. FinFET example. Results for ILU-BiCGstab: Average residual reduction factors, ρ.

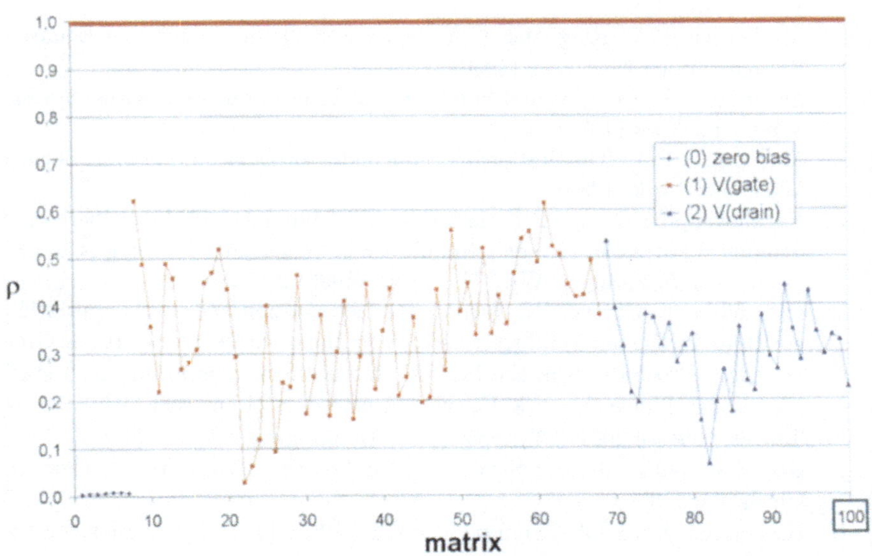

Fig. 10. FinFET example. Results for AMG-BiCGstab: Average residual reduction factors, ρ.

References

1. Ruge, J.W., Stüben, K.: Algebraic Multigrid (AMG). In: McCormick, S.F. (ed.): Multigrid Methods. Frontiers in Applied Mathematics, Vol. 3. SIAM, Philadelphia (1987) 73–130
2. Stüben, K.: An Introduction to Algebraic Multigrid. In: Trottenberg, U., Oosterlee, C.W., Schüller, A.: Multigrid. Academic Press, London San Diego (2001) 413–532
3. Stüben, K.: A review of algebraic multigrid. Journal of Computational and Applied Mathematics **128** (2001) 281–309
4. Stüben, K., Clees, T.: SAMG Release 21b, User's Manual. Fraunhofer SCAI, St. Augustin, Germany (2002)
5. Füllenbach, T., Stüben, K.: Algebraic Multigrid for Selected PDE Systems. In: Elliptic and Parabolic Problems, Rolduc and Gaeta 2001. Proceedings of the 4th European Conference. World Scientific, New Jersey London (2002) 399–410
6. Oeltz, D.: Algebraische Mehrgittermethoden für Systeme partieller Differentialgleichungen. Diplomarbeit, Rheinische Friedrich-Wilhelms-Universität, Bonn (2001)
7. Griebel, M., Oeltz, D., Schweitzer, M.A.: An algebraic multigrid method for linear elasticity. SIAM J. Sci. Comp., submitted
8. Füllenbach, T., Stüben, K., Mijalković, S.: Application of an algebraic multigrid solver to process simulation problems. In: Proceedings SISPAD 2000, International Conference on Simulation of Semiconductor Processes and Devices. IEEE, Piscataway, NJ (2000) 225–228
9. Markowich, P.A., Ringhofer, C.A., Schmeiser, C.: Semiconductor Equations. Springer, Wien New York (1990)
10. Selberherr, S.: Analysis and Simulation of Semiconductor Devices. Springer, Wien New York (1984)
11. Markowich, P.A.: The Stationary Semiconductor Device Equations. Springer, Wien New York (1986)
12. Kerkhoven, T.: On the Scharfetter-Gummel box method. In: Selberherr, S., Stippel, H., Strasser, E. (eds.): Simulation of Semiconductor Devices and Processes, Vol. 5. Springer, Wien New York (1993) 237–240
13. Fuhrmann, J., Gärtner, K.: Incomplete factorization and linear multigrid algorithms for the semiconductor device equations. In: Beauwens, R., de Groen, P. (eds.): Proceedings of the IMACS International Symposium on Iterative Methods in Linear Algebra. Elsevier, Amsterdam (1992) 493–503
14. Taurus, Release 2001.4. Synopsys Inc., Mountainview, CA (2002)
15. Sze, S.M. (ed.): Modern Semiconductor Device Physics. Wiley, New York Chichester (1998)
16. Huang, X., et al.: Sub-50 nm p-channel FinFET. IEEE Transactions on Electron Devices **48** (2001) 880–886

Simulation of Microwave and Semiconductor Laser Structures Including Absorbing Boundary Conditions

Georg Hebermehl[1], Friedrich-Karl Hübner[1], Rainer Schlundt[1], Thorsten Tischler[2], Horst Zscheile[2], and Wolfgang Heinrich[2]

[1] Weierstrass Institute for Applied Analysis and Stochastics, Mohrenstr. 39, 10117 Berlin, Germany
[2] Ferdinand-Braun-Institut für Höchstfrequenztechnik, Albert-Einstein-Str. 11, 12489 Berlin, Germany

Summary. The transmission properties of microwave and optical structures can be described in terms of their scattering matrix using a three-dimensional boundary value problem for Maxwell's equations. The computational domain is truncated by electric or magnetic walls, open structures are treated using the Perfectly Matched Layer (PML) Absorbing Boundary Condition. The boundary value problem is solved by a finite-volume scheme. This results in a two-step procedure: an eigenvalue problem for general complex matrices and the solution of a large-scale system of linear equations with indefinite symmetric complex matrices. The modes of smallest attenuation are located in a region bounded by two parabolas, and are found solving a sequence of eigenvalue problems of modified matrices. To reduce the execution times a coarse and a fine grid, and two levels of parallelization can be used. For the computation of the discrete grid equations, advanced preconditioning techniques are applied to reduce the dimension and the number of iterations solving the large-scale systems of linear algebraic equations. These matrix problems need to be solved repeatedly for different right-hand sides, but with the same coefficient matrix. The used block quasi-minimal residual algorithm is a block Krylov subspace iterative method that incorporates deflation to delete linearly and almost linearly dependent vectors in the block Krylov sequences. Special attention is paid to the PML which causes a significantly increased number of iterations within Krylov subspace methods.

1 Introduction

The fields of applications include mobile communications, radio links, automobile radar systems, optical communications, and material processing. The commercial applications of microwave circuits cover the frequency range between 1 GHz and about 100 GHz; special applications in radioastronomy use even higher frequencies up to 1 THz. For optoelectronic devices frequencies about several hundred THz are common.

The subject under investigation are passive structures of arbitrary geometry. They are connected to the remaining circuit by transmission lines. Ports p (see Fig. 1) are defined on the transmission lines outer terminations. In order to characterize their electrical behavior the transmission lines are assumed to be infinitely long and longitudinally homogeneous. Short parts of the transmission lines and the passive structure form the so-called discontinuity. This structure has to be surrounded with an enclosure.

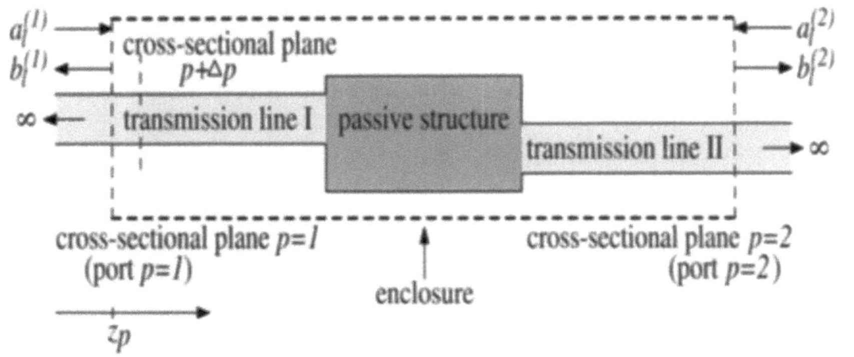

Fig. 1. Structure under investigation

The scattering matrix describes the structure in terms of wave modes on the transmission line sections at the ports. One can extract this matrix from the orthogonal decomposition of the electric field at two neighboring cross-sectional planes on each transmission line for a number of linearly independent excitations.

A three-dimensional boundary value problem can be formulated using the integral form of Maxwell's equations in the frequency domain in order to compute the electromagnetic field [1] and subsequently the scattering matrix. For numerical treatment, the computational domain has to be truncated by electric or magnetic walls or by a so-called Absorbing Boundary Condition (ABC) simulating open space. Among the ABC's, the Perfectly Matched Layer (PML) [13] technique represents the most powerful formulation. At the ports the transverse electric field is given by superposing transmission line modes. The transverse electric mode fields are the solutions of an eigenvalue problem for the transmission lines [2].

Maxwell's equations are discretized with staggered nonequidistant rectangular grids, using the Finite Integration Technique (FIT) [20], [1], [7], that transforms the continuous Maxwell equations into a set of discrete Maxwellian grid equations.

Assuming longitudinal homogeneity for the transmission line structure gives an eigenvalue problem for the computation of the transverse electric mode fields.

Only a few modes of smallest attenuation are able to propagate and have to be taken into consideration. Using a conformal mapping between the plane of propagation constants and the plane of eigenvalues the task is to compute all eigenmodes in a region, bounded by two parabolas. The region is covered by a number of overlapping circles. The eigenmodes in these circles are found solving a sequence of eigenvalue problems of modified matrices with the aid of the invert mode of the Arnoldi iteration [10], [17] using shifts.

The PML influences the mode spectrum. Modes that are related to the PML boundary can be detected, using the power part criterion given with [19]. This criterion is based on the comparison between the power concentration inside the PML region to the whole computational domain.

The integral over the power density vector, required for the calculation of the scattering matrix, of the power part criterion, and for the orthogonalization of eigenvectors, is computed for the used staggered grids.

In the case of multiple eigenvalues the eigenfunctions are orthogonalized according to the orthogonality relation for the scattering matrix using the algorithm of Gram-Schmidt [18].

The presented method, developed initially for a reliable calculation of all interesting complex eigenvalues from microwave structure computations [9], is expanded to meet the very special requirements of optoelectronic structure calculations. Relatively large cross sections and highest frequencies yield increased dimensions and numbers of eigenvalue problems. Using the results of a coarse grid calculation within the final fine grid calculation yields a remarkable reduced numerical effort. The use of two levels of parallelization results in an additional speed up of computation time.

After solution of the eigenvalue problem all boundary conditions are known and we can solve the boundary value problem to compute the three-dimensional electric field. The electromagnetic fields are calculated by the solution of large-scale systems of linear equations with indefinite complex symmetric coefficient matrices. In general, these matrix problems have to be solved repeatedly for different right-hand sides, but with the same coefficient matrix. The number of right-hand sides depends on the number of ports and modes. The systems of linear equations are solved using a block Krylov subspace iterative method. Independent set orderings, Jacobi and SSOR preconditioning techniques are applied to reduce the dimension and the number of iterations [16]. In comparison to the simple lossy case the number of iterations of Krylov subspace methods increases significantly in the presence of Perfectly Matched Layers. This growth is reduced by applying suitable grids and PML properties.

In general, the computation of the eigenvalue problem and of the system of linear algebraic equations have to be done for several frequencies.

2 Scattering Matrix

The scattering matrix

$$S = \begin{pmatrix} S_{11} & S_{12} & \cdots & S_{1m_s} \\ S_{21} & S_{22} & \cdots & S_{2m_s} \\ \cdots & \cdots & \cdots & \cdots \\ S_{m_s1} & S_{m_s2} & \cdots & S_{m_sm_s} \end{pmatrix} = (S_{\rho,\sigma}), \quad \rho,\sigma = 1(1)m_s, \tag{1}$$

$$\text{with} \quad m_s = \sum_{p=1}^{\overline{p}} m^{(p)}, \quad \rho = l + \sum_{q=1}^{p-1} m^{(q)}, \tag{2}$$

describes the energy exchange and phase relation between all outgoing and all incoming modes. $m^{(p)}$ denotes the number of modes which have to be taken into account at the port p. $\overline{p}$ is the number of ports. The modes on a port p are numbered with l, $l = 1(1)m^{(p)}$. That means, m_s is the number of all modes at all ports. To simplify matters, the cross-sectional planes and ports coincide in the structure under investigation (see Fig. 1). Generally, more than one port can be defined on one cross-sectional plane.

The computation and the explicit definition of the scattering matrix are described in the following.
The scattering matrix can be extracted from the orthogonal decomposition of the electric field (see (3)) at a pair of neighboring cross-sectional planes z_p and $z_{p+\Delta p}$ (see Fig. 1) on each waveguide for a number of linear independent excitations of the transmission lines. The electric fields at the cross-sectional planes z_p and $z_{p+\Delta p}$ are the solutions of an eigenvalue (see section 5) and of a boundary value problem (see section 3), respectively.

The transverse electric mode fields $\boldsymbol{E}_{t,l}(z)$ satisfy an orthogonality relation

$$\int_{\Omega} (\boldsymbol{E}_{t,l}(z) \times \boldsymbol{H}_{t,m}(z)) \cdot d\boldsymbol{\Omega} = \eta_m \delta_{l,m}. \tag{3}$$

Here $\boldsymbol{H}_{t,m}$ are the transverse magnetic mode fields.

In the case of degenerate modes, i.e., the algebraic multiplicity of the corresponding eigenvalues is greater than 1, we have to use first (3) in order to orthogonalize the modes (see section 9).

We consider all exciting modes with amplitudes a_l in positive z-direction and all outgoing modes with amplitudes b_l in negative z-direction. This assumption corresponds with the case, in which the cross section is located on the left-handed (x, y)-plane of the enclosure (see Fig. 1). The transverse mode field at a cross-sectional plane z is given by

$$\boldsymbol{E}_t(z) = \sum_{l=1}^{m^{(p)}} a_l \boldsymbol{E}_{t,l} e^{-\jmath k_{z_l} z} + \sum_{l=1}^{m^{(p)}} b_l \boldsymbol{E}_{t,l} e^{+\jmath k_{z_l} z} = \sum_{l=1}^{m^{(p)}} w_l(z) \boldsymbol{E}_{t,l} \tag{4}$$

with

$$w_l(z) = a_l e^{-\jmath k_{z_l} z} + b_l e^{+\jmath k_{z_l} z} = \tilde{a}_l(z) + \tilde{b}_l(z), \tag{5}$$

where k_{z_l} is the propagation constant. The application of (4) with (5) at a pair of neighboring cross-sectional planes z_p and $z_{p+\Delta p}$ gives because of $\boldsymbol{H}_{t,m}^{(p+\Delta p)} = \boldsymbol{H}_{t,m}^{(p)}$:

$$\begin{aligned} \frac{1}{\eta_m} \int_\Omega (\boldsymbol{E}_t^{(p)} \times \boldsymbol{H}_{t,m}^{(p)}) \cdot d\boldsymbol{\Omega} &= \tilde{a}_m^{(p)} + \tilde{b}_m^{(p)} &= w_m^{(p)}, \\ \frac{1}{\eta_m} \int_\Omega (\boldsymbol{E}_t^{(p+\Delta p)} \times \boldsymbol{H}_{t,m}^{(p)}) \cdot d\boldsymbol{\Omega} &= \tilde{a}_m^{(p+\Delta p)} + \tilde{b}_m^{(p+\Delta p)} &= w_m^{(p+\Delta p)} \end{aligned} \tag{6}$$

with

$$\begin{aligned} \boldsymbol{E}_t^{(p)} &= \boldsymbol{E}_t(z_p), & \boldsymbol{E}_t^{(p+\Delta p)} &= \boldsymbol{E}_t(z_{p+\Delta p}), \\ \boldsymbol{H}_{t,m}^{(p)} &= \boldsymbol{H}_{t,m}(z_p), & \boldsymbol{H}_{t,m}^{(p+\Delta p)} &= \boldsymbol{H}_{t,m}(z_{p+\Delta p}). \end{aligned} \tag{7}$$

We get $\boldsymbol{E}_t^{(p)}$ solving eigenvalue problems for the transmission lines (see section 5). $H_{t,m}^{(p)}$ can be computed from the known electric field $E_{t,m}$ of mode m (see section 8). The values of the weighted mode amplitude sums $w_m^{(p)}$ are given (see the discussion to follow). Thus, the normalization constant η_m can be computed by evaluating the orthogonality relation in the first equation of (6). $\boldsymbol{E}_t^{(p+\Delta p)}$ is computed solving a boundary value problem for the discontinuity (see section 3). Thus, the weighted mode amplitude sums $w_m^{(p+\Delta p)}$ can be calculated by using the second equation of (6).

The scattering matrix is defined with $\tilde{a}_m^{(p)}$ and $\tilde{b}_m^{(p)}$. Because of (5) we have

$$\tilde{a}_m^{(p+\Delta p)} = \tilde{a}_m^{(p)} e^{-\jmath k_{z_l}^{(p)} \Delta z_p}, \quad \tilde{b}_m^{(p+\Delta p)} = \tilde{b}_m^{(p)} e^{+\jmath k_{z_l}^{(p)} \Delta z_p}. \tag{8}$$

By using (8) we eliminate $\tilde{a}_m^{(p+\Delta p)}$ and $\tilde{b}_m^{(p+\Delta p)}$ in (6), and obtain

$$\tilde{a}_m^{(p)} = \frac{w_m^{(p)} e^{+\jmath k_{zm}^{(p)} \Delta z_p} - w_m^{(p+\Delta p)}}{e^{+\jmath k_{zm}^{(p)} \Delta z_p} - e^{-\jmath k_{zm}^{(p)} \Delta z_p}}, \quad \tilde{b}_m^{(p)} = \frac{w_m^{(p+\Delta p)} - w_m^{(p)} e^{-\jmath k_{zm}^{(p)} \Delta z_p}}{e^{+\jmath k_{zm}^{(p)} \Delta z_p} - e^{-\jmath k_{zm}^{(p)} \Delta z_p}}. \tag{9}$$

By using (9) reflection coefficients

$$r_m^{(p)} = \frac{\tilde{b}_m^{(p)}}{\tilde{a}_m^{(p)}} = \frac{e^{-jk_{zm}^{(p)}\Delta z_p} - \frac{w_m^{(p+\Delta p)}}{w_m^{(p)}}}{\frac{w_m^{(p+\Delta p)}}{w_m^{(p)}} - e^{+jk_{zm}^{(p)}\Delta z_p}} \tag{10}$$

are computed for all modes $\rho = 1(1)m_s$ and all excitations $\nu = 1(1)m_s$.

The excitations $w_m^{(p)}$ are given and can be described by the vectors

$$\bar{\boldsymbol{w}}_\nu = (\bar{w}_{1,\nu}, \ldots, \bar{w}_{\rho,\nu}, \ldots, \bar{w}_{m_s,\nu})^T, \quad \nu = 1(1)m_s. \tag{11}$$

The vectors (11) have to be linear independent. That is achieved here by choosing the components of $\bar{\boldsymbol{w}}_\nu$ in the following way:

$$\bar{w}_{\rho,\nu} = \begin{cases} |w_m^{(p)}| \text{ for } 1 \leq \rho \leq m_s + 1 - \nu \\ -|w_m^{(p)}| \text{ for } m_s + 2 - \nu \leq \rho \leq m_s \end{cases}, \quad \rho = m + \sum_{q=1}^{p-1} m^{(q)}, \tag{12}$$

with

$$w_m^{(p)} = 1.0, \quad m = 1(1)m^{(p)}, \quad p = 1(1)\bar{p}. \tag{13}$$

With this choice of $\bar{\boldsymbol{w}}_\nu$ (see (11), (12) and (13)) the vectors $\bar{\boldsymbol{r}}_\nu$, $\bar{\boldsymbol{a}}_\nu$ and $\bar{\boldsymbol{b}}_\nu$ are built up analogously (see (9) and (10)):

$$\begin{aligned} \bar{\boldsymbol{r}}_\nu &= (\bar{r}_{1,\nu}, \ldots, \bar{r}_{\rho,\nu}, \ldots, \bar{r}_{m_s,\nu})^T, & \bar{r}_{\rho,\nu} &= r_m^{(p)}, \\ \bar{\boldsymbol{a}}_\nu &= (\bar{a}_{1,\nu}, \ldots, \bar{a}_{\rho,\nu}, \ldots, \bar{a}_{m_s,\nu})^T, & \bar{a}_{\rho,\nu} &= \tilde{a}_m^{(p)}, \\ \bar{\boldsymbol{b}}_\nu &= (\bar{b}_{1,\nu}, \ldots, \bar{b}_{\rho,\nu}, \ldots, \bar{b}_{m_s,\nu})^T, & \bar{b}_{\rho,\nu} &= \tilde{b}_m^{(p)}. \end{aligned} \tag{14}$$

The relation between (ρ, ν) on the one hand and $(m, (p))$ on the other hand is given by (12) and (13). The choice of $\bar{\boldsymbol{w}}_\nu$ and the relations between the indices are demonstrated by a small example at the end of this section.

That means, we have to solve m_s boundary value problems with the boundary condition

$$\boldsymbol{E}_{t,\nu} = \sum_{\rho=1}^{m_s} \bar{w}_{\rho,\nu} \boldsymbol{E}_{t,l}^{(p)}, \quad \rho = l + \sum_{q=1}^{p-1} m^{(q)}, \quad p = 1(1)\bar{p}, \quad \nu = 1(1)m_s, \tag{15}$$

in order to compute $w_m^{(p+\Delta p)}$ and subsequently $\bar{\boldsymbol{r}}_\nu$.

The scattering matrix S (see (1)) is defined by

$$\bar{\boldsymbol{b}}_\nu = S\bar{\boldsymbol{a}}_\nu, \quad \nu = 1(1)m_s, \tag{16}$$

or (see (14))

$$\bar{b}_{\rho,\nu} = \sum_{\sigma=1}^{m_s} S_{\rho,\sigma} \cdot \bar{a}_{\sigma,\nu}, \quad \rho, \nu = 1(1)m_s. \tag{17}$$

Because of (6) and (10) we have

$$\begin{array}{ll} \bar{w}_{\rho,\nu} = \bar{a}_{\rho,\nu} + \bar{b}_{\rho,\nu}, & \\ 0 \;\;\;\;\, = \bar{r}_{\rho,\nu}\bar{a}_{\rho,\nu} - \bar{b}_{\rho,\nu}, \end{array} \quad \text{or} \quad \begin{array}{l} \bar{a}_{\rho,\nu}(1+\bar{r}_{\rho,\nu}) = \bar{w}_{\rho,\nu}, \\ \bar{b}_{\rho,\nu}(1+\bar{r}_{\rho,\nu}) = \bar{r}_{\rho,\nu}\bar{w}_{\rho,\nu}, \end{array} \quad \rho,\nu = 1(1)m_s. \quad (18)$$

Multiplying Equation (17) with the product $\prod_{\mu=1}^{m_s}(1+\bar{r}_{\mu,\nu})$ gives

$$\bar{b}_{\rho,\nu}\prod_{\mu=1}^{m_s}(1+\bar{r}_{\mu,\nu}) = \sum_{\sigma=1}^{m_s} S_{\rho,\sigma}\bar{a}_{\sigma,\nu}\prod_{\mu=1}^{m_s}(1+\bar{r}_{\mu,\nu}), \quad \rho,\nu = 1(1)m_s. \quad (19)$$

Substitution of (18) into the relation (19) gives

$$\bar{r}_{\rho,\nu}\bar{w}_{\rho,\nu}\prod_{\substack{\mu=1\\ \mu\neq\rho}}^{m_s}(1+\bar{r}_{\mu,\nu}) = \sum_{\sigma=1}^{m_s} S_{\rho,\sigma}\bar{w}_{\sigma,\nu}\prod_{\substack{\mu=1\\ \mu\neq\rho}}^{m_s}(1+\bar{r}_{\mu,\nu}), \quad \rho,\nu = 1(1)m_s, \quad (20)$$

or

$$R_{\rho,\nu} = \sum_{\sigma=1}^{m_s} S_{\rho,\sigma}W_{\sigma,\nu} \text{ with } W_{\rho,\nu} = \bar{w}_{\rho,\nu}\prod_{\substack{\mu=1\\ \mu\neq\rho}}^{m_s}(1+\bar{r}_{\mu,\nu}),\ R_{\rho,\nu} = \bar{r}_{\rho,\nu}W_{\rho,\nu}. \quad (21)$$

(21) can be written as matrix equation:

$$R = SW. \quad (22)$$

That means, we have to solve m_s linear algebraic equations in order to compute the $(m_s)^2$ coefficients of S:

$$W^T(S_{\rho,1},\ldots,S_{\rho,m_s})^T = (R_{\rho,1},\ldots,R_{\rho,m_s})^T, \quad \rho = 1(1)m_s. \quad (23)$$

Example

The choice of the excitations (11) and the relations between the indices in (14) are demonstrated by an example:

Number of ports: $\overline{p} = 2,\ p = 1,2.$
Number of modes at the port $p = 1$: $m^{(1)} = 2.$
Number of modes at the port $p = 2$: $m^{(2)} = 3.$

Sum of all modes:

$$\begin{array}{l} m_s = m^{(1)} + m^{(2)} = 5 \text{ (see (2))}, \\ p = 1:\ \ l = m = 1,2,\ \ \rho = 1,2, \\ p = 2:\ \ l = m = 1,2,3,\ \ \rho = 3,4,5. \end{array}$$

Number of excitations: $m_s = 5,\ \nu = 1(1)5,$ (see (11) - (14)):

$$\begin{aligned}
\bar{\boldsymbol{w}}_\nu &= (\bar{w}_{1,\nu}, \ldots, \bar{w}_{\rho,\nu}, \ldots, \bar{w}_{5,\nu})^T, \\
\bar{\boldsymbol{r}}_\nu &= (\bar{r}_{1,\nu}, \ldots, \bar{r}_{\rho,\nu}, \ldots, \bar{r}_{5,\nu})^T, \\
\bar{\boldsymbol{a}}_\nu &= (\bar{a}_{1,\nu}, \ldots, \bar{a}_{\rho,\nu}, \ldots, \bar{a}_{5,\nu})^T, \\
\bar{\boldsymbol{b}}_\nu &= (\bar{b}_{1,\nu}, \ldots, \bar{b}_{\rho,\nu}, \ldots, \bar{b}_{5,\nu})^T.
\end{aligned}$$

The $\bar{w}_{\rho,\nu}$, $\bar{r}_{\rho,\nu}$, $\bar{a}_{\rho,\nu}$, and $\bar{b}_{\rho,\nu}$ are given for $\nu = 1$:

$$\begin{array}{llllll}
\bar{w}_{1,1} = |\bar{w}_1^{(1)}| = 1.0, & \bar{r}_{1,1} = r_1^{(1)}, & \bar{a}_{1,1} = \tilde{a}_1^{(1)}, & \bar{b}_{1,1} = \tilde{b}_1^{(1)}, & \\
\rho = 1, & p = 1, & m = 1, & & \\
\bar{w}_{2,1} = |\bar{w}_2^{(1)}| = 1.0, & \bar{r}_{2,1} = r_2^{(1)}, & \bar{a}_{2,1} = \tilde{a}_2^{(1)}, & \bar{b}_{2,1} = \tilde{b}_2^{(1)}, & \\
\rho = 2, & p = 1, & m = 2, & & \\
\bar{w}_{3,1} = |\bar{w}_1^{(2)}| = 1.0, & \bar{r}_{3,1} = r_1^{(2)}, & \bar{a}_{3,1} = \tilde{a}_1^{(2)}, & \bar{b}_{3,1} = \tilde{b}_1^{(2)}, & \\
\rho = 3, & p = 2, & m = 1, & m^{(1)} = 2, & \\
\bar{w}_{4,1} = |\bar{w}_2^{(2)}| = 1.0, & \bar{r}_{4,1} = r_2^{(2)}, & \bar{a}_{4,1} = \tilde{a}_2^{(2)}, & \bar{b}_{4,1} = \tilde{b}_2^{(2)}, & \\
\rho = 4, & p = 2, & m = 2, & m^{(1)} = 2, & \\
\bar{w}_{5,1} = |\bar{w}_3^{(2)}| = 1.0, & \bar{r}_{5,1} = r_3^{(2)}, & \bar{a}_{5,1} = \tilde{a}_3^{(2)}, & \bar{b}_{5,1} = \tilde{b}_3^{(2)}, & \\
\rho = 5, & p = 2, & m = 3, & m^{(1)} = 2, &
\end{array}$$

and for $\nu = 3$:

$$\begin{array}{llllll}
\bar{w}_{1,3} = |\bar{w}_1^{(1)}| = 1.0, & \bar{r}_{1,3} = r_1^{(1)}, & \bar{a}_{1,3} = \tilde{a}_1^{(1)}, & \bar{b}_{1,3} = \tilde{b}_1^{(1)}, & \\
\rho = 1, & p = 1, & m = 1, & & \\
\bar{w}_{2,3} = |\bar{w}_2^{(1)}| = 1.0, & \bar{r}_{2,3} = r_2^{(1)}, & \bar{a}_{2,3} = \tilde{a}_2^{(1)}, & \bar{b}_{2,3} = \tilde{b}_2^{(1)}, & \\
\rho = 2, & p = 1, & m = 2, & & \\
\bar{w}_{3,3} = |\bar{w}_1^{(2)}| = 1.0, & \bar{r}_{3,3} = r_1^{(2)}, & \bar{a}_{3,3} = \tilde{a}_1^{(2)}, & \bar{b}_{3,3} = \tilde{b}_1^{(2)}, & \\
\rho = 3, & p = 2, & m = 1, & m^{(1)} = 2, & \\
\bar{w}_{4,3} = -|\bar{w}_2^{(2)}| = -1.0, & \bar{r}_{4,3} = r_2^{(2)}, & \bar{a}_{4,3} = \tilde{a}_2^{(2)}, & \bar{b}_{4,3} = \tilde{b}_2^{(2)}, & \\
\rho = 4, & p = 2, & m = 2, & m^{(1)} = 2, & \\
\bar{w}_{5,3} = -|\bar{w}_3^{(2)}| = -1.0, & \bar{r}_{5,3} = r_3^{(2)}, & \bar{a}_{5,3} = \tilde{a}_3^{(2)}, & \bar{b}_{5,3} = \tilde{b}_3^{(2)}, & \\
\rho = 5, & p = 2, & m = 3, & m^{(1)} = 2. &
\end{array}$$

3 Boundary Value Problem

A three-dimensional boundary value problem can be formulated using the integral form of Maxwell's equations in the frequency domain in order to compute the electromagnetic field:

$$\oint_{\partial\Omega} \boldsymbol{H} \cdot d\boldsymbol{s} = \int_{\Omega} \jmath\omega[\epsilon]\boldsymbol{E} \cdot d\boldsymbol{\Omega}, \qquad \oint_{\cup\Omega} ([\epsilon]\boldsymbol{E}) \cdot d\boldsymbol{\Omega} = 0, \tag{24}$$

$$\oint_{\partial\Omega} \boldsymbol{E} \cdot d\boldsymbol{s} = -\int_{\Omega} \jmath\omega[\mu]\boldsymbol{H} \cdot d\boldsymbol{\Omega}, \qquad \oint_{\cup\Omega} ([\mu]\boldsymbol{H}) \cdot d\boldsymbol{\Omega} = 0, \tag{25}$$

$$\boldsymbol{D} = [\epsilon]\boldsymbol{E}, \;\; \boldsymbol{B} = [\mu]\boldsymbol{H}, \;\; [\epsilon] = \mathrm{diag}\,(\epsilon_x, \epsilon_y, \epsilon_z)\,, \;\; [\mu] = \mathrm{diag}\,(\mu_x, \mu_y, \mu_z)\,. \tag{26}$$

The electric and magnetic flux densities $\boldsymbol{D}$ and $\boldsymbol{B}$ are complex functions of the spatial coordinates. ω is the circular frequency of the sinusoidal excitation, and $\jmath^2 = -1$. In the left-hand sides of formulae (24) and (25) Ω is an open surface surrounded by a closed contour $\partial\Omega$, while in the right-hand sides of (24) and (25) $\cup\Omega$ is a closed surface with an interior volume. The direction of the element $\boldsymbol{ds}$ of the contour $\partial\Omega$ is such that when a right-handed screw is turned in that direction, it will advance in the direction of the vector element $\boldsymbol{d\Omega}$.

At the ports p the transverse electric field $\boldsymbol{E}_t(z_p)$ is given by superposing transmission line modes $\boldsymbol{E}_{t,l}(z_p)$ (see (15)):

$$\boldsymbol{E}_t(z_p) = \sum_{l=1}^{m^{(p)}} w_l(z_p)\boldsymbol{E}_{t,l}(z_p). \tag{27}$$

The transverse electric mode fields have to be computed solving an eigenvalue problem for the transmission lines (see section 5). All other parts of the surface of the computation domain are assumed to be an electric or a magnetic wall:

$$\boldsymbol{E} \times \boldsymbol{n} = 0 \quad \text{or} \quad \boldsymbol{H} \times \boldsymbol{n} = 0. \tag{28}$$

At the material boundaries the tangential components of the electric and the magnetic field, and the normal components of the electric and the magnetic flux density have to be continuous.

In order to simulate open structures, the Perfectly Matched Layer absorbing boundary condition is implemented, namely the uniaxial PML formulation according to [13]. The PML region is filled with an artificial material with complex anisotropic material properties. A complex permittivity $[\epsilon] = [\epsilon(x, y, z)]$ and a complex permeability $[\mu] = [\mu(x, y, z)]$ diagonal tensor are introduced, resulting in a reflection free interface between the computational area and the lossy PML region.

4 Grid Equations

Maxwell's equations are discretized using staggered nonequidistant rectangular grids. Using the Finite Integration Technique (FIT) [20], [1], [7] with the lowest order integration formulae

$$\oint_{\partial\Omega} \boldsymbol{f} \cdot \boldsymbol{ds} \approx \sum(\pm f_i s_i), \qquad \int_{\Omega} \boldsymbol{f} \cdot \boldsymbol{d\Omega} \approx f\Omega \tag{29}$$

Equations (24), (25) are transformed into a set of grid equations:

$$A^T D_{s/\mu}\boldsymbol{b} = \jmath\omega\epsilon_0\mu_0 D_{A_\epsilon}\boldsymbol{e}, \qquad B D_{A_\epsilon}\boldsymbol{e} = 0, \tag{30}$$

$$A D_s\boldsymbol{e} = -\jmath\omega D_A\boldsymbol{b}, \qquad \tilde{B} D_A\boldsymbol{b} = 0. \tag{31}$$

The vectors $\boldsymbol{e}$ and $\boldsymbol{b}$ contain the components of the electric field intensity and the magnetic flux density of the elementary cells, respectively. The diagonal matrices $D_{s/\mu}$, D_{A_ϵ}, D_s, and D_A contain the information on cell dimension and material. A, B, and $\tilde{B}$ are sparse.

By eliminating the components of the magnetic flux density from the two equations on the left-hand sides of (30) and (31), we get the system of linear algebraic equations

$$(A^T D_{s/\mu} D_A^{-1} A D_s - k_0^2 D_{A_\epsilon})\boldsymbol{e} = 0, \quad k_0 = \omega\sqrt{\epsilon_0\mu_0}, \tag{32}$$

which have to be solved using the boundary conditions (27) and (28), possibly supplemented by PML. k_0 is the wavenumber in vacuum.

5 Eigenvalue Problem

The field distribution at the ports is computed assuming longitudinal homogeneity for the transmission line structure. Thus, any field can be expanded into a sum of so-called modal fields which vary exponentially in the longitudinal direction:

$$\boldsymbol{E}(x, y, z \pm 2h) = \boldsymbol{E}(x, y, z)e^{\mp \jmath k_z 2h}. \tag{33}$$

k_z is the propagation constant. $2h$ is the length of an elementary cell in z-direction. We consider the field components in three consecutive elementary cells. The electric field components of the vector $\boldsymbol{e}$ (see (32)) $E_{x_{i,j,k+1}}$, $E_{x_{i,j,k-1}}$, $E_{y_{i,j,k+1}}$, $E_{y_{i,j,k-1}}$, $E_{z_{i,j,k-1}}$, $E_{z_{i+1,j,k-1}}$, and $E_{z_{i,j+1,k-1}}$ are expressed by the values of cell k using ansatz (33). The longitudinal electric field components E_z can be eliminated by means of the electric-field divergence equation $BD_{A_\epsilon}\boldsymbol{e} = 0$ (see (30)). Thus, we get an eigenvalue problem for the transverse electric field $\underline{\boldsymbol{e}}$ on the transmission line region:

$$C\underline{\boldsymbol{e}} = \gamma\underline{\boldsymbol{e}}, \quad \gamma = e^{-\jmath k_z 2h} + e^{+\jmath k_z 2h} - 2 = -4\sin^2(hk_z). \tag{34}$$

The sparse matrix C is in general nonsymmetric complex. The order of C is $n = 2n_x n_y - n_b$. $n_x n_y$ is the number of elementary cells at the port. The size n_b depends on the number of cells with perfectly conducting material. The relation between the propagation constants k_z and the eigenvalues γ is nonlinear, and can be expressed as

$$k_z = \frac{\jmath}{2h}\ln\left(\frac{\gamma}{2} + 1 + \sqrt{\frac{\gamma}{2}\left(\frac{\gamma}{2} + 2\right)}\right) = \beta - \jmath\alpha. \tag{35}$$

We are interested only in a few modes with the smallest attenuation. These are the modes with the smallest magnitude of imaginary part, but possibly with large real part of their propagation constant. The computation of all eigenvalues in order to find a few propagation constants must be avoided for the high-dimensional problem. For numerical treatment we have to limit the search for propagation constants by setting a maximum value k_f for their real parts. A reasonable estimation of this maximum value is derived for the lossy case including PML for inhomogeneously filled waveguides in [8]:

$$\mathbb{R}(k_z) \leq k_f = \omega \mathbb{R}(\sqrt{\epsilon_m \mu_m}). \tag{36}$$

ϵ_m and μ_m are properties of the material that yields the largest value of the right-hand side of equation (36). Using the limited k_f and a preset maximum value α_m of the imaginary part of the propagation constants the region containing the interesting constants is defined as a rectangle $\hat{F}$ bounded by the lines (see Fig. 3)

$$\beta = \pm k_f \quad \text{and} \quad \alpha = \pm \alpha_m. \tag{37}$$

We can use the approximation $\sin(x) \approx x$ in (34) if we choose h to be small enough, which is necessary in any case in order to get small discretization errors:

$$\gamma = -4\sin^2(hk_z) \approx -4(hk_z)^2 = u + \jmath v. \tag{38}$$

With the aid of the approximation (38) the conformal mapping (35) between the plane of eigenvalues (γ-plane, see Fig. 2) and the plane of propagation constants (k_z-plane, see Fig. 3) simplifies to:

$$u = -4h^2(\beta^2 - \alpha^2), \quad v = 8h^2\alpha\beta. \tag{39}$$

Under this mapping the rectangle $\hat{F}$ of the k_z-plane is transformed into a region F of the γ-plane bounded by the two parabolas

$$v = \pm 4hk_f\sqrt{u + 4h^2{k_f}^2} \quad \text{and} \quad v = \pm 4h\alpha_m\sqrt{-u + 4h^2{\alpha_m}^2}. \tag{40}$$

This means, we have to find all eigenvalues of the region bounded by the parabolas.

6 Computation of Eigenmodes

We need an algorithm that computes just a few selected eigenvalues and eigenvectors of a complex sparse matrix. A state-of-the-art algorithm for such problems is the Arnoldi method. We use the implicitly restarted Arnoldi iteration, implemented in the package ARPACK of the authors of [10] and [17]. In general, the method converges for our high-dimensional nonsymmetric eigenvalue problems only using the invert mode and looking for eigenvalues of largest magnitude. Thus, a simple way to find the eigenvalues located in the region F would be to look for all eigenvalues of smallest magnitude, which are located in a circle centered on the origin and covering the region F. Because of the

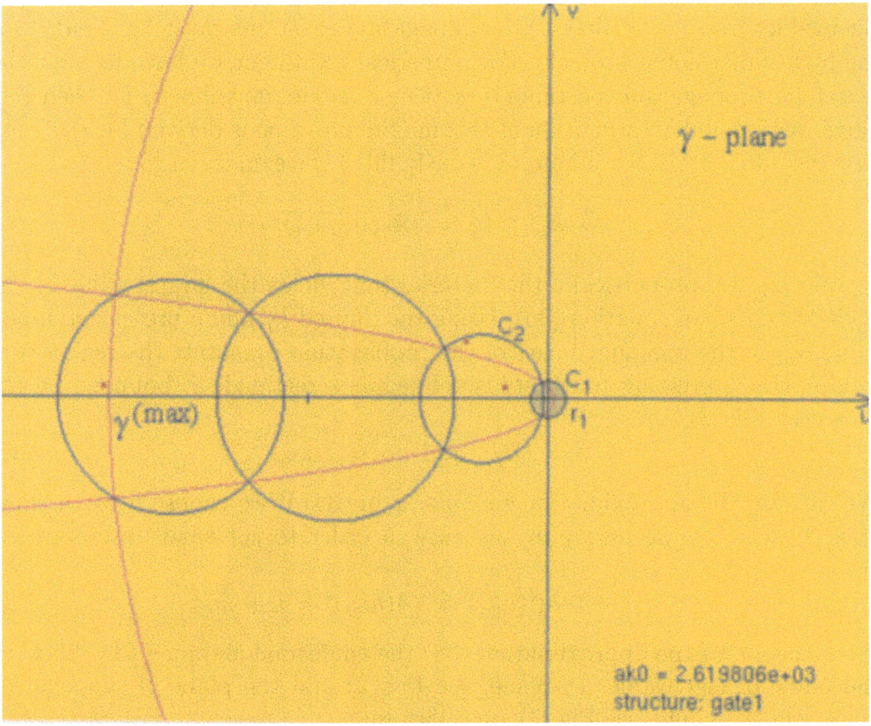

Fig. 2. γ-plane. The region bounded by two parabolas contains all eigenvalues of interest. For numerical treatment the region is covered by a number of smaller overlapping circles. $\gamma^{(max)} = -4\sin^2(hk_f)$.

high wavenumber k_f, the number of eigenvalues located in this circle is too large in general for a feasible computation using an iterative method. We can solve this problem covering the region F with $s \geq 1$ circles C_i, $i = 1(1)s$, centered on the u-axis and calculating the eigenvalues located in these circles. That is done in the following way. First s points

$$\hat{P}_i(\beta_i, \alpha_m),\ i = 1(1)s,\ \beta_1 = \frac{k_f}{s} \geq \bar{\beta},\ \beta_s = k_f, \quad \text{with} \quad \bar{\beta} = \sqrt{3}\alpha_m \qquad (41)$$

are defined on the interval $[0, k_f]$ of the line $\alpha = \alpha_m$. The distance between the points is controlled as shown below. Even the meaning of the distance $\bar{\beta}$ is discussed below. The points $\hat{P}_i$ are transformed into the points P_i of the γ-plane. They are located on the parabola ((40), right formula). The s circles C_i of the γ-plane (see Fig. 2)

$$(u - m_i)^2 + v^2 = {r_i}^2,\ r_i = \sqrt{(\mathbb{R}(P_i) - m_i)^2 + (\Im(P_i))^2}, \quad i = 1(1)s, \quad (42)$$

with

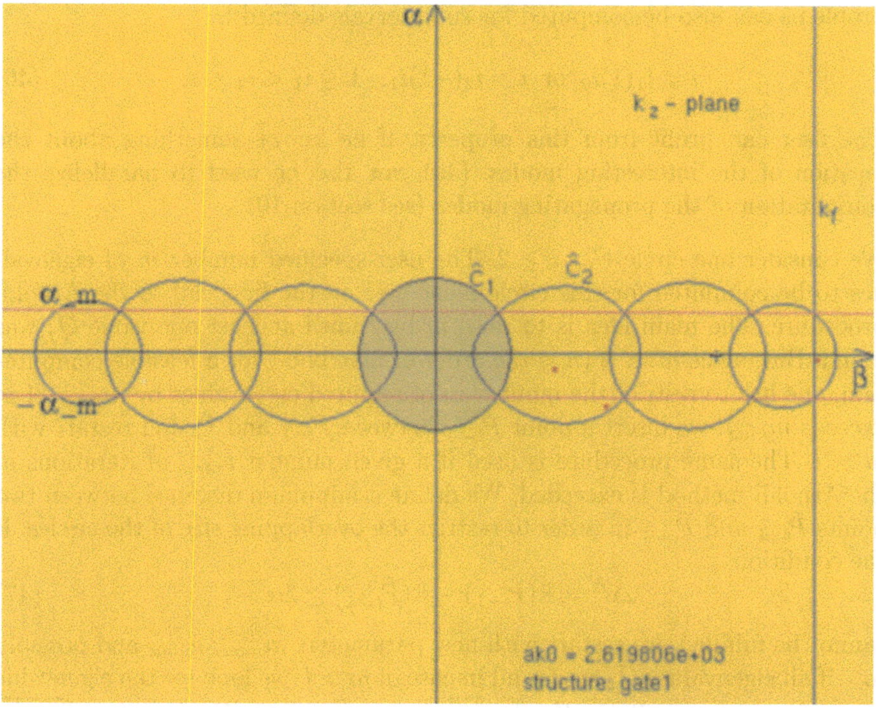

Fig. 3. k_z-plane. The region containing the interesting propagation constants is defined as a rectangle. The rectangle is covered by Cassinian curves which corresponds with the circles of Fig. 2. One mode was identified as guided wave mode. The corresponding propagation constant is chararacterized by a +.

$$m_1 = 0, \;\; m_i = \frac{(\mathbb{R}(P_i))^2 - (\mathbb{R}(P_{i-1}))^2 + (\Im(P_i))^2 - (\Im(P_{i-1}))^2}{2|\mathbb{R}(P_i) - \mathbb{R}(P_{i-1})|} \tag{43}$$

are centered on the u-axis, covering the region bounded by the parabolas.

In order to find all eigenvalues, located in the circle C_i, l points Q_j are defined on the periphery of C_i. The matrix C is extended by the diagonal matrix Q. The diagonal elements of Q are the l complex elements Q_j:

$$\bar{C} = \begin{pmatrix} Q & \\ & C \end{pmatrix}, \quad Q = diag(Q_1, ..., Q_l). \tag{44}$$

The complex elements Q_j are eigenvalues of $\bar{C}$.

The s eigenvalue problems

$$(\bar{C} - m_i I)\underline{e} = (\gamma - m_i)\underline{e}, \quad i = 1(1)s, \tag{45}$$

are solved with the aid of the implicitly restarted Arnoldi method using the invert mode. The eigenvalue problems can be solved separately. The eigenvalue problems can also be computed for subintervals defined by

$$i = i_1(1)i_2 \text{ or } i = i_2(-1)i_1, \;\; 1 \le i_1 < i_2 \le s. \tag{46}$$

The user can profit from this property, if he knows something about the location of the interesting modes. (46) can also be used to parallelize the computation of the propagating modes (see section 10).

We consider one circle $C_i, \;\; i \ge 2$. The user-specified number m of eigenvalues to be computed for this circle must be l on the first call to the Arnoldi procedure. The main idea is to raise m by l until at least one value Q_j was found. But, since $m \ll \bar{n}$ ($\bar{n}$ is the order of matrix $\bar{C}$) for a feasible computation, one has to restrict the number m of required eigenvalues by m_{max}. If m exceeds m_{max}, we insert a point $\hat{P}_{i-\frac{1}{2}}$ between $\hat{P}_{i-1}$ and $\hat{P}_i$ and restart with $m = l$. The same procedure is used if a given number ν_{max} of iterations in the Arnoldi method is exceeded. We define a minimum distance between two points $\hat{P}_{i-1}$ and $\hat{P}_{i-\frac{1}{2}}$ in order to restrict the overlapping size of the circles. If the condition

$$\Delta\hat{P} = \mathbb{R}(\hat{P}_{i-1}) - \mathbb{R}(\hat{P}_{i-\frac{1}{2}}) \ge 2\alpha_m \tag{47}$$

cannot be fulfilled, we restart with new parameters m_{max}, ν_{max} and possibly α_m. If all eigenvalues Q_j are found in case of $m > l$, we look for the eigenvalue γ_{max} of largest magnitude. If $\sqrt{|\gamma_{max}|} > r_i$, a new circle $\tilde{C}_i$ of radius $\sqrt{|\gamma_{max}|}$ with the same center as C_i is defined. The left intersection point of this circle with the parabola ((40), right formula) is used as new point P_i, and $\Delta\hat{P} = \mathbb{R}(\hat{P}_i) - \mathbb{R}(\hat{P}_{i-1})$ as distance for the next step. m is reduced by the number of eigenvalues with $\sqrt{|\gamma|} > r_i$ for the next circle.

Because in general the Arnoldi method does not converge using the regular mode for our eigenvalue problem the invert mode with shifting (see (45)) is applied. A time and memory consuming system of partly ill-conditioned non-symmetric complex linear algebraic equations (see also [8]) has to be solved on each iteration step in this case. The linear sparse solver PARDISO [14], [15] is applied in order to fulfill the high accuracy requirements of the eigenvalue problem. The algorithm is split into three phases: symbolic factorization, numerical factorization, and forward and backward solve. The symbolic factorization can be used for all modified matrices of our problem. The numerical factorization has to be repeated for every new shift. The factorization is applied to matrix C. The diagonal matrix Q (see (44)) is considered in the forward and backward solve phase.

The typical ratio of factorization time to solution time on a single CPU can be used to define ν_{max} in the subinterval control process. This ratio amounts on the average to 20 using the linear sparse solver PARDISO. That means, the costs using $\nu_{max} = 60$ Arnoldi iterations for the computation of m eigenmodes in a circle C_i defined by the points P_{i-1}, P_i are comparable with the costs, defined by the costs for two circles defined by the points $P_{i-1}, P_{i-\frac{1}{2}}$

and $P_{i-\frac{1}{2}}, P_i$ using $\nu_{max} = 20$ iterations. On the other hand the time is lost, interrupting the computation of m eigenmodes after $\nu_{max} = 60$ iterations and starting a new iteration process for two reduced circles. Thus, we use a greater ν_{max} (see section 11 for an example). Moreover, due to the significant difference between the length and the height of the rectangular region $\hat{F}$ in the k_z-plane we have to solve a large number s of eigenvalue problems (see section 10). In order to diminish this number we use large intervals in the k_z-plane, i.e., circles with relatively large diameters in the γ-plane. That means, a number of non desired eigenvalues outside of the area F has to be calculated. In general the computation of a large number m of eigenvalues in one circle needs more iterations than a small number.

Separating the new values on each eigenvalue problem i, we are sure to have found all eigenvalues which are located in the corresponding circles C_i. Applying the mapping (39) the circles C_i (see (42)) are transformed into Cassinian curves $\hat{C}_i$ (see Fig. 3)

$$(\beta^2 + \alpha^2)^2 - \frac{m_i}{2h^2}(\beta^2 - \alpha^2) = \frac{{r_i}^2}{16h^4} - \frac{{m_i}^2}{16h^4}, \tag{48}$$

which cover the rectangle $\hat{F}$ containing all desired propagation constants. Propagation constants outside of $\hat{F}$ and PML-modes (see section 7) are eliminated.

The Cassinian curves $\hat{C}_i, i = 2(1)s$, consist of two disjoint ovals [3], if

$$r_i < m_i. \tag{49}$$

That means, the circles C_i may neither cut nor be tangent to the v-axis. (49) can be fulfilled by a special choice of $\hat{P}_1$:
Because of $m_1 = 0$ (see (43)) the Cassinian curve $\hat{C}_1$ (see (48)) is a circle:

$$\beta^2 + \alpha^2 = \hat{r}_1^2 = \frac{r_1}{4h^2}. \tag{50}$$

The radius r_1 of C_1 may be defined by $\hat{P}_1(\bar{\beta}, \alpha_m)$ (see (41) and (50)). A circle

$$C_2: \quad (u + r_1)^2 + v^2 = {r_1}^2 \tag{51}$$

of radius $r_2 = r_1$ with the center $M_2(-r_1, 0)$ is evidently tangent to the v-axis. Because of (39) and (50) the point $P_1(u_1 = -\frac{r_1}{2}, v_1 = \frac{\sqrt{3}}{2}r_1)$ is an intersection point of C_2 with the parabola ((40), right formula), if $\bar{\beta} = \sqrt{3}\alpha_m$. All circles C_i with a center left of M_2 and the intersection point P_1 cut the u-axis left of the v-axis. Thus, using $\bar{\beta}$ as minimum distance between the origin and $\hat{P}_1$ (see (41)) other shapes of Cassinian curves (e.g. waisted ovals, see Fig. 4 and 5), which would lead to higher execution times, are avoided.

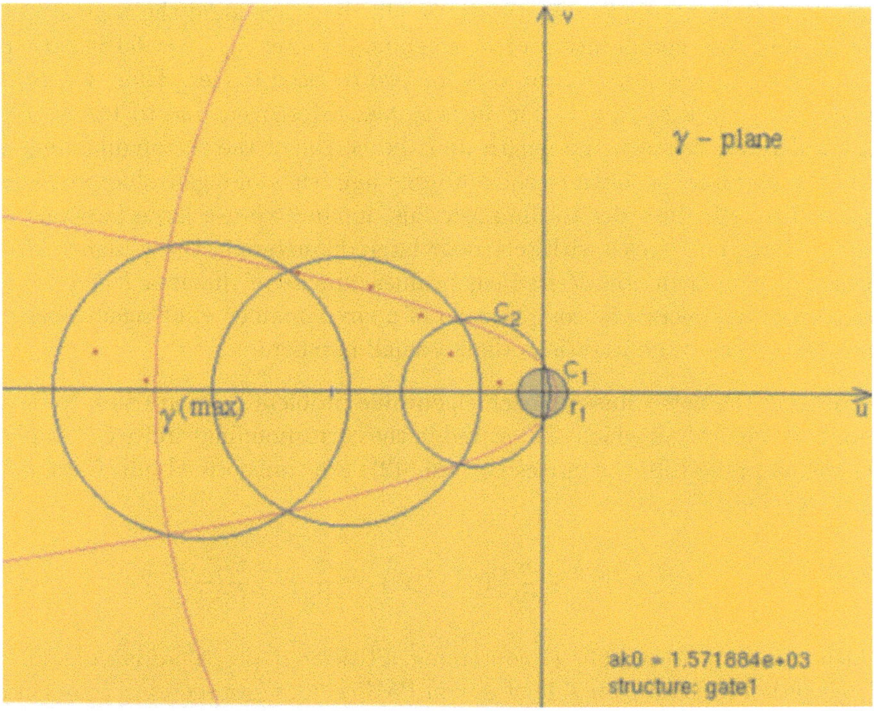

Fig. 4. γ-plane. The circle C_2 cut the v-axis in contrast to Fig. 2.

7 Power Part Criterion

The PML is applied in order to compute the eigen modes of open waveguide structures simulating the infinite space. Introducing the PML permits the calculation of radiation effects. Additionally the absorbing boundary suppresses the interaction between the waveguide modes and higher order modes, caused by the finite simulation domain. However, undesired so-called PML-modes are generated, due to the electric or magnetic walls behind these absorbing boundary layers, and the PML shifts only gently these box modes within the area $\hat{F}$ (see (37)). Therefore we need an additional tool to distinguish these PML-modes from the desired ones. As a result of our numerical calculations we found that examination of the eigenfunctions provides a useful criterion to select the modes of interest. As a criterion to distinguish between waveguide modes and undesired modes we use the power concentration of all modes by way of comparison. Undesired modes are characterized by a high power concentration inside the PML region, while propagating modes are concentrated in the waveguide area. Thus, to eliminate the PML-modes we calculate the magnitude of the power flow (see section 8) of each computed mode in the PML ($P^{(P)}$), in the waveguide region ($P^{(W)}$), and in the total computational domain (P):

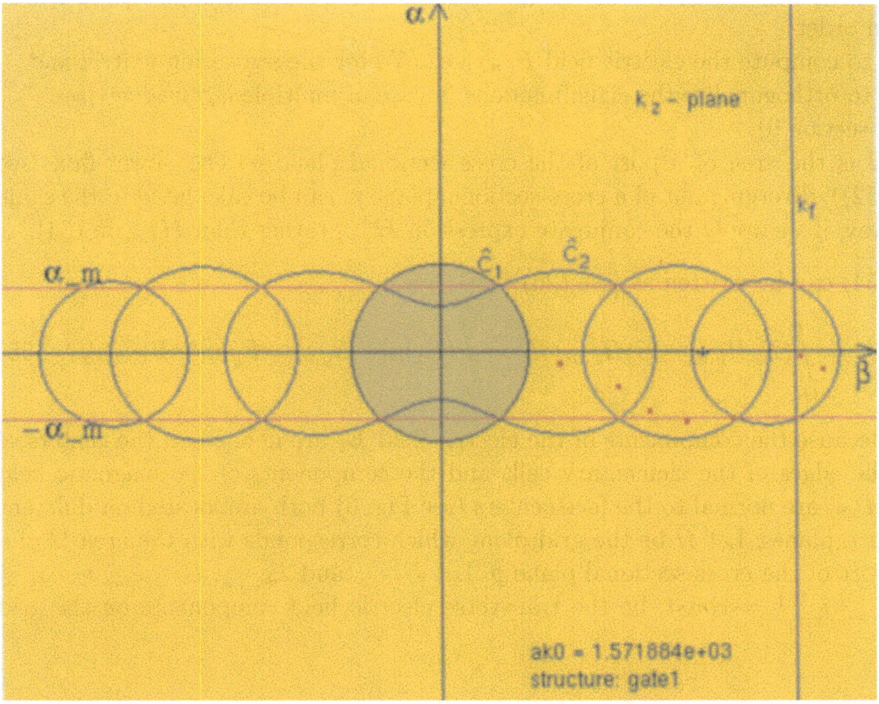

Fig. 5. k_z-plane. The Cassinian curve $\hat{C}_2$, which corresponds with the circle C_2 of Fig. 4, is a waisted oval. Propagations constants, which are located in a region, covered by $\hat{C}_1$ and $\hat{C}_2$, would be computed twice.

$$P = P^{(P)} + P^{(W)} = \int\limits_{\Omega^{(P)}} \left(\boldsymbol{E}_t \times \boldsymbol{H}^*_{t,m}\right) \cdot d\boldsymbol{\Omega} + \int\limits_{\Omega^{(W)}} \left(\boldsymbol{E}_t \times \boldsymbol{H}^*_{t,m}\right) \cdot d\boldsymbol{\Omega} . \quad (52)$$

A mode is specified as PML-mode if the ratio

$$r^{(P)} = \frac{P^{(P)}}{P} > \xi , \quad (53)$$

with values $\xi = 0.2, \ldots, 0.6$, found empirically, is satisfied.

8 Computation of the Integral

We have to calculate numerically the surface integral (3) for the mode m

$$\Im = \int\limits_{\Omega} (\boldsymbol{E}_t \times \boldsymbol{H}_{t,m}) \cdot d\boldsymbol{\Omega} = \int\limits_{\Omega} (E_x H_{y,m} - E_y H_{x,m}) d\Omega , \quad (54)$$

in order
- to compute the electric field $E_{t,m}$ (see (6)) for the scattering matrix and
- to orthogonalize the eigenfunctions in case of multiple eigenvalues (see section 9).

Ω is the area of a port of the cross-sectional plane p. The power flux (see (52)), diverging out of a cross-sectional plane p, can be calculated in the same way, if we apply the conjugate expression $\boldsymbol{H}^*_{t,m}$ rather than $\boldsymbol{H}_{t,m}$ in (54).

(54) can be written as (see (26)):

$$\Im = \int_{\Omega} (E_x H_{y,m} - E_y H_{x,m}) d\Omega = \int_{\Omega} (E_x [\mu]^{-1} B_{y,m} - E_y [\mu]^{-1} B_{x,m}) d\Omega. \quad (55)$$

Because the components of the electric field $\boldsymbol{E}_t$ are defined on the centers of the edges of the elementary cells and the components of the magnetic field $\boldsymbol{H}_{t,m}$ are normal to the face centers (see Fig. 6) both are located on different grid planes. Let $\overline{\Omega}$ be the grid plane which corresponds with the area Ω of a port of the cross-sectional plane p. Let $E_{x_{i,j,k}}$ and $E_{y_{i,j,k}}$, $i_1 \leq i \leq i_2$, $j_1 \leq j \leq j_2$, $k = const$, be the transverse electric field components on the grid

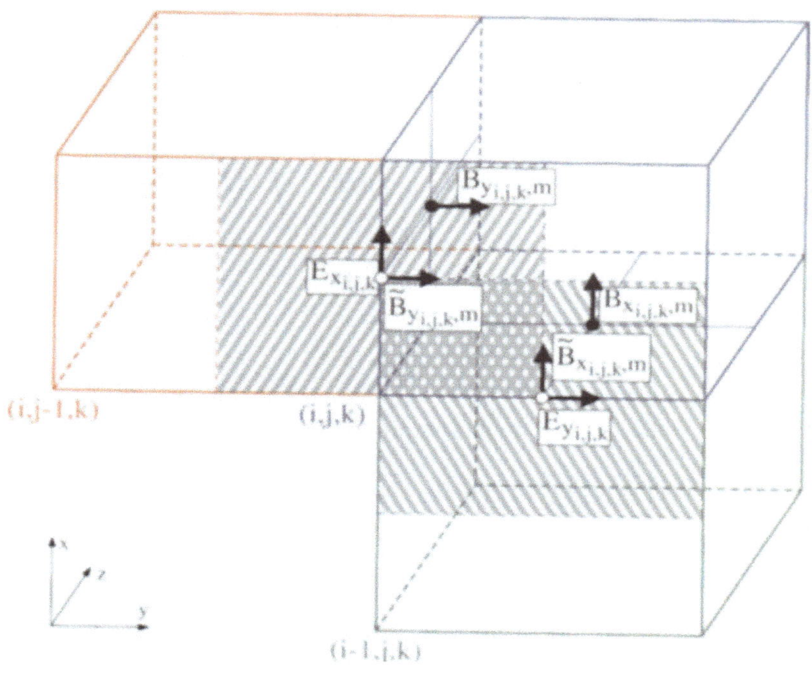

Fig. 6. Integration domain

plane $\overline{\Omega}$, and let $\tilde{B}_{x_{i,j,k},m}$ and $\tilde{B}_{y_{i,j,k},m}$, $i_1 \leq i \leq i_2$, $j_1 \leq j \leq j_2$, $k = const$, be the transverse magnetic flux density on the same grid plane (see Fig. 6). We use the lowest order integration formula (29) to approximate the integral (55). The material constants can be different between two different elementary cells of the primary grid. Thus, we have to divide the integration domain. The grid plane $\overline{\Omega}$ consists of $(i_2 - i_1 + 1)(j_2 - j_1 + 1)$ partial planes, and we get the following approximation of the integral:

$$\begin{aligned}\Im = & \sum_{\substack{i_1 \leq i \leq i_2 \\ j_1 \leq j \leq j_2}} E_{x_{i,j,k}} \left(\frac{x_{i,j,k} y_{i,j-1,k}}{2\mu_{y,i,j-1,k}} \tilde{B}_{y_{i,j,k},m} + \frac{x_{i,j,k} y_{i,j,k}}{2\mu_{y,i,j,k}} \tilde{B}_{y_{i,j,k},m} \right) \\ & - \sum_{\substack{i_1 \leq i \leq i_2 \\ j_1 \leq j \leq j_2}} E_{y_{i,j,k}} \left(\frac{x_{i-1,j,k} y_{i,j,k}}{2\mu_{x,i-1,j,k}} \tilde{B}_{x_{i,j,k},m} + \frac{x_{i,j,k} y_{i,j,k}}{2\mu_{x,i,j,k}} \tilde{B}_{x_{i,j,k},m} \right)\end{aligned} \tag{56}$$

or

$$\Im = \sum_{\substack{i_1 \leq i \leq i_2 \\ j_1 \leq j \leq j_2}} E_{x_{i,j,k}} G_{x_{i,j,k},m} + \sum_{\substack{i_1 \leq i \leq i_2 \\ j_1 \leq j \leq j_2}} E_{y_{i,j,k}} G_{y_{i,j,k},m} \tag{57}$$

with

$$\begin{aligned} G_{x_{i,j,k},m} &= \frac{x_{i,j,k}}{2} \left(\frac{y_{i,j-1,k}}{\mu_{y,i,j-1,k}} + \frac{y_{i,j,k}}{\mu_{y,i,j,k}} \right) \tilde{B}_{y_{i,j,k},m} \ , \\ G_{y_{i,j,k},m} &= -\frac{y_{i,j,k}}{2} \left(\frac{x_{i-1,j,k}}{\mu_{x,i-1,j,k}} + \frac{x_{i,j,k}}{\mu_{x,i,j,k}} \right) \tilde{B}_{x_{i,j,k},m} \end{aligned} \tag{58}$$

or

$$\Im = \boldsymbol{E}_t \cdot \boldsymbol{G}_m \quad \text{with} \quad \boldsymbol{G}_m = (\boldsymbol{G}_{x,m}, \boldsymbol{G}_{y,m})^T . \tag{59}$$

The components $G_{x_{i,j,k},m}$ and $G_{y_{i,j,k},m}$ (see (58)) contain the unknown quantities $\tilde{B}_{x_{i,j,k},m}$ and $\tilde{B}_{y_{i,j,k},m}$. They can be computed using the known electric field as follows. Because of the longitudinal homogeneity of the transmission lines we determine similar to (33)

$$B_{y_{i,j,k},m} = \tilde{B}_{y_{i,j,k},m} e^{-jk_{z_m} h}, \quad B_{y_{i,j,k-1},m} = \tilde{B}_{y_{i,j,k},m} e^{+jk_{z_m} h} \tag{60}$$

with

$$h = \frac{z_{i,j,k}}{2} = \frac{z_{i,j,k-1}}{2} \ . \tag{61}$$

Using this ansatz and the left-hand sides of the formulae (30) and (31) the quantities $\tilde{B}_{x_{i,j,k},m}$ and $\tilde{B}_{y_{i,j,k},m}$ can be expressed by the electric field components.

9 Orthogonalization

The orthogonality relation (3) is valid for nondegenerate modes. However, for degenerate modes we may choose a suitable linear combination of the

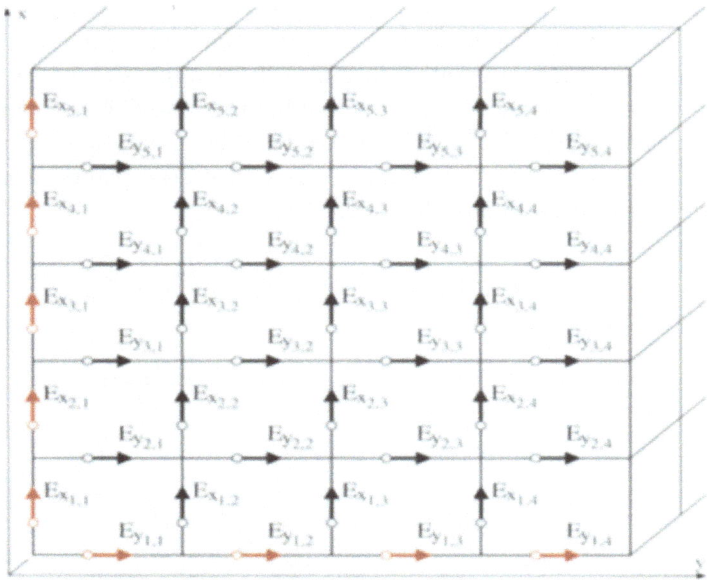

Fig. 7. Boundary conditions on the port of the structure

degenerate modes such that this subset of modes is an orthogonal set. We consider the eigenvalue problem for a selected port as an example (see Fig. 7). The vector

$$\begin{aligned} \boldsymbol{a}_\kappa = \underline{e}_l{}^{(\kappa)} = \underline{e} = (\underline{e}_x, \underline{e}_y)^T \,, \, \underline{e}_x &= \left(\underline{e}_{x_1}, \underline{e}_{x_2}, \dots, \underline{e}_{x_{n_{xy}}}\right) \,, \, \underline{e}_{x_\ell} = E_{x_{i,j,k}} \,, \\ \underline{e}_y &= \left(\underline{e}_{y_1}, \underline{e}_{y_2}, \dots, \underline{e}_{y_{n_{xy}}}\right) \,, \, \underline{e}_{y_\ell} = E_{y_{i,j,k}} \,, \end{aligned} \tag{62}$$

with

$$\begin{aligned} &\ell = (j-1)n_x + i, \quad i = 1(1)n_x, \quad j = 1(1)n_y, \\ &n_{xy} = n_x n_y, \quad \text{and} \quad k = 1 \quad \text{or} \quad k = n_z, \end{aligned} \tag{63}$$

consists of the electric field intensity components of a port. The cross-sectional planes (see Fig. 1) can be located on the 6 different planes of the three-dimensional rectangular structure. The assumption $k = 1$ corresponds with the case, in which the cross-sectional plane (see Fig. 1) is located on the left-handed (x, y)-plane of the enclosure. We consider the boundary condition $\boldsymbol{E} \times \boldsymbol{n} = 0$ (see (28)) on the left-hand side and at the bottom of the port, that is,

$$\underline{e}_{x_\ell} = E_{x_{i,1,1}} = 0, \quad \ell = i, \quad i = 1(1)n_x, \tag{64}$$

and

$$\underline{e}_{y_\ell} = E_{y_{1,j,1}} = 0, \quad \ell = (j-1)n_x + 1, \quad j = 1(1)n_y. \tag{65}$$

That means, these components of the eigenvector are known, and the dimension of the eigenvalue problem (34) to be computed is reduced to

$$n = n_x n_y - n_b, \quad n_b = n_x + n_y. \tag{66}$$

Let λ_l be an eigenvalue of the matrix C (see (34) and (35)), and let ν be the algebraic multiplicity of λ_l. Then the eigenvectors $\boldsymbol{a}_\kappa = \underline{\boldsymbol{e}}_l{}^{(\kappa)}, \kappa = 1(1)\nu$, associated with λ_l, form a n-by-ν matrix A with $n \geq \nu$. The corresponding vectors $\boldsymbol{b}_\kappa = \underline{\boldsymbol{h}}_l^{(\kappa)}$ of the magnetic flux density components (see (67)) can be calculated as described in section 8.

The eigenvectors $\boldsymbol{a}_\kappa = \underline{\boldsymbol{e}}_l{}^{(\kappa)}, \kappa = 1(1)\nu$, generally do not fulfill the orthogonality relation (see (3), and (62) - (66))

$$\int_\Omega (\underline{\boldsymbol{e}}_l{}^{(\kappa_1)} \times \underline{\boldsymbol{h}}_l{}^{(\kappa_2)}) \cdot d\boldsymbol{\Omega} = \delta_{\kappa_1,\kappa_2}, \quad \kappa_1, \kappa_2 = 1(1)\nu, \tag{67}$$

but, the $\boldsymbol{a}_\kappa$ are linear independent. That means, that A has full column rank. Then there exists a unique n-by-ν matrix Q of vectors (modes), which fulfills the orthogonality relation (67), and an unique ν-by-ν upper triangular matrix U with positive diagonals $u_{\kappa,\kappa} > 0$ such that $A = QU$.

We can use the Gram-Schmidt algorithm [18] to orthogonalize each vector of A against the previously computed vectors according the orthogonality relation (67). The Gram-Schmidt process determines the columns of Q and U recursively as follows:
We begin with

$$u_{11} = \sqrt{\int_\Omega (\boldsymbol{a}_1 \times \boldsymbol{b}_1) \cdot d\boldsymbol{\Omega}}, \quad \boldsymbol{q}_1 = \frac{\boldsymbol{a}_1}{u_{11}} \quad \text{with} \quad \boldsymbol{b}_1 = \underline{\boldsymbol{h}}_l^{(1)}. \tag{68}$$

We assume that the vectors $\boldsymbol{q}_1, ..., \boldsymbol{q}_{\kappa-1}$ of Q have been already calculated by applying the Gram-Schmidt algorithm to the linear independent vectors $\boldsymbol{a}_1, ..., \boldsymbol{a}_{\kappa-1}$ of A such that the orthogonality relation

$$\int_\Omega (\boldsymbol{q}_i \times \boldsymbol{b}_j) \cdot d\boldsymbol{\Omega} = \delta_{i,j}, \quad i, j = 1(1)\kappa - 1, \tag{69}$$

is fulfilled, and that the corresponding vectors $\boldsymbol{b}_j$ of the magnetic flux density components have been computed as described in section 8. In order to find $\boldsymbol{q}_\kappa$ with aid of $\boldsymbol{a}_\kappa$ we construct a vector

$$\boldsymbol{x} = \boldsymbol{a}_\kappa - \sum_{j=1}^{\kappa-1} u_{j,\kappa} \boldsymbol{q}_j, \tag{70}$$

which fulfills the orthogonality relation (69) for the vectors $\boldsymbol{q}_1, ..., \boldsymbol{q}_{\kappa-1}$. That gives

$$u_{j,\kappa} = \int_\Omega (\boldsymbol{q}_j \times \boldsymbol{b}_\kappa) \cdot d\boldsymbol{\Omega}, \quad j = 1(1)\kappa - 1. \tag{71}$$

We get $\boldsymbol{q}_\kappa$ by normalizing the vector $\boldsymbol{x}$ (see (69) and (70)):

$$\boldsymbol{q}_k = \boldsymbol{x}/u_{\kappa,\kappa} \tag{72}$$

with

$$u_{\kappa,\kappa} = \sqrt{\int_\Omega (\boldsymbol{x} \times \boldsymbol{b}_\kappa) \cdot d\boldsymbol{\Omega}}. \tag{73}$$

After reformulation, (70) and (72) can be written as matrix equation: $A = QU$.

10 Optoelectronic Devices

For optoelectronic devices frequencies about several hundred THz are common. The region containing potentially propagating modes grows substantially. A significant higher number of eigenvalue problems have to be solved within our algorithm. Additionally, the maximum cell size of the discretization should be less than $\frac{\lambda}{10}$, where λ denotes the wavelength in the material with the highest $\mathbb{R}(\epsilon)$. Additional mesh refinements have to be used for structure regions with highly varying fields. Besides, large cross sections are common for the waveguides under investigation. Thus, high dimensional problems have to be handled.

Due to electric and magnetic walls terminating the PML regions, undesired modes are generated inside the computation domain. These non-physical modes can be detected by examining the eigenfunctions (see section 7). Anyway, the number of eigen modes to be calculated increases because of the shifted modes.

Because of the significant difference between the magnitude of the real and imaginary part of the propagation constant a high computational accuracy is required. That means, the numerical effort increases significantly.

To overcome these problems new strategies have been realized.
The few interesting modes are located in a partial region of the rectangle $\hat{F}$. By covering the corresponding area F in the plane of eigenvalues with circles (see (42) and (43)) a set of eigenpairs is scanned, which include the interesting modes. PML modes as well as propagating constants, that are located in the corresponding Cassinian curves (see (48)), but outside of $\hat{F}$, are eliminated. Additionally, the user might give an additional limit of mode numbers to be calculated. Therefore, only the remaining set of eigenpairs have to be computed with high accuracy.

Thus, to reduce the execution times, in a first step the problem is solved using a coarse grid with lower accuracy requirements in order to find approximately the locations of the interesting propagation constants. Anyway, the number of modified eigenvalue problems to be solved is high. Thus, we split the interval $[0, k_f]$ (see (41)) into subintervals (see (46)), and compute the corresponding eigenpairs independently and in parallel, for instance on different workstations or shared memory multiprocessors.

Finally the modes of interest are calculated in a second step for an essentially reduced region using a fine grid, that fulfills higher accuracy requirements. The parallel CPU mode of PARDISO provides the additional possibility to reduce the computing times for this high dimensional problem on shared memory multiprocessors without essential additional memory requirements.

11 Laser Application

As an example we have calculated the guided mode of an optoelectronic device which leads to a more moderate dimension of the corresponding eigenvalue problem. A so-called self aligned stripe (SAS) laser with a discretized large cross section of (4050 × 7750) nm is investigated (see Fig. 8). Thin layers of 100 nm with complex material properties have to be taken into account. The laser structure contains an additional, antiguided layer (marked with yellow color in Figure 8) outside the emitting stripe (marked with red color). This high power laser diode excites only the fundamental mode. The frequency is fixed to $300 * 10^{12}$ Hz, which corresponds to a vacuum wavelengths of 1000 nm.

A graded mesh of 121 times 127 elementary cells, including 10-cell PML regions, is used as a coarse grid. The maximum cell size amounts 80 nm, and the minimum cell size 4 nm. Maximum cell size is scaled down exponentially in the vertical direction near the emitting zone and in the horizontal direction near the material cut at the end of the active zone. The dimension of the eigenvalue problem is 29 625. 84 Cassinian curves have been used to cover the long small region of the complex plane ($\alpha_m = 2500\,\mathrm{m}^{-1}$, $k_f = 21\,765\,592\,\mathrm{m}^{-1}$, see (37)) containing potential guided modes. The eigenpairs have been computed with the relative accuracy $tol = 10^{-7}$, and with $m_{max} = 16$, $l = 5$ (see section 6). A maximum number $\nu_{max} = 120$ of Arnoldi iteration has been used. The total computational time amounts approximately 1 145 s. One guided mode according to the fundamental mode of the laser, was found. The circle that contains the guided mode is known after this step. The computed complex propagation constant is given by $k_z = (20\,818\,302 + \jmath 1\,401)\,\mathrm{m}^{-1}$ using this coarse grid.

A graded mesh of 283 times 345 elementary cells, including 10-cell PML regions, is used as a fine grid. The maximum cell size amounts $\frac{\lambda}{12} = 25$nm, where λ denotes the wavelength in the material with the highest $\mathbb{R}(\epsilon)$. The

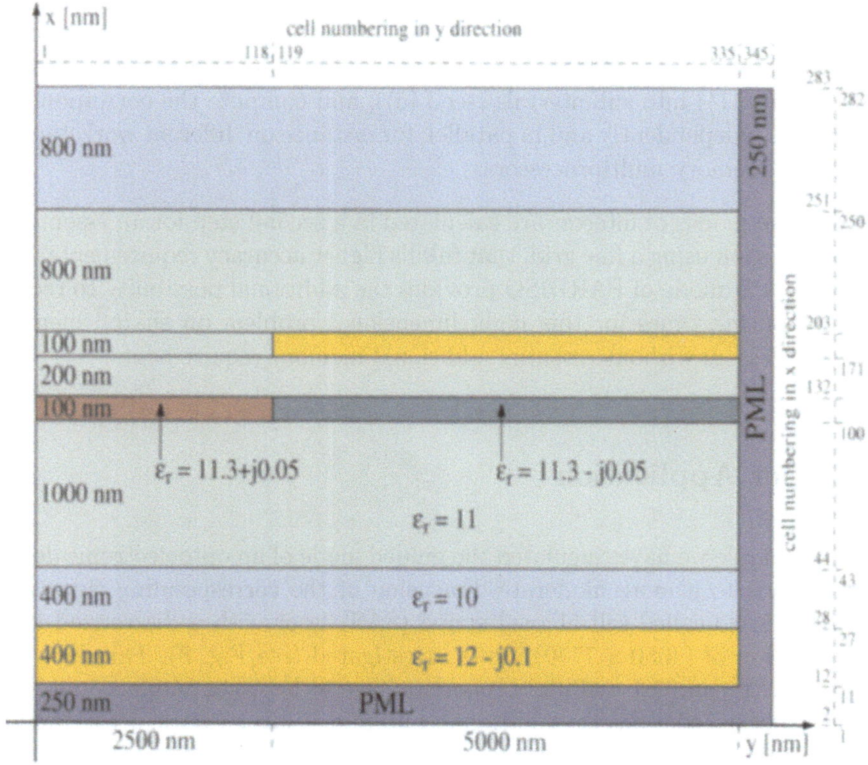

Fig. 8. Laser (amplifier). The complex permittivity $\epsilon = \epsilon_0 \epsilon_r$ and the permeability $\mu = \mu_0 \mu_r$ are isotrop ($\epsilon_r = \epsilon_x = \epsilon_y = \epsilon_z, \mu_r = \mu_x = \mu_y = \mu_z$) in the non-PML region. ϵ_0 and μ_0 denote the permittivity and permeability in vacuum, respectively. ϵ_r - relative permittivity. μ_r - relative permeability. $\mu_r = 1$. The thickness of the layers is given in nm.

minimum cell size is 1 nm. The dimension of the eigenvalue problem is 192 423. The eigenpairs have been calculated with the relative accuracy $tol = 10^{-10}$. The time to find the accurate value of the guided mode in the reduced region using the fine grid amounts only 70 s. The computed complex propagation constant is given by $k_z = (20\,817\,578 + \jmath\, 1\,488)\,\mathrm{m}^{-1}$ using the fine grid.

The calculation of all eigenvalues within the long small region $\hat{F}$ using the fine grid, needs a total computational time amount of approximately 3h and 23 minutes. Thus, the computational time is reduced by a factor of 10 using a coarse and a fine grid. A Compaq Professional Workstation with processor XP1000 alpha 667 MHz has been used for the computations.

Splitting the interval $[0, k_f]$ into q subintervals (see (46)) and solving the corresponding eigenvalue problems in parallel the time for the coarse grid computation is reduced by a factor of q. Additionally, applying the parallel

CPU mode of PARDISO to the fine grid computation in the reduced region the computing times for LU decomposition and for the solve phase could be reduced to 68% and 86%, respectively, using two processors.

As the technical realization of such semiconductor lasers implies significant material uncertainties, measurements of the field intensity are in such a coarse range that a comparison with our numerical result is ruled out.

A calculation of the given structure with the Effective Index Method (EIM) [11] results in a complex propagation constant $k_z = (20\,816\,706 + \jmath 1\,929)$ which is in relatively good agreement with our result. The deviation in the imaginary part of both results may be related to the fact that the EIM does not guaranty continuity at the transition between areas of different vertical properties.

12 System of Linear Algebraic Equations

Multiplying (32) by $D_s^{1/2}$ yields a symmetric form of linear algebraic equations:

$$\bar{A}\boldsymbol{x} = 0, \quad \bar{A} = (D_s^{1/2} A^T D_{s/\mu} D_A^{-1} A D_s^{1/2} - k_0^2 D_{A_\epsilon}) \tag{74}$$

with $\boldsymbol{x} = D_s^{1/2}\boldsymbol{e}$. Four kinds of preconditioning and a block quasi-minimal residual algorithm are applied to solve the large scale systems of linear algebraic equations. Details are given with [16].

The gradient of the electric field divergence (see (24), right formula)

$$[\epsilon]\nabla([\epsilon]^{-2}\nabla \cdot [\epsilon]\boldsymbol{E}) = 0 \tag{75}$$

is equivalent to the matrix equation

$$\bar{B}\boldsymbol{x} = 0, \quad \bar{B} = D_s^{-1/2} D_{A_\epsilon} B^T D_{V_{\epsilon\epsilon}}^{-1} B D_{A_\epsilon} D_s^{-1/2}. \tag{76}$$

The diagonal matrix $D_{V_{\epsilon\epsilon}}$ is a volume matrix for the 8 partial volumes of the dual elementary cell. Taking into account the boundary conditions Equations (74) and (76) yields the form

$$\tilde{A}\boldsymbol{x} = \boldsymbol{b}, \quad \tilde{B}\boldsymbol{x} = 0. \tag{77}$$

The effect of the combination of Equations (77) is described for the used Finite Integration Technique in [21] and [1]. Analytically, taking into account the electric-field divergence equation (see (24), right formula) irrotational solutions are avoided. By using the FIT-method the grid equations (30) and (31) are consistent with Maxwell's equations (24) and (25), respectively, in the sense that the basic properties of the analytical fields are maintained. That means, fields that are source-free in the real space also are source-free in the grid space. Thus, the matrix defined by (78) has non-vanishing eigenvalues, and this linear system of equations can be solved numerical faster than (74).

If we assume that ${k_0}^2$ is not an eigenvalue of $D_{A_\epsilon}^{-1} D_s^{1/2} A^T D_{s/\mu} D_A^{-1} A D_s^{1/2}$ (see Equation (74)), the effect of the addition of Equations (77) also can be interpreted as preconditioning [16] with the preconditioner $(I + \tilde{B}\tilde{A}^{-1})^{-1}$ for system ((77), left equation):

$$(\tilde{A} + \tilde{B})\boldsymbol{x} = \boldsymbol{b}. \tag{78}$$

$(\tilde{A} + \tilde{B})$ is a complex indefinite symmetric matrix.

In addition, independent set orderings [12] , Jacobi and SSOR preconditioning using Eisenstat's trick [5] are applied to accelerate the speed of convergence of the used block Krylov subspace method [6] for the system of linear algebraic equations (78) that has to be solved with the same coefficient matrix, but multiple right-hand sides. The number m_s (see (2)) of right-hand sides depends on the number of ports and guided modes.

Independent set orderings are permutations P_i (see Fig. 9) to transform the matrix A_i with $A_0 = \tilde{A} + \tilde{B}$ in the form

$$A_i \quad \longrightarrow \quad P_i A_i P_i^T = \begin{pmatrix} D_i & E_i^T \\ E_i & H_i \end{pmatrix}, \tag{79}$$

where D_i is a diagonal, E_i, and H_i are sparse matrices. The unknowns of the independent set D_i are eliminated to get the next reduced matrix

$$A_{i+1} = H_i - E_i D_i^{-1} E_i^T. \tag{80}$$

We get a system of linear equations

$$P_i A_i P_i^T P_i \boldsymbol{x}_i = P_i \boldsymbol{b}_i \tag{81}$$

with $\boldsymbol{y}_i = P_i \boldsymbol{x}_i = (\boldsymbol{y}_{i,1}, \boldsymbol{y}_{i,2})^T$ and $\boldsymbol{c}_i = P_i \boldsymbol{b}_i = (\boldsymbol{c}_{i,1}, \boldsymbol{c}_{i,2})^T$ and have to solve the reduced system of linear equations

$$A_{i+1}\boldsymbol{x}_{i+1} = \boldsymbol{b}_{i+1}, \quad \boldsymbol{x}_{i+1} = \boldsymbol{y}_{i,2}, \quad \boldsymbol{b}_{i+1} = \boldsymbol{c}_{i,2} - E_i D_i^{-1} \boldsymbol{c}_{i,1} \tag{82}$$

for $\boldsymbol{y}_{i,2}$, and then we get

$$\boldsymbol{y}_{i,1} = D_i^{-1}(\boldsymbol{c}_{i,1} - E_i^T \boldsymbol{y}_{i,2}). \tag{83}$$

Then we have to permute the solution vector $\boldsymbol{y}_i$ back to the vector $\boldsymbol{x}_i$.

In comparison to the simple lossy case the number of iterations of Krylov subspace methods increases significantly in the presence of Perfectly Matched Layers (see Tab. 1). The speed of convergence depends on the relations of the

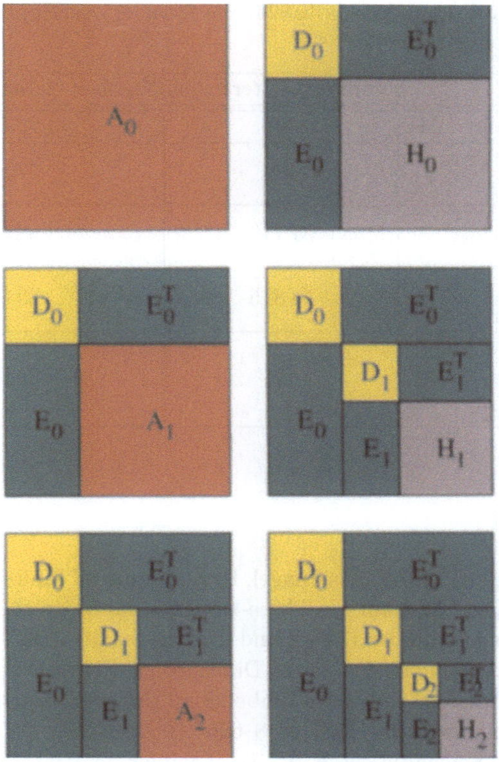

Fig. 9. Independent set ordering

edges in a elementary cell of the nonequidistant rectangular grid in this case. The best results can be obtained using nearly cubic cells. Moreover, overlapping PML conditions on the corner regions of the computational domain lead to an increase of the magnitude of the corresponding off-diagonal elements in comparison to the diagonal of the coefficient matrix. That downgrades the properties of the matrix. Thus, overlapping PML should be avoided. Otherwise, eigenvalues of the matrix are shifted into the negative half plane. That means, the Krylov subspace methods need more iterations.

Acknowledgments

The authors would like to thank Dr. H. Wenzel and Dr. R. Güther from the Optoelectronics Department, Ferdinand-Braun-Institut, Berlin, for helpful discussions concerning the investigated laser structure.

Table 1. Influence of the PML setup on computational effort. The order of the system of linear algebraic equations is 40 824. We consider overlapping PML conditions on the corner regions.

	Number of Iterations	Number of Nonzero Elements
Structure without PML	55	235 696
Structure with PML in z-direction	489	379 220
Structure with PML in yz-direction	37 719	387 806
Structure with PML in xyz-direction	54 815	394 378
Structure with PML in xyz-direction (nonoverlapping)	458	394 378

References

1. Beilenhoff, K., Heinrich, W., Hartnagel, H. L.: Improved Finite-Difference Formulation in Frequency Domain for Three-Dimensional Scattering Problems. IEEE Transactions on Microwave Theory and Techniques **40** (1992) 540-546
2. Christ, A., Hartnagel, H. L.: Three-Dimensional Finite-Difference Method for the Analysis of Microwave-Device Embedding. IEEE Transactions on Microwave Theory and Techniques **35** (1987) 688–696
3. Bronstein, I. N., Semendjajew, K.A.: Taschenbuch der Mathematik. B. G. Teubner Verlagsgesellschaft (1965)
4. Davis, T. A., Duff, I. S.: A Combined Unifrontal/Multifrontal Method for Unsymmetric Sparse Matrices. University of Florida, Technical Report **16** (1997) 1–18
5. Eisenstat, S. C.: Efficient implementation of a class of preconditioned conjugate gradient methods. SIAM J. Sci Statist. Comput. **2** (1981) 1–4
6. Freund, R. W., Malhotra, W.: A Block-QMR Algorithm for Non-Hermitian Linear Systems with Multiple Right-Hand Sides. Linear Algebra and Its Applications, **254** (1997) 119–157
7. Hebermehl, G., Schlundt, R., Zscheile, H., Heinrich W.: Improved Numerical Methods for the Simulation of Microwave Circuits, Surveys on Mathematics for Industry, **9** (1999) 117–129
8. Hebermehl, G., Hübner, F.-K., Schlundt, R., Tischler, T., Zscheile, H., Heinrich, W.: On the Computation of Eigen Modes for Lossy Microwave Transmission Lines Including Perfectly Matched Layer Boundary Conditions. The International Journal for Computation and Mathematics in Electrical and Electronic Engineering **20** (2001) 948–964
9. Hebermehl, G., Hübner, F.-K., Schlundt, R., Tischler, T., Zscheile, H., Heinrich, W.: Numerical Simulation of Lossy Microwave Transmission Lines Including PML. In: Scientific Computing in Electrical Engineering (ed. U. van Rienen, M. Günther, D. Hecht), Lecture Notes in Computational Science and Engineering, Springer Verlag, (2001) 267–275

10. Lehoucq, R. B.: Analysis and Implementation of an Implicitly Restarted Arnoldi Iteration. Rice University, Department of Computational and Applied Mathematics, Technical Report **13** (1995) 1-135
11. März, R.: Integrated Optics - Design and Modeling. Artech House, (1995)
12. Saad, Y.: Iterative methods for sparse linear systems. PWS Publishing Company (1996)
13. Sacks, Z. S., Kingsland, D. M., Lee, R., Lee, J.-F.: A Perfectly Matched Anisotropic Absorber for Use as an Absorbing Boundary Condition. IEEE Transactions on Antennas and Propagation **43** (1995) 1460–1463
14. Schenk, O., Gärtner, K., Fichtner, W.: Efficient Sparse LU Factorization with Left-Right Looking Strategy on Shared Memory Multiprocessors. BIT **40** (2000) 158-176
15. Schenk, O., Gärtner, K.: Two-Level Dynamic Scheduling in PARDISO: Improved Scalability on Shared Memory Multiprocessing Systems. Parallel Computing **28** (2002) 187–197
16. Schlundt, R., Hebermehl, G., Hübner, F.-K., Heinrich, W., Zscheile, H.: Iterative Solution of Systems of Linear Equations in Microwave Circuits Using a Block Quasi-Minimal Residual Algorithm. In: Scientific Computing in Electrical Engineering (ed. U. van Rienen, M. Günther, D. Hecht), Lecture Notes in Computational Science and Engineering, Springer Verlag, (2001) 325–333
17. Sorensen, D. C.: Implicit Application of Polynomial Filters in a k-Step Arnoldi Method. SIAM J. Matr. Anal. Apps. **13** (1992) 357–385
18. Stoer, J., Bulirsch, R.: Introduction to Numerical Analysis, Second Edition, Springer-Verlag (1993)
19. Tischler, T., Heinrich, W.: The Perfectly Matched Layer as Lateral Boundary in Finite-Difference Transmission-Line Analsysis. IEEE Transactions on Microwave Theory and Techniques **48** (2000) 2249-2253
20. Weiland, T.: A Discretization Method for the Solution of Maxwell's Equations for Six-Component Fields. Electronics and Communication (AEÜ) **31** (1977) 116–120
21. Weiland, T.: On the Unique Numerical Solution of Maxwellian Eigenvalue Problems in Three Dimensions. Particle Accerelators **17** (1985) 227–242

Finite Element Methods for Coupled Problems in Ferrohydrodynamics

Olga Lavrova[1], Gunar Matthies[1], Teodora Mitkova[1], Viktor Polevikov[2], and Lutz Tobiska[1]

[1] Institut für Analysis und Numerik, Otto-von-Guericke-Universität Magdeburg, PF4120, 39016 Magdeburg, Germany
[2] Belarusian State University, 4 F. Skaryna Ave., 220050 Minsk, Belarus

Summary. The interaction between a magnetisable fluid and an external magnetic field gives rise to several interesting phenomena in ferrohydrodynamics. Our mathematical models and solution strategies are mainly focused on cases in which the magnetic liquid exhibits a free surface which is not known a-priori. In particular, two special problems are taken into consideration: the behaviour of a ferrofluid drop in a rotary shaft seal and the generation of an ordered pattern of surface protuberances when the applied field exceeds a critical value.

1 Ferrofluids: Properties and Applications

Ferrohydrodynamics (FHD) deals with the mechanics of fluid motion influenced by strong forces of magnetic polarisation. In the well-known discipline of magnetohydrodynamics (MHD) the body force acting on the fluid is the Lorentz force that arises when electric current flows at an angle to the direction of an imposed magnetic field. However, in FHD the fluid is nonconducting and the body force is due to polarisation force. The principal type of a magnetic fluid arising in FHD is a ferrofluid.

Ferrofluids are colloidal suspensions of finely divided ferromagnetic (Fe, Co, Ni) or ferrimagnetic (Fe_3O_4) particles (of size 3-15 nm) in a carrier liquid (water, oil, bio-compatible liquids). In order to avoid agglomeration the monodomain particles are coated with a molecular layer of a dispersant. A schematic representation is shown in Fig. 1 taken from [13]. A colloidal ferrofluid has do be synthesised, since it is not found in nature. A typical ferrofluid contains about 10^{14} particles per cubic millimetre and is photoresist ('black').

If the applied magnetic field is increased then the following states can be observed:

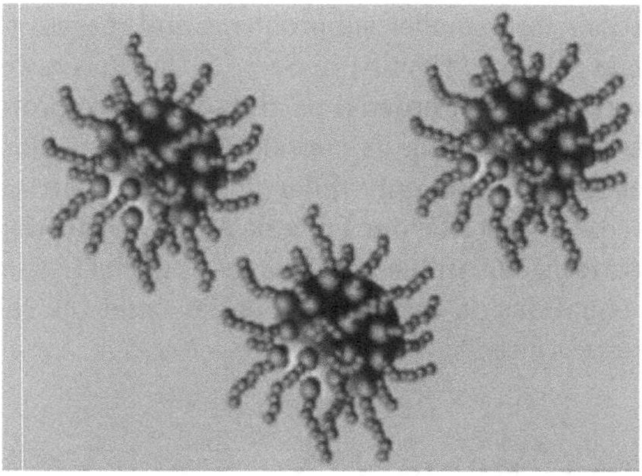

Fig. 1. Magnetic particles with dispersant layer in a ferrofluid

- a random orientation of the particles with no net magnetisation,
- a tendency for alignment, partially overcome by thermal agitation,
- the particles become more and more aligned with the field direction,
- the particles are completely aligned and the magnetisation achieves its saturation value.

The possibility to influence the hydrodynamic parameters of the fluid by an external magnetic field opens the way for a variety of technical and medical applications, like zero-leakage rotary shaft seals, liquid-cooled loudspeakers and magnetic drug targeting. For example, magnetic fluid seals are used to admit both a rotary motion and a hermetic isolation of a high from a low pressure environment. That way sensitive electronical components, like hard disks, can be protected from outside contaminants. In liquid-cooled loudspeakers the lubrication and damping properties can be controlled by the viscosity of the carrier liquid. Further, the heat transfer from the coil can be improved since the thermal conductivity is larger than that of air. In the field of medical drug targeting a magnetic field can pilot the path of a drop of ferrofluid in the body bringing and retaining drugs to a target site. Localised treatment of a disease becomes possible and unwanted side effects can be considerably decreased. For further interesting technical applications of ferrofluids we refer to [3, 18].

The paper is organised as follows. Section 2 is devoted to modelling aspects for describing the interaction between magnetic fields and magnetic fluids. The basic model consists of the incompressible Navier-Stokes equations coupled with the Maxwell equations for a nonconducting fluid. Then, in Sect. 3 the general model is simplified to describe the behaviour of a ferrofluid drop in a rotary shaft seal. We propose a splitting of the coupled nonlinear partial differential equations into smaller subproblems and discuss their robust and accurate numerical solution. Finally, in Sect. 4, the Rosensweig instability of spontaneous generation of an ordered pattern of surface protuberances when the field exceeds a critical value is considered. In particular two geometric models are discussed which mainly differ in the parametrisation of the unknown free surface. We describe our numerical solution strategies and present numerical tests showing the influence of the used geometric model and different constitutive equations on the shape of surface and the calculated critical magnetic field, respectively.

2 Basic Model in Ferrohydrodynamics

We start with the Maxwell equations for a nonconducting media in the stationary case given by

$$\operatorname{curl} \mathbf{H} = \mathbf{0}, \qquad \operatorname{div} \mathbf{B} = 0 \qquad \text{in } \Omega \ , \tag{1}$$

where $\mathbf{H}$ and $\mathbf{B}$ denote the magnetic field strength and magnetic induction, respectively. The first equation guarantees the existence of a magnetostatic potential. Let us suppose that the ferrofluid occupies the domain Ω_F, with $\overline{\Omega}_F \subset \Omega \subset \mathbb{R}^d$, $d = 2$ or $d = 3$, and that the media outside of Ω_F is non-magnetisable. Then, we complete (1) by the constitutive relation

$$\mathbf{B} = \begin{cases} \mu_0(\mathbf{M} + \mathbf{H}) & \text{in } \Omega_F \ , \\ \mu_0 \mathbf{H} & \text{in } \Omega \backslash \Omega_F \ , \end{cases} \tag{2}$$

where $\mathbf{M}$ denotes the magnetisation and $\mu_0 = 4\pi \cdot 10^{-7}\text{Vs/Am}$ is the permeability constant. The magnetisation $\mathbf{M}$ of a ferrofluid is parallel to the magnetic field $\mathbf{H}$ and the modulus $M = |\mathbf{M}|$ follows the nonlinear magnetisation law

$$M(H) = M_S \, L(\gamma H) = M_S \left(\coth(\gamma H) - \frac{1}{\gamma H} \right) \tag{3}$$

with $H = |\mathbf{H}|$, the saturation magnetisation M_S, the Langevin function L, the Langevin parameter $\gamma = 3\chi_0/M_S$, and the initial susceptibility $\chi_0 = \partial_H M(0)$. Consequently, we have

$$\mathbf{M} = M(H)\frac{\mathbf{H}}{H} = \frac{M_S}{H} \left(\coth(\gamma H) - \frac{1}{\gamma H} \right) \mathbf{H} \ . \tag{4}$$

The hydrodynamical properties of the magnetic liquid are described by the time-dependent incompressible Navier-Stokes equations

$$\varrho\,(\mathbf{u}_t + (\mathbf{u} \cdot \nabla)\mathbf{u}) = \operatorname{div} \boldsymbol{\sigma}(\mathbf{u}, p, \mathbf{H}) + \mathbf{f} \qquad \text{in } \Omega_F(t),\ t > 0 \ , \tag{5}$$

$$\operatorname{div} \mathbf{u} = 0 \qquad \text{in } \Omega_F(t),\ t > 0 \ , \tag{6}$$

where $\boldsymbol{\sigma}(\mathbf{u}, p, \mathbf{H})$ is the stress tensor

$$\boldsymbol{\sigma}_{ij}(\mathbf{u}, p, \mathbf{H}) = 2\eta \mathcal{D}(\mathbf{u})_{ij} - \left(p + \frac{\mu_0}{2} H^2\right) \delta_{ij} + B_i H_j, \qquad i, j = 1, \ldots d \ , \tag{7}$$

consisting of a hydrodynamic and magnetic part. Here, ϱ denote the fluid density, $\mathbf{f}$ the volume force, η the dynamic viscosity, $\mathbf{u}$ the velocity, $p = p_f + p_m$ the (composite) pressure with the thermodynamic pressure p_f in the fluid and the fluid-magnetic pressure

$$p_m = \mu_0 \int_0^H M(H')\, dH' \ , \tag{8}$$

and $\mathcal{D}(\mathbf{u})$ the velocity deformation tensor defined by

$$\mathcal{D}(\mathbf{u})_{ij} = \frac{1}{2} \left(\frac{\partial u_i}{\partial x_j} + \frac{\partial u_j}{\partial x_i} \right), \qquad i, j = 1, \ldots d \ . \tag{9}$$

Using (7) and (9) we can rewrite (5), (6) into

$$\varrho \left(\mathbf{u}_t + (\mathbf{u} \cdot \nabla)\mathbf{u}\right) - \eta \Delta \mathbf{u} + \boldsymbol{\nabla}\, p = \mathbf{f} + \mu_0 M \nabla H \quad \text{in } \Omega_F(t),\ t > 0 \ , \tag{10}$$

$$\operatorname{div} \mathbf{u} = 0 \quad \text{in } \Omega_F(t),\ t > 0 \ . \tag{11}$$

Let us consider the force balance

$$[|\boldsymbol{\sigma} \mathbf{n}|] = \alpha \mathcal{K} \mathbf{n} \qquad \text{on } \Gamma_F(t),\ t > 0 \ , \tag{12}$$

where $\Gamma_F(t)$ is the free surface between a magnetisable medium 1 (ferrofluid) and a nonmagnetisable medium 2 (vacuum, air). Here, $\mathcal{K}$ denote the sum of principal curvatures, $\mathbf{n}$ the unit normal vector, and α the coefficient of surface tension. Multiplying with the unit normal vector we obtain the Young-Laplace equation at the interface $\Gamma_F(t)$

$$-2\eta \mathbf{n} \cdot \mathcal{D}(\mathbf{u})\mathbf{n} + p_f + p_m + \frac{\mu_0}{2} (\mathbf{M} \cdot \mathbf{n})^2 = p^{(2)} + \alpha \mathcal{K} \tag{13}$$

where $p^{(2)}$ is the pressure in medium 2.

Finally, we add the kinematic condition

$$\mathbf{u} \cdot \mathbf{n} = V_\Gamma, \qquad \text{on } \Gamma_F(t),\ t > 0 \ , \tag{14}$$

with the normal velocity of the free boundary V_Γ. Note that (1)-(3), (10)-(12), (14) are a coupled system of nonlinear partial differential equations with a free boundary $\Gamma_F(t)$, $t > 0$. In the following sections we consider two examples with a time-independent free boundary $\Gamma_F(t) = \Gamma_F$.

3 Rotary Shaft Seals

Let us consider a rotary shaft seal as shown in Fig. 2 where 1 denotes the magnet, 2 the core, 3 the concentrator, 4 the shaft (rotor), and 5 the magnetic fluid bridge. A complete simulation of the behaviour of the ferrofluid drop in the seal would require

- the solution of the Maxwell equations between the rotary shaft and the concentrator in the fluid and the surrounding media,
- the solution of the incompressible Navier-Stokes equations in the fluid,
- the solution of the heat equation in the fluid, shaft, and concentrator.

Furthermore, we have to take into consideration the interface conditions at the free surfaces. Finally, we get a coupled system of nonlinear partial differential equations for the velocity $\mathbf{u}$, the pressure p, the magnetic field strength $\mathbf{H}$, the temperature T, and the positions of the free boundaries.

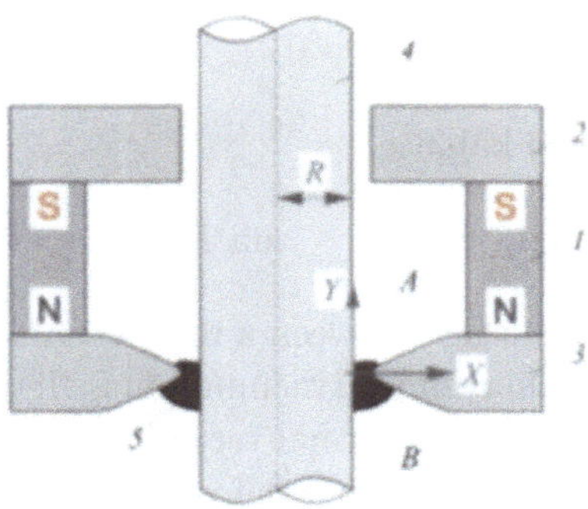

Fig. 2. Rotary shaft seal

3.1 Simplified Model

Under stationary operating conditions the fluid domain can be considered to be rotationally symmetric. Thus, using cylindrical coordinates $x = r\cos\varphi$, $y = r\sin\varphi$, $z = z$ and assuming that the velocity $\mathbf{u} = (u_r, u_\varphi, u_z)$, the thermodynamic pressure p_f and the temperature T depend on r and z only, the incompressible Navier-Stokes equations (10), (11) and the heat equation can be represented in the form

$$\left(u_r\frac{\partial u_r}{\partial r}-\frac{u_\varphi^2}{r}+u_z\frac{\partial u_r}{\partial z}\right)+\frac{\partial p_f}{\partial r}=\nu\left(\frac{\partial^2 u_r}{\partial r^2}+\frac{1}{r}\frac{\partial u_r}{\partial r}-\frac{u_r}{r^2}+\frac{\partial^2 u_r}{\partial z^2}\right),$$

$$\left(u_r\frac{\partial u_\varphi}{\partial r}+\frac{u_r u_\varphi}{r}+u_z\frac{\partial u_\varphi}{\partial z}\right)=\nu\left(\frac{\partial^2 u_\varphi}{\partial r^2}+\frac{1}{r}\frac{\partial u_\varphi}{\partial r}-\frac{u_\varphi}{r^2}+\frac{\partial^2 u_\varphi}{\partial z^2}\right),$$

$$\left(u_r\frac{\partial u_z}{\partial r}+u_z\frac{\partial u_z}{\partial z}\right)+\frac{\partial p_f}{\partial z}=\nu\left(\frac{\partial^2 u_z}{\partial r^2}+\frac{1}{r}\frac{\partial u_z}{\partial r}+\frac{\partial^2 u_z}{\partial z^2}\right),$$

$$\left(\frac{\partial u_r}{\partial r}+\frac{1}{r}u_r+\frac{\partial u_z}{\partial z}\right)=0\,,$$

$$u_r\frac{\partial T}{\partial r}+u_z\frac{\partial T}{\partial z}-\frac{\nu}{c_p}\Phi(\mathbf{u})=\lambda\left(\frac{\partial^2 T}{\partial r^2}+\frac{1}{r}\frac{\partial T}{\partial r}+\frac{\partial^2 T}{\partial z^2}\right).$$

Here, the gravitational force has been neglected in the momentum equation and λ denotes the thermal conduction, c_p the specific heat, $\nu=\eta/\varrho$ the kinematic viscosity, and Φ is the dissipation function given by

$$\Phi(\mathbf{u})=2\left[\left(\frac{\partial u_r}{\partial r}\right)^2+\left(\frac{1}{r}u_r\right)^2+\left(\frac{\partial u_z}{\partial z}\right)^2\right]+\left(\frac{\partial u_r}{\partial z}+\frac{\partial u_z}{\partial r}\right)^2$$
$$+\left(\frac{\partial u_\varphi}{\partial z}\right)^2+\left(\frac{\partial u_\varphi}{\partial r}-\frac{1}{r}u_\varphi\right)^2.$$

Let R be the radius of the rotor, $a\ll R$ the gap width, u_0 the circumferential velocity of the rotor at $r=R$, and T_0 the characteristic temperature. We introduce dimensionless variables

$$X=\frac{r-R}{a},\ Y=\frac{z}{a},\ v_1=\frac{u_r}{u_0},\ v_2=\frac{u_z}{u_0},\ w=\frac{u_\varphi}{u_0},\ p=\frac{p_f}{\varrho u_0^2},\ \vartheta=\frac{T}{T_0}$$

and approximate the term representing centrifugal forces in the momentum equation

$$-\frac{u_\varphi^2}{r}=-\frac{u_0^2}{a}\,\frac{a}{R+aX}\,w^2=-\frac{u_0^2}{a}\,\frac{\delta}{1+\delta X}\,w^2\approx-\frac{u_0^2}{a}\,\delta\,w^2$$

by taking into consideration that the relative width of the gap $\delta=a/R\ll 1$. All other terms with the factor

$$\frac{a}{r}=\frac{a}{R+aX}=\frac{\delta}{1+\delta X}\ll 1$$

are neglected. That way we end up with a problem for the azimuthal velocity w, the secondary flow $\mathbf{v}=(v_1,v_2)$, the pressure p, and the temperature ϑ in the cross-section $\Omega_F\subset\mathbb{R}^2$:

$$-\frac{1}{\mathrm{Re}}\Delta w + \mathbf{v}\cdot\nabla w = 0 \qquad \text{in } \Omega_F \ , \tag{15}$$

$$-\frac{1}{\mathrm{Re}}\Delta \mathbf{v} + \mathbf{v}\cdot\nabla \mathbf{v} + \nabla p = \delta\, w^2\, \mathbf{e}_X \qquad \text{in } \Omega_F \ , \tag{16}$$

$$\nabla\cdot\mathbf{v} = 0 \qquad \text{in } \Omega_F \ , \tag{17}$$

$$-\frac{1}{\mathrm{Pr}\ \mathrm{Re}}\Delta\vartheta + \mathbf{v}\cdot\nabla\vartheta = \frac{\Lambda}{\mathrm{Re}}\frac{a^2}{u_0^2}\Phi(\mathbf{u}) \qquad \text{in } \Omega_F \ . \tag{18}$$

Here, we have used the notations $\mathrm{Re} = u_0 a/\nu$ for the Reynolds number, $\mathrm{Pr} = \nu/\lambda$ for the Prandtl number, $\mathbf{e}_X$ for the unit vector in the X-direction, and $\Lambda = u_0^2/c_p T_0$.

The derivation of weak formulations of (15)-(17) can be found in [12], where also the solvability of the problem is analysed. Both existence and uniqueness (for small δ) of the weak solution follow from the general theory of saddle point problems, the standard Galerkin method and a weak maximum principle.

We can achieve a further simplification by taking into consideration that the magnitude of the azimuthal velocity w is much larger than the secondary flow $\mathbf{v}$. Under this assumption we can neglect $\mathbf{v}\cdot\nabla w \approx 0$ in (15) and use the simpler form of the dimensionless dissipation function

$$\frac{a^2}{u_0^2}\Phi(\mathbf{u}) \approx \tilde{\Phi}(w) = \left(\frac{\partial w}{\partial X}\right)^2 + \left(\frac{\partial w}{\partial Y}\right)^2$$

in (18). Thus, given the azimuthal velocity we can determine the secondary flow and the temperature by solving the two-dimensional problem (16)-(18).

Let us consider the case of a hyperbolic concentrator shape and assume that the permeability of the shaft and concentrator is much higher than the permeability of the fluid. Moreover, let $M \ll H$ which implies that the shape of the free surface has practically no influence on the local magnetic field strength. Under these assumptions we are able to give an explicit analytical expression for the magnetic field strength by transforming the domain between the shaft and the concentrator onto an infinite strip [17].

Concerning the Young-Laplace equation we neglect the influence of gravity, surface tension, the magnetic normal pressure $(\mathbf{M}\cdot\mathbf{n})^2\mu_0/2$, and the flow on the free surface (cf. [17]). Thus, (13) reduces finally to the balance $p_f + p_m = p^{(2)}$ of fluid and fluid-magnetic pressure with the outer pressure $p^{(2)}$ at the free surface. Since in our case $M(H) \approx M_S$ we have $p_m \approx \mu_0 M_S H$. The fluid pressure p_f can be approximated up to a constant by integrating the approximated momentum equation

$$\nabla p \approx \delta\, w^2\, \mathbf{e}_X \ .$$

The constant can be calculated from the condition that the volume of fluid enclosed by the two arcs of free surfaces is given. Together with the analytical expression for the magnetic field strength the coupled problem for determing the azimuthal velocity w and the shape of the two arcs consists of a boundary value problem for a linear elliptic equation and algebraic equations for the free surfaces [17].

3.2 Solution Strategy

Our simplified model admits the solution strategy shown in Fig. 3.

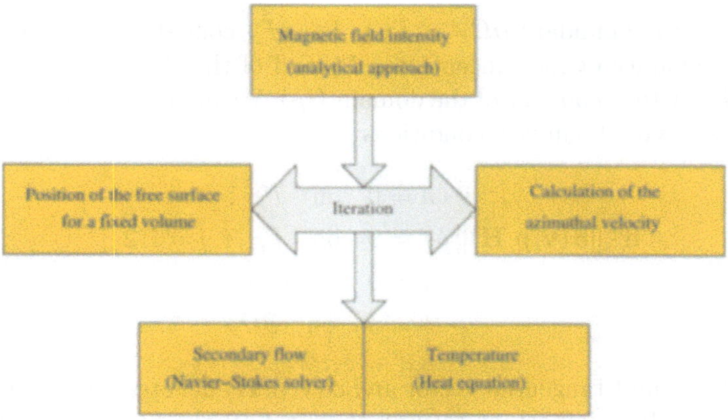

Fig. 3. Iterative decoupling strategy

We start with the analytical expression for the magnetic field and a given cross-section of the fluid domain Ω_F^0. Next, we use an inner iteration to determine w and the free surfaces Γ_F. To this end we calculate the solution of the elliptic equation

$$-\Delta w^k = 0 \quad \text{in } \Omega_F^{k-1}$$

subjected to the boundary conditions

$$w^k = 0 \;\text{ on } \Gamma_C^{k-1}, \quad w^k = 1 \;\text{ on } \Gamma_S^{k-1}, \quad \frac{\partial w^k}{\partial n} = 0 \;\text{ on } \Gamma_F^{k-1} \; ,$$

where the closed contour $\partial\Omega_F^{k-1} = \Gamma_C^{k-1} \cup \Gamma_S^{k-1} \cup \Gamma_F^{k-1}$ consists of the axial cross section of the solid (concentrator and shaft) and free surfaces. Then, to find the new update of the free surfaces $Y_i^k = Y_i^k(X)$, $i = 1, 2$ we solve algebraic equations of the form

$$\mu_0 M_s H(X, Y_i^k(X)) = p_i^{(2)} - p_f(w)(X, Y_i^{k-1}), \quad i = 1, 2 \; ,$$

subjected to a fixed volume between the two arcs. The two arcs $(X, Y_i^k(X))$ define the new fluid domain Ω_F^k. We set $k = k + 1$ and repeat the iteration unless the changes between Ω_F^{k-1} and Ω_F^k are small. Finally, for the results of this iteration w and Ω_F the secondary flow $\mathbf{v}$ and the temperature field ϑ are calculated from (16)-(18).

3.3 Numerical Results

For details on the calculation of the azimuthal velocity w and the free surfaces Γ_F see [17]. We calculate the secondary flow $\mathbf{v}$ and the temperature field ϑ in a fixed domain Ω_F for a given azimuthal velocity w.

In our case the boundary $\partial\Omega_F = \Gamma_F \cup \Gamma_C \cup \Gamma_S$ consist of two free surfaces Γ_F, the boundary of the concentrator Γ_C and of the shaft Γ_S (Fig. 4 shows schematically the geometry of the domain Ω_F). We apply for completing (16)-(18) the following boundary conditions:

$$\mathbf{v}\cdot\mathbf{n} = 0 \quad \text{on} \quad \Gamma_F \ , \tag{19}$$

$$\mathbf{n}\cdot[|\widetilde{\boldsymbol{\sigma}}(\mathbf{v},p,\mathbf{H})|]\cdot\tau = 0 \quad \text{on} \quad \Gamma_F\, , i,j = 1,2 \ , \tag{20}$$

$$\mathbf{v} = 0 \quad \text{on} \quad \Gamma_C \cup \Gamma_S \ , \tag{21}$$

$$\vartheta = \vartheta_0 \quad \text{on} \quad \partial\Omega_F \ . \tag{22}$$

Here, τ is the unit tangential vector and $\widetilde{\boldsymbol{\sigma}}(\mathbf{v},p,\mathbf{H})$ denotes the dimensionless stress tensor. It is important to note that in (20) the magnetic part of the stress tensor does not play any role since the jumps of $\mathbf{H}\cdot\tau$ and $\mathbf{B}\cdot\mathbf{n}$ vanish. Consequently, a slip boundary condition (19) and a condition on the tangential stresses (20) are imposed on the free boundary Γ_F whereas a Dirichlet (or no-slip) boundary condition (21) is prescribed for $\mathbf{u}$ on the remaining part $\partial\Omega_F \backslash \Gamma_F = \Gamma_C \cup \Gamma_S$. The slip and no-slip boundary condition describe different physical situations. This is also reflected in the mathematical treatment of the problem and of the finite element discretisation.

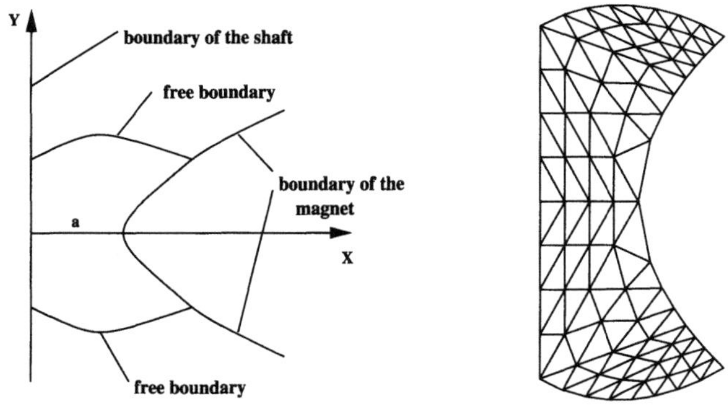

Fig. 4. Geometry of the fluid domain Ω_F (left) and the triangulation for refinement level 2 (right)

All results which we present here were obtained by using the conforming P_2/P_1 finite element pair, i.e., the velocity and the temperature are approximated by continuous, piecewise quadratic functions while the pressure is approximated by continuous, piecewise linear functions. The conforming finite element pair P_2/P_1 satisfies the Babuška-Brezzi stability condition [7], which guarantees a stable discretisation of the saddle point problem (16)-(17). We consider a discretisation where the slip boundary condition (19) is incorporated in the ansatz and the finite element space. It is possible to treat the slip boundary condition (19) as a Lagrange multiplier, see e.g. [2]. Numerically, however, our treatment of the slip boundary condition is less complicated and cheaper.

The discretisation of (16)-(18) corresponds to a linear algebraic system of equations for (18) and a nonlinear algebraic system for the saddle-point problem (16)-(17). The nonlinearity within the problem (16)-(17) is resolved by using a fixed-point iteration (linearisation of the convective term). The nonlinear fixed-point iteration is stopped if the l_2-norm of the residual is less than 10^{-8}. Thus, it remains to solve efficiently a large linear systems in each nonlinear iteration step. We use the geometric multigrid method as an efficient solver. For a detailed discussion of the design of multigrid methods and their efficiency we refer to [8]. The triangulation of the refinement multigrid level 2 is shown in Fig. 4. All finer meshes are constructed by uniform refinement and boundary adaption.

Within the multigrid method for solving the saddle-point problem (16)-(17) we use a V-cycle with one pre-smoothing and one post-smoothing step. As a smoother, we use the block Gauss-Seidel method (Vanka-type smoother, see [20]). The solution of a linear problem (obtained from the linearisation of the convective term) is stopped after a residual reduction of a factor 1/10 or the residual is smaller than 10^{-10}.

The linear algebraic system corresponding to the discretisation of (18) is solved also by applying a geometric multilevel approach. As a smoother on the coarsest level we consider the ILU method while the Jacobi methods is applied on all finer levels. For details see [19]. The stopping criterion is the same as above.

As a testexample we consider a rotary shaft seal with a minimum gap width $a = 1$ and a concrete geometry, which is presented in Fig. 4. We carried out calculations for ferrofluid EMG 905 (for more details on the physical parameters of ferrofluids see [1, 18]. We obtain $\mathrm{Re} = 63.2456$, $\mathrm{Pr} = 0.7$, $\delta = 0.01$, $\Lambda = 1.2295$ as dimensionless parameters. The temperature ϑ_0 on the boundary is given as the room temperature ($\vartheta_0 = 20$). All presented numerical results we have obtained using MooNMD [9].

Table 1 gives the number of cells and degrees of freedom for the P_2/P_1 pair on the different refinement levels. Since we consider a two-dimensional problem the number of unknowns increases by a factor of about 4 from one level to next finer level.

Table 1. Number of degrees of freedom in each level

Level	2	3	4	5	6	7
number of cells	32	128	512	2048	8192	32768
velocity	170	594	2210	8514	33410	132354
pressure	27	85	297	1105	4257	16705
temperature	85	297	1105	4257	16705	66177

Fig. 5 shows that the multigrid method is an efficient solver with respect to the computing times. The computing time to determine the secondary flow increases like $\mathcal{O}(N)$ with the number N of degrees of freedom.

The computed streamlines of the secondary flow on the level 5 and level 7 are visualised in Fig. 6. Figure 7 illustrates the isolines of the temperature on level 5 and level 7. Considering finer triangulations we obtain really reliable results.

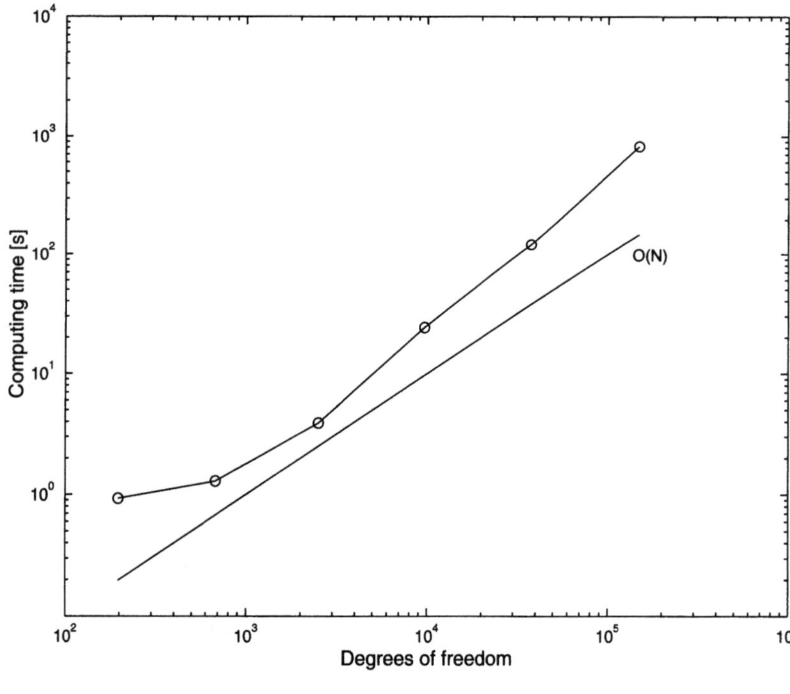

Fig. 5. Computing times (in sec.) to determine the secondary flow for different refinement levels

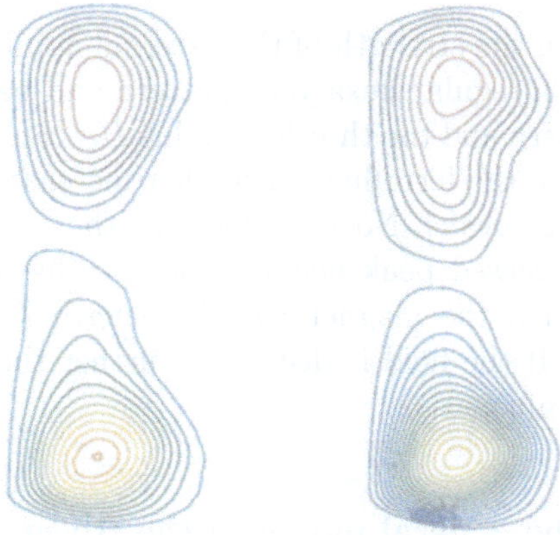

Fig. 6. Streamlines of the secondary flow on level 5 (left) and level 7(right)

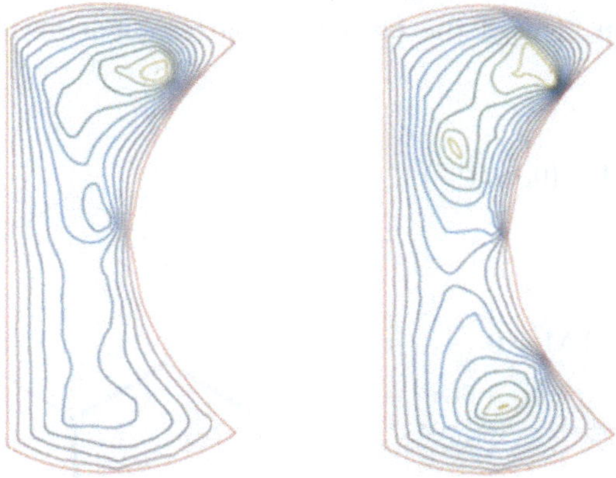

Fig. 7. Isolines of the temperature on level 5 (left) and level 7(right)

4 Rosensweig Instability

4.1 Physical Observation

A perpendicular, uniform magnetic field applied to a layer of ferrofluid generates a spontaneous surface deformation if the field strength exceeds a critical value. This stationary wave pattern on the surface was first observed by Cowley and Rosensweig [4]. This surface instability is caused by the interaction of gravitational, capillary and magnetic forces.

First we will describe in detail a physical experiment which demonstrates the behaviour of a ferrofluid layer in the Rosensweig instability. We start with an undisturbed and flat surface. For small value of the magnetic field strength the interface between ferrofluid and surrounding air remains flat. At a certain threshold value H_C of the strength of the external applied field the ferrofluid surface deforms into a regular hexagonal pattern and peaks occur. The wave length λ_C of the pattern and the threshold value H_C can be determined by the physical parameters of the ferrofluid. A further increase of the magnetic field strength causes higher peaks. Now we decrease the magnetic field strength. This results in a decreased peak height. Due to a hysteresis the pattern of peaks will exist even if the magnetic field strength H is smaller than the threshold value H_C. If the field is decreased further then the peaks collapse at a second critical value.

4.2 Adaption of the general model to the Rosensweig instability

We consider a horizontally unbounded ferrofluid layer of infinite depth. Figure 8 displays a two-dimensional cut through the three-dimensional surface pattern and a top view of the hexagonal pattern. Furthermore, the critical

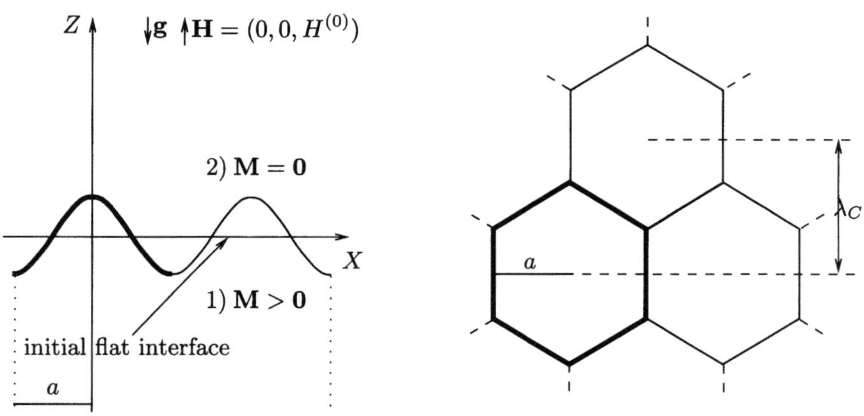

Fig. 8. Illustration of surface deformation, two-dimensional cut (left) and top view with hexagonal structure (right)

wave length λ_C is shown. The value of λ_C can be obtained by

$$\lambda_C = 2\pi\sqrt{\frac{\alpha}{\varrho g}} \ ,$$

see [18]. The critical magnetic field $H_C^{(1)}$ inside the fluid can be calculated from the nonlinear equations

$$M^2(H_C^{(1)}) = \frac{2}{\mu_0}\left(1 + \frac{1}{r_0\left(H_C^{(1)}\right)}\right)\sqrt{\varrho g \alpha}$$

with

$$r_0\left(H_C^{(1)}\right) = \frac{1}{\mu_0}\sqrt{\mu_c\left(H_C^{(1)}\right)\mu_t\left(H_C^{(1)}\right)}$$

and

$$\mu_c\left(H_C^{(1)}\right) = \frac{B^{(1)}\left(H_C^{(1)}\right)}{H_C^{(1)}}, \qquad \mu_t\left(H_C^{(1)}\right) = \frac{\partial B^{(1)}}{\partial H}\left(H_C^{(1)}\right) .$$

The critical value $H_C^{(2)}$ for the outer applied field is given by

$$H_C^{(2)} = H_C^{(1)} + M\left(H_C^{(1)}\right) ,$$

see [18].

Since the surface protuberances form a regular pattern we restrict our considerations to a domain which is bounded in both horizontal directions, i.e., we consider $\Omega = \Omega' \times \mathbb{R}$ with a bounded domain $\Omega' \in \mathbb{R}^2$. The magnetostatic problem (1)-(2) can be written by using the magnetostatic potential φ as

$$-\text{div}\left(\mu(x, |\nabla\varphi|)\nabla\varphi\right) = 0 \qquad \text{in } \Omega , \tag{23}$$

where

$$\mu(x, |\nabla\varphi|) = \begin{cases} \mu_0 & x \in \Omega \setminus \Omega_F , \\ \mu_0\left(\frac{M(|\nabla\varphi|)}{|\nabla\varphi|} + 1\right) & x \in \Omega_F . \end{cases}$$

In order to apply boundary conditions for the magnetostatic problem we restrict the z-coordinate to the interval $(-Z, Z)$, i.e., we consider the magnetostatic problem in the bounded domain $\Omega' \times (-Z, Z)$. At the bottom and the top boundary we use

$$\varphi = \begin{cases} -ZH^{(1)} & \text{on } \Omega' \times \{-Z\} \\ ZH^{(2)} & \text{on } \Omega' \times \{Z\} \end{cases} \tag{24}$$

as boundary conditions where $H^{(1)}$ and $H^{(2)}$ are the magnetic field strength for the case of an undisturbed interface inside and outside the ferrofluid, respectively. Using this boundary conditions makes sense since we assume that the magnetic field far away from the interface coincide with the field for the case of an undisturbed, flat interface. Due to the regularity of the surface pattern it is reasonable to apply

$$\frac{\partial\varphi}{\partial n} = 0 \qquad \text{on } \partial\Omega' \times (-Z, Z) . \tag{25}$$

For the Navier–Stokes equations (10) and (11) we assume that the ferrofluid is at rest, i.e., $\mathbf{u} = \mathbf{0}$. If we take only the gravity as acting force $\mathbf{f}$ in (10) we get

$$\nabla p = -\varrho\mathbf{g} + \mu_0 M \nabla H$$

and obtain

$$p_f + p_m = -\varrho g z + \mu_0 \int_0^H M(H')\, dH' \ .$$

Taking this we get the Young–Laplace equation (13) in the form

$$p^{(2)} + \alpha\mathcal{K} = -\varrho g z + \mu_0 \int_0^H M(H')\, dH' + \frac{\mu_0}{2}(\mathbf{M}\cdot\mathbf{n})^2 + C \qquad (26)$$

where the constant C will be fixed by using a volume constraint.

4.3 Decoupling Strategy

The problem is a coupled system of nonlinear partial differential equations which is splitted into two subproblems. The first one is the magnetostatic problem in a given domain and the second one is the Young–Laplace equation for a given magnetic field. By using a parametrisation of the unknown interface we have a representation of the curvature $\mathcal{K}$ in terms of the unknowns of the interface parametrisation.

For solving the whole system we apply the following decoupling strategy.

S1. Construct an initial mesh for the two-dimensional interface.

S2. Generate a three-dimensional mesh for the magnetostatic potential from the two-dimensional interface mesh.

S3. Solve the magnetostatic problem for a given interface position.

S4. Calculate the unknowns of the surface parametrisation from the Young–Laplace equation using the magnetic field obtained in S3 and get the new interface mesh.

S5. If the norm of the surface change within one iteration is greater than a prescribed threshold then continue with S2 otherwise STOP.

The step S1 has to provided an initial surface. This can be for instance a surface which is slightly disturbed flat interface or a computed surface for a parameter set close to the new one.

4.4 Magnetostatic problem

The magnetostatic problem is nonlinear due to the nonlinearity of the permeability function μ. For solving this problem we apply a fixed-point iteration, i.e., we solve

$$-\mathrm{div}\left(\mu(x, |\nabla\varphi^{k-1}|)\nabla\varphi^k\right) = 0, \qquad k \geq 1 \ ,$$

with the boundary conditions (24) and (25). In the very first calculation for obtaining φ^1 we start with a magnetic field which corresponds to the case of the undisturbed interface. The magnetostatic problem is uniformly elliptic and has slightly jumping coefficients.

4.5 Parametric free surface equations

We suppose that Γ_F can be considered as a surface of revolution and formulate the problem in the cylindrical coordinates R, Z so that $Z = 0$ is the equation of the unperturbed plane surface. The equilibrium shape is described by the parametric functions $R = R(S)$, $Z = Z(S)$, where S is the arc length of the unknown equilibrium line, $S = 0$ at the peak top on the OZ axis, $S = \ell$ at the peak foot where $R = a$ is the radius of the circle Ω'.

The surface curvature is calculated by the formula $\mathcal{K} = -(RZ')' / (RR')$ and can be written as $\mathcal{K} = -Z''/R' - Z'/R = R''/Z' - Z'/R$ according to the natural property $(Z')^2 + (R')^2 \equiv 1$ of such a type of parametrisation.

We will use the Vislovich approximation of the magnetisation law [3]

$$M(H) \approx M_S H/(H + H_T)$$

where H_T is the magnetic field strength at which the magnetisation of the fluid is equal to $M_S/2$.

The Young-Laplace equation (26) in axisymmetrical case has the form

$$z'' = r'(f + C), \quad r'' = -z'(f + C); \quad 0 < s < L = \ell/a \ , \tag{27}$$
$$f = \lambda^2 z - \frac{2\lambda Si}{\chi}(H - \ln(1 + H)) - \lambda Si \left(\frac{\mathbf{H} \cdot \mathbf{n}}{1 + H}\right)^2 - z'/r \ ,$$

where $r = R/a$, $z = Z/a$, $s = S/a$ are the dimensionless variables, C the unknown constant, $\lambda = a\sqrt{\rho g/\alpha}$, $Si = \mu_0 M_S^2/(2\sqrt{\rho g \alpha})$, $\chi = M_S/H_T$, and H the field in the magnetic fluid dimensionless on H_T.

Boundary conditions follow from the shape symmetry which in dimensionless formulation take the form

$$r(0) = z'(0) = z'(L) = 0, \quad r(L) = r'(0) = r'(L) = 1 \ , \tag{28}$$

and are supplemented by the condition of volume conservation

$$\int_0^L zrr' \, ds = 0 \ . \tag{29}$$

We reformulate the problem (27)-(29) in new dimensionless variables $\tilde{s} = s/L, \tilde{r} = r/L, \tilde{z} = z/L$, what allows us to make computation on a fixed domain $(0, 1)$. Then we construct a second order difference scheme for this problem on a uniform grid and solve the corresponding nonlinear difference problem by an iterative method elaborated in [15, 16].

4.6 Surface as a graph

In this section we abandon the assumption that the free surface can be considered as a revolution surface. We will consider the hexagonal surface pattern and choose a hexagon as domain Ω'. Furthermore, we assume that the interface Γ_F is given as a graph of a function z, i.e.,

$$\Gamma_F = \left\{(x_1, x_2, z) \,:\, z = z(x_1, x_2),\, (x_1, x_2) \in \Omega'\right\} .$$

Then, the curvature $\mathcal{K}$ is given by

$$\mathcal{K} = -\operatorname{div} \frac{\nabla z}{\sqrt{1 + |\nabla z|^2}} .$$

Using a characteristic length A and the strength $H^{(0)}$ of the outer applied field to translate the Young-Laplace equation (26) into dimensionless quantities we obtain after multiply (26) by A/α the following equation

$$\begin{aligned} -\operatorname{div} \frac{\nabla z}{\sqrt{1 + |\nabla z|^2}} + \Lambda^2 z = & \\ \Lambda Si \ln\left(\frac{\sinh(\gamma H)}{\gamma H}\right) + \frac{3\chi_0 \Lambda Si}{2} L^2(\gamma H) \left(\frac{\mathbf{H}\cdot\mathbf{n}}{H}\right)^2 + C & \end{aligned} \tag{30}$$

with the dimensionless parameters

$$\Lambda = A\sqrt{\frac{\varrho g}{\alpha}}, \qquad Si = \frac{\mu_0 M_S^2}{3\chi_0\sqrt{\varrho g \alpha}}, \qquad \gamma = \frac{3\chi_0 H^{(0)}}{M_S}, \qquad \chi_0 = \frac{\partial M}{\partial H}(0) .$$

Due to the symmetry of the surface pattern we equip this equation with homogeneous Neumann boundary conditions.

Let F contain all forces on the right-hand side of (30). In the following, we will consider a slightly more general problem in which Λ is not longer a positive constant but a uniformly positive and bounded function on Ω', i.e., $\Lambda \in L^\infty(\Omega')$ and $\Lambda(x_1, x_2) \geq \Lambda_0 > 0$ for $(x, y) \in \Omega'$. This means that we consider the problem

$$-\operatorname{div} \frac{\nabla z}{\sqrt{1 + |\nabla z|^2}} + \Lambda^2 z = F \qquad \text{in } \Omega' , \tag{31}$$

$$\frac{\partial z}{\partial n} = 0 \qquad \text{on } \partial\Omega' . \tag{32}$$

This equation is elliptic but not uniformly elliptic. Moreover, it is nonlinear. We solve this problem by a fixed-point iteration, i.e., we solve the equation

$$-\operatorname{div} \frac{\nabla z^k}{\sqrt{1 + |\nabla z^{k-1}|^2}} + \Lambda^2 z^k = F \qquad \text{in } \Omega', \qquad k \geq 1 ,$$

with homogenous Neumann boundary conditions on $\partial\Omega'$. As initial guess z^0 we choose the interface position which was used during the calculation of the magnetostatic potential.

Since we want to solve the Young–Laplace equation (31) by a finite element method we have to derive a weak formulation of (31). In order to do so we multiply (31) by a test function $v \in H^1(\Omega')$ and integrate over Ω'. After an integration by parts of the first term we obtain as a weak formulation

Find $z \in H^1(\Omega')$ such that

$$\int_{\Omega'} \frac{\nabla z}{\sqrt{1+|\nabla z|^2}} \cdot \nabla v \, dx + \int_{\Omega'} \Lambda^2 \, zv \, dx = \int_{\Omega'} Fv \, dx \qquad \forall v \in H^1(\Omega') \ . \tag{33}$$

In the following we assume that the problem (33) has a unique solution z which satisfies $z \in W^{1,\infty}(\Omega')$.

Let $V_h \subset H^1(\Omega')$ an arbitrary finite element space. The discretisation of (33) is given by

Find $z_h \in V_h$ such that

$$\int_{\Omega'} \frac{\nabla z_h}{\sqrt{1+|\nabla z_h|^2}} \cdot \nabla v_h \, dx + \int_{\Omega'} \Lambda^2 \, z_h v_h \, dx = \int_{\Omega'} F v_h \, dx \qquad \forall v_h \in V_h \ . \tag{34}$$

Now we will estimate the error between the solutions z and z_h of (33) and (34), respectively. As usual in finite element analysis we get an error estimate which is bounded by the interpolation error only.

Theorem 1 (Abstract error estimate). *Let z and z_h be the solutions of (33) and (34), respectively. Furthermore, we assume that $z \in W^{1,\infty}(\Omega')$. Then we get*

$$\frac{K(z)}{2}\Theta^2 + \frac{3}{4}\Lambda_0^2 \|z - z_h\|_{0,2,\Omega'}^2 \le C(z) \inf_{v_h \in V_h} \|z - v_h\|_{1,2,\Omega'}^2 \tag{35}$$

where

$$\Theta := \left(\int_{\Omega'} \frac{|\nabla(z - z_h)|^2}{\sqrt{1+|\nabla z_h|^2}} \, dx \right)^{1/2} .$$

The quantities $C(z)$ and $K(z)$ are positive constants which are independent of V_h.

The statement of the previous theorem leads together with standard approximation results for finite element spaces to the following corollary.

Corollary 1 (Error estimates). *Let $z \in H^{k+1}(\Omega') \cap W^{1,\infty}(\Omega')$ be the solution of (33). The space V_h consists of finite elements of order k. Then we have*

$$\|z - z_h\|_{0,2,\Omega'} \le C(z)\, h^k, \qquad \textit{and} \qquad |z - z_h|_{1,2,\Omega'} \le C(z)\, h^k \ ,$$

where the constant $C(z)$ depends only on z but not on h.

The proof of Theorem 1 and Corollary 1 can be found in [10].

4.7 Numerical Results

Calculated peak shapes.

As a first result we will show in Fig. 9 the peak shapes which we have obtained by the method describe in Sect. 4.6 using MooNMD [9]. As expected, the

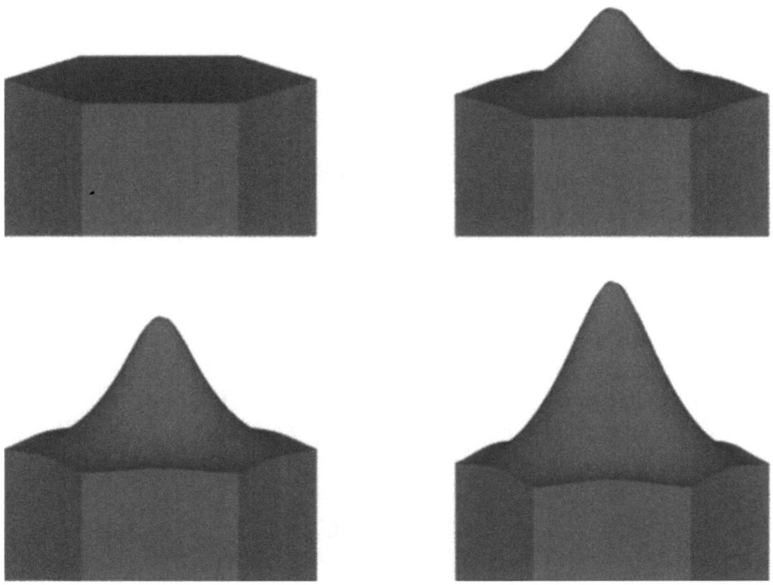

Fig. 9. Peak shape for different applied fields

interface remains flat if the outer applied field doesn't exceed the critical threshold (upper left subplot). If the applied field is larger than the critical value H_C then we obtain a peak (upper right). The lower row in Fig. 9 shows that an increasing magnetic field will lead to an increased peak height.

Hysteresis.

Ferrofluids show an hysteretic behaviour if the magnetic field strength is decreased from a supercritical value to a subcritical value. Figure 10 shows the peak height for the ferrofluid APG 512A. The marker $\circ$ indicates the path with an increasing field while the marker $*$ labels the path of a decreasing

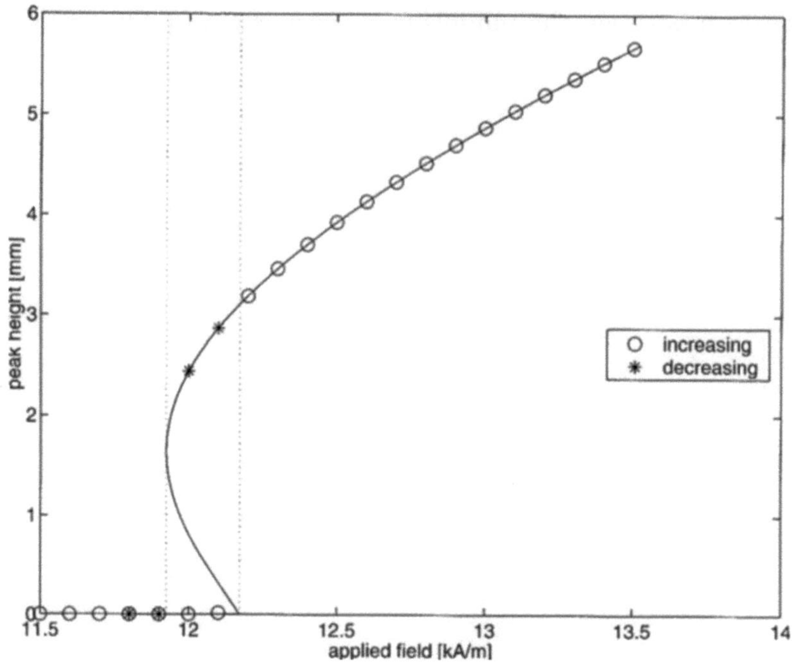

Fig. 10. Hysteresis for ferrofluid APG 512A together with a fit

field. There is a region for the magnetic field strength, between the two vertical lines in Fig. 10, below the critical value where two surface shapes are possible, a flat interface or a developed peak. Which shape occurs in the experiment depends on the way in which the applied field strength was achieved. If the magnetic field strength is increased from a lower value then the surface remains flat. If the magnetic field strength is decreased from a value above the critical value then a pattern of peaks is obtained. A classical result for the dependence of the peak height on the strength of the applied field can be found in [6] where the magnetisation M was approximated by $M = \chi H$ with a constant susceptibility χ. Recently, the pattern and wavenumber selection in ferrofluid layers of finite and infinite depth was investigated in [5]. The physical theory predicts that the dimensionless quantity

$$\varepsilon = \frac{H^2 - H_C^2}{H_C^2}$$

is a polynomial in dimensionless peak height h. Figure 10 shows a fit with a polynomial of degree 6 through the data which were obtained by our simulations. The agreement between the fit and the data is quite good.

Comparison between the models from Sect. 4.5 and 4.6.

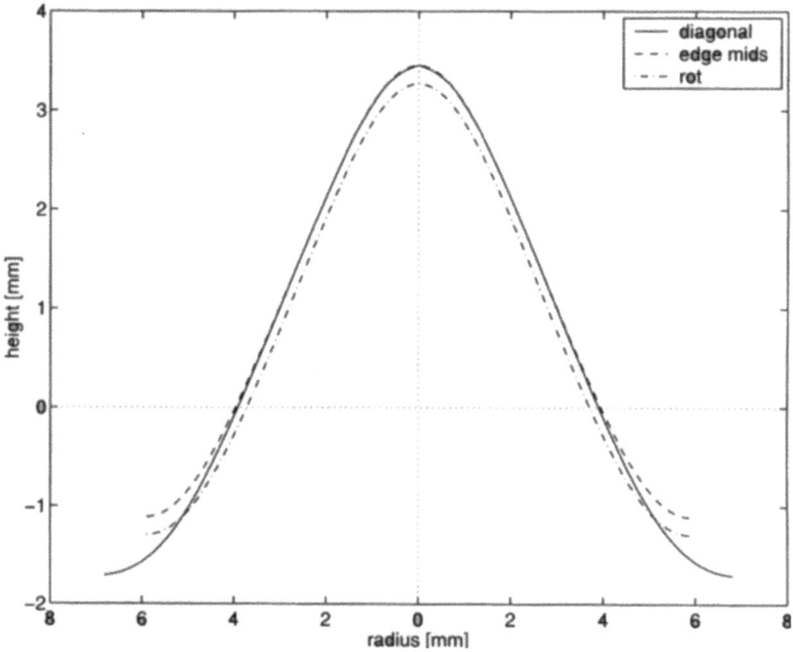

Fig. 11. Cuts through a peak along diagonal, line connecting midpoints of two opposite edges, and shape obtained by the parametric surface model

The following results were obtained for the ferrofluid EMG 909. Figure 11 shows the peak shape obtained by the parametric surface model and two cuts through a peak obtained by the graph model. One cut goes along a diagonal from one vertex to the opposite one. The second cut connects the midpoints of two opposite edges of the hexagon. We see that for the centre of the peak the assumption of rotational symmetry is reasonable. The difference occur only in the regions which have a greater distance from the middle of the peak. The results from the calculation with assumed rotational symmetry produces a peak which is a little bit smaller but the shape is very similar to that obtained by the fully three-dimensional calculations.

Now we will investigate how the assumption of rotational symmetry influences the peak height and the value for the critical field. Figure 12 shows the peak top and foot z-coordinates in dependence on the outer applied field. We see that the calculations which were made with the assumption of rotational symmetry produces peaks which are smaller than those obtained by the fully 3d model. Note that for both cases the Vislovich approximation was used which

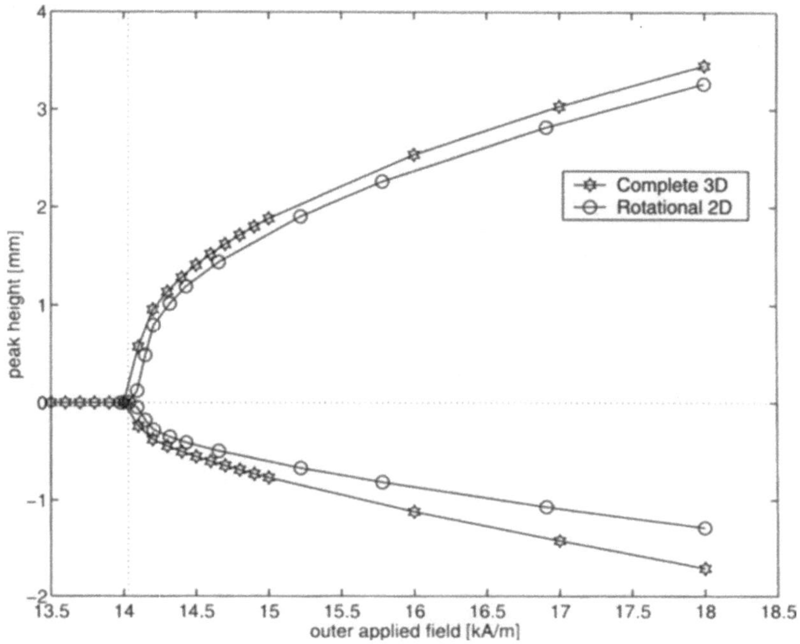

Fig. 12. Comparison of rotational symmetric and complete 3d calculation

results in a theoretical value $H_C = 14.03$ kA/m for the critical magnetic field strength. This value is shown in Fig. 12 as a vertical line. The fully 3d calculations produce, compared to the 2d rotational calculations, a value for the critical magnetic field strength which is closer to the theoretical one.

Comparison of different magnetisation laws in the 3d model.

Now we will consider the influence of the used constitutive equation on the peak height. We compare calculations with the model describe in Sect. 4.6 for the Langevin function and its Vislovich approximation. Due to the different magnetisation laws we obtain also different values for the critical magnetic field H_C. For the Langevin function this value is given by 14.85 kA/m while the Vislovich approximation results in a value of 14.03 kA/m. Both values are given in Fig. 13 as vertical lines. Figure 13 shows that both critical values are reproduced very well by the finite element approximation. Furthermore, we see that the peak height for the Langevin function increases more rapidly compared to the Vislovich approximation.

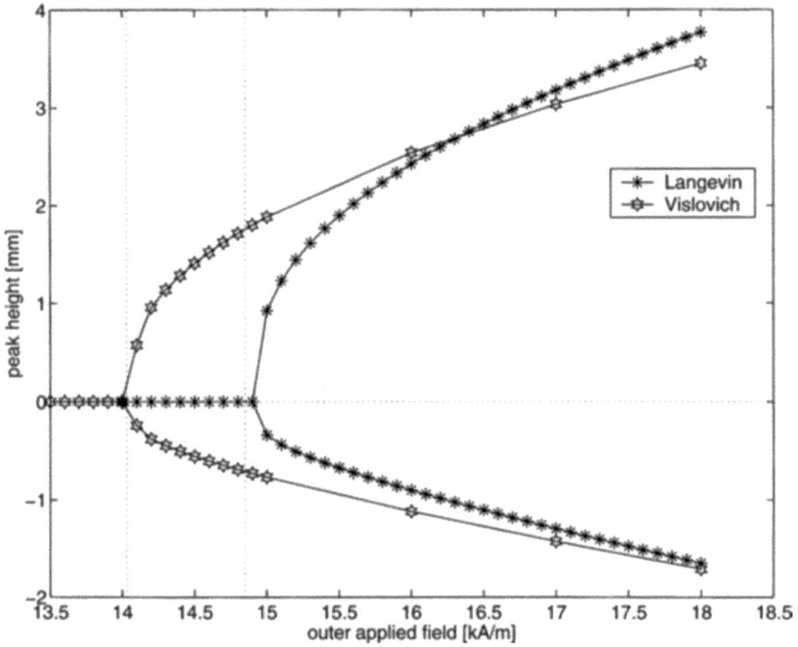

Fig. 13. Comparison of Langevin function and Vislovich approximation

Acknowledgement

For supporting partially the research presented in this paper, the authors like to thanks the German Research Foundation (DFG) under grants within the research group FOR 301 and within the high priority research program SPP 1104, as well as the German Academic Exchange Service (DAAD) and the State Sachsen-Anhalt. Several pictures were produced using the visualisation package GRAPE which was developed at the universities of Bonn and Freiburg i.Br. by the SFB 256 of the DFG.

References

1. Auernhammer, G. K., Brand, H. R.: Thermal convection in a rotating layer of a magnetic fluid. Eur. Phys. J. B **16** (2000) 157–168
2. Bänsch, E., Höhn, B.: Numerical treatment of the Navier-Stokes equations with slip boundary condition. SIAM J. Sci. Comp. **20** (2000) 2144–2162
3. Berkovsky, B. M., Bashtovoy, V.: Magnetic Fluids and Applications Handbook. Begell house, Inc., New York, Wallingford (UK), 1996
4. Cowley, M. D., Rosensweig, R. E.: The interfacial stability of a ferromagnetic fluid. J. Fluid Mech. **30** (1967) 671–688

5. Friedrichs, R., Engel, A.: Pattern and wavenumber selection in magnetic fluids. Phys. Rev. E **64** (2001) 021406
6. Gailitis, A.: Formation of hexagonal pattern on the surface of a ferromagnetic fluid in an applied magnetic field. J. Fluid Mech. **82** (1977) 401–413
7. Girault, V., Raviart, P.: Finite element methods for Navier-Stokes equations. Theory and algorithms. Springer-Verlag, Berlin, 1986
8. John, V., Knobloch, P., Matthies, G., Tobiska, L.: Non-nested multi-level solvers for finite element discretizations of mixed problems. Computing **68** (2002) 343–373
9. John, V., Matthies, G.: MooNMD - a program package based on mapped finite element methods. Technical Report 02-01, Fakultät für Mathematik, Otto-von-Guericke-Universität Magdeburg (2002)
10. Matthies, G.: Finite element methods for free boundary value problems with capillary surface. Shaker Verlag, Aachen, 2002, PhD thesis, Otto-von-Guericke-Universität Magdeburg
11. Matthies, G., Tobiska, L.: Simulation of free surfaces of magnetic liquids. In: Algoritmy 2000, 15th Conf. Scientific Comput., ed. by Handlovičová, A., et. all, Slovak Technical University, Bratislava (2000) 1–10
12. Mitkova, T.: Lösbarkeit eines mathematischen Modells für Dichtungen mit magnetischen Flüssigkeiten. Technische Mechanik **20** (2000) 283–293
13. Odenbach, S.: Magnetoviscous effects in ferrofluids. Lect. Notes Physics, m71, 2002
14. Polevikov, V.: Stability of a static magnetic-fluid seal under the action of an external pressure drop. Fluid Dynamics **32** (1997) 457–461
15. Polevikov, V.: Numerical modelling of equilibrium capillary surfaces: some methods and results. Preprint 98-10, Fakultät für Mathematik, Otto-von-Guericke-Universität Magdeburg (1998)
16. Polevikov, V.: Methods of numerical modeling of equilibrium capillary surfaces. Differents. Uravnenia **35** (1999) 975–981
17. Polevikov, V., Tobiska, L.: Modeling of a dynamic magneto-fluid seal in the presence of a pressure drop. Fluid Dynamics **36** (2001) 890–898
18. Rosensweig, R. E.: Ferrohydrodynamics. Dover Publications, Inc. Mineola, New York, 1997
19. Saad, Y.: Iterative method for sparse linear systems. PWS Publishing Company, 1996
20. Vanka, S.: Block-implicit multigrid calculation of two-dimensional recirculating flows. Comp. Meth. Appl. Mech. Eng. **59** (1986) 29–48

Iterative Substructuring Methods for Advection – Diffusion Problems in Heterogeneous Media

Paolo Zunino

Modeling and Scientific Computing Laboratory (MOCS), Institute of Mathematics, Ecole Polytéchnique Fédérale de Lausanne (EPFL), 1015 Ecublens, Switzerland.

Summary. This work is devoted to the numerical approximation of a system of advection-diffusion equations set in adjacent domains and coupled with non-standard matching conditions. The specific application of the model at hand is the study of the transfer of chemicals through media of heterogeneous nature, for instance a free fluid and a porous medium.

After a brief description of the model, we focus our attention on its numerical treatment. In particular, since our model couples subproblems in different media, we study an iterative procedure where the solutions provided separately on each subdomain are suitably matched (a so called *iterative substructuring method*). More precisely, we consider a strategy based on Robin interface conditions for both subdomains. The convergence of this iterative strategy is analyzed at both the continuous and the discrete level. Moreover, an algebraic reinterpretation of this technique is provided, leading to the definition of optimal preconditioners for the linear system arising from the discretization of the global problem. Finally, numerical results are presented, in order to assess the computational efficiency of the numerical methods proposed.

1 Introduction

In this work we address the problem of the transport of chemicals through heterogeneous media. This problem has several practical applications, for instance the study of the motion of a pollutant from a basin to the adjacent soil or the study of the transport of chemicals from the blood flowing into an artery and the wall of the vessel. In all these situations, the physical unknown of interest is the concentration of a chemical. Because of the heterogeneity of the media we consider, the concentration may feature large variations from one medium to the other. For example, if a pollutant is released into a basin, its concentration in the free fluid is reasonably higher than the one in the surrounding soil. In the framework of classical physics, the concentration is a continuous variable, consequently it may feature an extremely steep boundary layer at the interface of the heterogeneous media. However, the scale of the phenomena of interest is often much larger than the one of the interface boundary layers. Consequently, the numerical simulation of such problems could be an extremely challenging task. To this aim, we propose a particular choice of the matching conditions between the advection-diffusion problems on each medium, which allow the discontinuity of the concentration across the interface. In other words, we apply a lumping approach in space and we represent the region in the neighborhood of the interface, where large concentration gradients take place, as a *permeable membrane* separating the heterogeneous media and leading to a jump of the concentration on its sides. In order to describe the behavior of membranes with respect to mass transfer we use the so called Kedem-Katchalsky equations. These equations describe the diffusion and transport of chemicals across the membrane with respect to the concentrations on both sides, and will be introduced later on. For an extensive discussion of the Kedem-Katchalsky equations in the framework of biomechanics we refer to [4, 11] and references therein. Finally, we observe that the modelling of the considered phenomena can be split in two parts: the set up of models for the fluid dynamics and the set up of models for the dynamics of chemicals. As regards the flow in the free fluid, one may consider the incompressible Navier-Stokes equations while for the flow through porous media one may use the Darcy law of filtration. The mathematical modelling and the numerical approximation of the coupled flow in a free fluid and in a porous medium is a very demanding task that requires effective numerical methods and intensive computational effort. The numerical approximation of these equations in the case of environmental applications is discussed for example in [3]. However, in this work we focus on the solute dynamics and we assume that the transport fields for the chemicals, namely the velocities in each medium are given quantities.

To start the description of our reference problem, we introduce the following notation. We denote with the subscript 1 the quantities in one medium (typically a free fluid) and with the subscript 2 the quantities in the other (typically a porous medium). We refer to figure 1 for a description of the domains and for the partition of the boundaries. As regards the physical quantities of interest, let c_1, c_2 be the concentrations of the considered chemical. These are the unknowns of our reference problem. As regards the data, let $\mathbf{v}_1, \mathbf{v}_2$ be the velocities of the fluid in each medium. According to the mass conservation principle we assume $\mathrm{Div}\,\mathbf{v}_i = 0$ in Ω_i, $i = 1,2$ and $\mathbf{v}_1 \cdot \mathbf{n}_1 = -\mathbf{v}_2 \cdot \mathbf{n}_2$ on Γ. Moreover, we assume that $\mathbf{v}_i \cdot \mathbf{n}_i \geq 0$ on $\Gamma_{N,i}, i = 1,2$. This means that there is no reverse flow in the outflow regions of the domains, which is in general verified in our applications. Furthermore, we denote with $J_v = \mathbf{v}_1 \cdot \mathbf{n}_1|_\Gamma$ the normal velocity across the membrane. Finally, let D_1, D_2 be the diffusivities of the considered chemical and r_1, r_2 its reaction coefficients account-

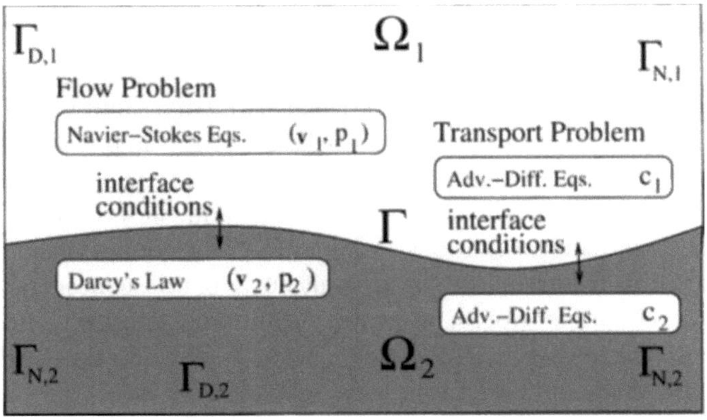

Fig. 1. Synoptic description of the reference problem.

ing for the consumption of the chemical due to possible reactions with the medium. Moreover, the forcing terms f_1, f_2 represent other possible consumption or production processes, independent of the concentration of the considered chemical. Regarding the parameters to characterize the membrane that represents the interface Γ, we introduce the permeability P and the average concentration within the membrane, denoted with $f(c_1, c_2) = w_1 c_1 + w_2 c_2$. We assume that the weights w_1, w_2 do not depend on the concentration and satisfy $w_1 + w_2 = 1$. With this notation, the flux of chemicals across the interface Γ can be defined as $J_s(c_1, c_2) = P(c_1 - c_2) + J_v f(c_1, c_2)$, that is indeed the aforementioned Kedem-Katchalsky model. This equation features a diffusive term, $P(c_1 - c_2)$, and a transport term, $J_v f(c_1, c_2)$. When the considered solute is a large molecule, the osmotic phenomena should be taken into account as well. This requires the addition of a third term in the definition of J_s. This term is of the form $\zeta f(c_1, c_2)(c_1 - c_2)$ and is thus nonlinear. The analysis of iterative substructuring methods for this case is much more delicate than in the linear case and is not completely understood yet. However, convergence results for simple iterative substructuring methods applied to nonlinear interface conditions have been derived in [11]. In what follows, we assume all the physical parameters to be bounded functions. More precisely, we set $D_i > 0$, $D_i \in L^\infty(\Omega_i)$, $r_i \geq 0$ with $r_i \in L^\infty(\Omega_i)$ as well as $f_i \in L^\infty(\Omega_i)$ and we define $\mu_i = 1/\, \|1/D_i\|\, L^\infty(\Omega_i), i = 1, 2$. Furthermore, on the interface Γ we assume $P > 0$ and $P, \mathbf{v}_i \cdot \mathbf{n}_i \in L^\infty(\Gamma)$.

Using this notation, our reference problem takes the form of the following system of advection-diffusion equations.

Problem 1 *Find the concentrations* c_1, c_2 *such that,*

$$\begin{cases} \partial_t c_1 + \mathrm{Div}\,(-D_1\,\mathrm{Grad}\,c_1 + \mathbf{v}_1 c_1) + r_1 c_1 = f_1, \ \ in\ \Omega_1\ t > 0, \\ c_1(0) = c_{1,0}\ \ in\ \Omega_1, \\ c_1 = c_{D,1}\ \ on\ \Gamma_{D,1},\ \ t > 0 \\ \mathrm{Grad}\,c_1 \cdot \mathbf{n}_1 = 0\ \ on\ \Gamma_{N,1},\ \ t > 0, \end{cases} \tag{1}$$

$$\begin{cases} \partial_t c_2 + \mathrm{Div}\,(-D_2\,\mathrm{Grad}\,c_2 + \mathbf{v}_2 c_2) + r_2 c_2 = f_2, \ \ in\ \Omega_2\ t > 0, \\ c_2(0) = c_{2,0}\ \ in\ \Omega_2, \\ c_2 = c_{D,2}\ \ on\ \Gamma_{D,2},\ \ t > 0 \\ \mathrm{Grad}\,c_2 \cdot \mathbf{n}_2 = 0\ \ on\ \Gamma_{N,2},\ \ t > 0, \end{cases} \tag{2}$$

with the following matching conditions on Γ,

$$-D_1\,\mathrm{Grad}\,c_1 \cdot \mathbf{n}_1 + \mathbf{v}_1 \cdot \mathbf{n}_1 c_1 = -[-D_2\,\mathrm{Grad}\,c_2 \cdot \mathbf{n}_2 + \mathbf{v}_2 \cdot \mathbf{n}_2 c_2]\ \ on\ \Gamma,\ \ t > 0, \tag{3}$$

$$-D_1\,\mathrm{Grad}\,c_1 \cdot \mathbf{n}_1 + \mathbf{v}_1 \cdot \mathbf{n}_1 c_1 = J_s(c_1, c_2)\ \ on\ \Gamma,\ \ t > 0. \tag{4}$$

The well posedness of this problem follows from the classical theory of linear parabolic problems (see for example [5]). In the particular case with $\mathbf{v}_2 = 0$, under standard regularity assumptions on the data there exists a unique weak solution $[c_1, c_2]$ of the coupled problem in the following functional space $[c_1, c_2] \in L^\infty(0,T; L^2(\Omega_1 \cup \Omega_2)) \cap L^2(0,T; H^1(\Omega_1) \times H^1(\Omega_2))$ (see [9]). In the case of a nonlinear flux $J_s(c_1, c_2)$ the proof of existence and uniqueness of the solution is a more challenging task. Under regularity assumptions and restrictions on the data, which are however physically acceptable, existence and uniqueness have been proven in [11] using fixed point techniques together with maximum principle properties.

2 A Robin-Robin iterative substructuring method

We observe that problem 1 can not be reformualted as a problem governed by a unique differential operator on the global domain $\Omega_1 \cup \Omega_2 \cup \Gamma$. For this reason, we focus our attention on *iterative substructuring methods* to split problem 1 into subproblems. A general theory of domain decomposition methods for symmetric operators is given in Quarteroni and Valli [7], but the advection-diffusion equations introduced in the latter section are not symmetric. Moreover, iterative substructuring methods for unsymmetric but coercive problems have been proposed for example in [1, 2]. However, problem 1 can not be casted into the latter theory because the matching conditions that we propose do not prescribe the continuity of the solution across the interface. For this reasons, we aim to introduce a suitable splitting strategy for problem 1 and to pursue the convergence analysis for this specific case. Finally, the numerical approximation of the problem at hand suggests a way to define an *optimal* preconditioner for the original coupled problem. This observation give rise to the discussion of efficient computational techniques to solve problem

1. Actually, this work is a generalization of the one by Quarteroni, Veneziani and the author in [8]. In that work, the authors proposed and analyzed an iterative substructuring method for problem 1 in the particular case $\mathbf{v}_2 = 0$ and $J_s(c_1, c_2) = P(c_1 - c_2)$.

To make our notation more simple, we define $\zeta_1 := P + w_1 J_v, \zeta_2 := P - w_2 J_v$ and rewrite the flux of chemicals across Γ as $J_s(c_1, c_2) = \zeta_1 c_1 - \zeta_2 c_2$. Moreover, we define the co-normal derivative on Γ, $\hat{\partial}_{\mathbf{n}_i} c_i := -D_i \partial_{\mathbf{n}_i} c_i + \mathbf{v}_i \cdot \mathbf{n}_i c_i$ where $\partial_{\mathbf{n}_i} c_i$ is equivalent to $\operatorname{Grad} c_i \cdot \mathbf{n}_i$. For the time discretization, we subdivide the time interval $[0, T]$ in N time steps $t^n = n\Delta t$ with $\Delta t > 0$ and $n = 1, \ldots, N$, and we use backward Euler finite difference schemes. We denote with $\mathbf{v}_1^n, \mathbf{v}_2^n$ the velocity field evaluated at time t^n. For the sake of simplicity, from now on we consider the particular case of homogeneous Dirichlet boundary conditions, namely we set $c_{D,i} = 0$ on $\Gamma_{D,i}, i = 1, 2$. Indeed, the general case can be reduced to the homogeneous one by considering the unknown $c_i' = c_i - g_i$, where $g_i \in H^1(\Omega_i)$ is a suitable lifting of the Dirichlet boundary data. The time discrete counterpart of problem 1 reads as follows.

Problem 2 *Given c_1^0, c_2^0, for all $n = 1, \ldots, N$, find c_1^n, c_2^n such that,*

$$\begin{cases} (a)\ \frac{1}{\Delta t} c_i^n + \operatorname{Div}(-D_i \operatorname{Grad} c_i^n + \mathbf{v}_i^n c_i^n) + r_i c_i^n = \\ \qquad\qquad f_i + (1/\Delta t) c_i^{n-1}, \ \text{in } \Omega_i, \ i = 1, 2, \\ (b)\ c_i^n = 0 \ \text{on } \Gamma_{D,i}, \ i = 1, 2, \\ (c)\ \partial_{\mathbf{n}_i} c_i^n = 0 \ \text{on } \Gamma_{N,i}, \ i = 1, 2, \\ (d)\ \hat{\partial}_{\mathbf{n}_1} c_1^n = \zeta_1 c_1^n - \zeta_2 c_2^n \ \text{on } \Gamma, \\ (e)\ \hat{\partial}_{\mathbf{n}_2} c_2^n = \zeta_2 c_2^n - \zeta_1 c_1^n \ \text{on } \Gamma. \end{cases} \tag{5}$$

Since from now on all the relevant equations deal only with unknowns evaluated at the time step t^n, for notational convenience we drop the index n. The time index will be explicitly indicated only when referring to a time step different than t^n. To split 2 into subproblems, we introduce an iterative procedure where the concentration c_2, in the advection diffusion equation in Ω_1, is evaluated at the previous iterative step. This leads to the following iterative procedure.

Problem 3 *Given an initial guess λ^0 and a relaxation parameter $\theta > 0$, for $k = 1, 2, \ldots$ find a sequence $[c_1^k, c_2^k]$ such that*

$$\begin{cases} \frac{1}{\Delta t} c_1^k + \operatorname{Div}(-D_1 \operatorname{Grad} c_1^k + \mathbf{v}_1 c_1^k) + r_1 c_1^k = f_1 + (1/\Delta t) c_1^{n-1} \ \text{in } \Omega_1, \\ c_1^k = 0, \ \text{on } \Gamma_{D,1}, \\ \partial_{\mathbf{n}_1} c_1^k = 0 \ \text{on } \Gamma_{N,1}, \\ \hat{\partial}_{\mathbf{n}_1} c_1^k = \zeta_1 c_1^k - \zeta_2 \lambda^{k-1}, \ \text{on } \Gamma, \end{cases}$$

$$\begin{cases} \frac{1}{\Delta t} c_2^k + \operatorname{Div}(-D_2 \operatorname{Grad} c_2^k + \mathbf{v}_2 c_2^k) + r_2 c_2^k = f_2 + (1/\Delta t) c_2^{n-1} \ \text{in } \Omega_2, \\ c_2^k = 0, \ \text{on } \Gamma_{D,2}, \\ \partial_{\mathbf{n}_2} c_2^k = 0 \ \text{on } \Gamma_{N,2}, \\ \hat{\partial}_{\mathbf{n}_2} c_2^k = \zeta_2 c_2^k - \zeta_1 c_1^k, \ \text{on } \Gamma \times (0, T), \end{cases}$$

$$\lambda^k = \theta c_2^k + (1-\theta)\lambda^{k-1}.$$

In the sequel we will refer to this method as *Robin-Robin* iterative method, because the interaction between the subdomains is achieved by Robin matching conditions.

To conclude this section, we introduce the variational formulation of problems 2 and 3, giving a correct framework for the convergence analysis of the Robin-Robin method. In order to simplify the exposition we introduce the following notation. We denote with $c = [c_1, c_2]$ the couple of unknown functions. A similar notation will be adopted for the coefficients in Ω_1, Ω_2. For example, D denotes the couple of the diffusivities $D := [D_1, D_2]$. Then, in order to treat indices, we introduce the following product, $ab = [a_1 b_1, a_2 b_2]$. Furthermore, if $a, b \in L^2(\Omega_1) \times L^2(\Omega_2)$ we define $(a, b) = (a_1, b_1) + (a_2, b_2)$ where (a_i, b_i) denotes the inner product in $L^2(\Omega_i)$. No implicit summation on the repeated indices will be adopted in the sequel. Moreover, let us introduce the following functional spaces: $V := V_1 \times V_2 := H^1_{\Gamma_{D,1}}(\Omega_1) \times H^1_{\Gamma_{D,2}}(\Omega_2)$ endowed with the norm $\|v\| V := (\|v_1\| H^{1^2} + \|v_2\| H^{1^2})^{1/2}$, $L^2(\Omega_1) \times L^2(\Omega_2)$ endowed with the norm $\|v\| L^2 := (\|v_1\| L^{2^2} + \|v_2\| L^{2^2})^{1/2}$, and $L^\infty(\Omega_1) \times L^\infty(\Omega_2)$ endowed with the norm $\|v\| \infty := \max_{i=1,2}(\|v_i\| \infty)$. Finally, defining $\mathcal{F}_i = f_i + \frac{1}{\Delta t} c_i^{n-1}, i = 1, 2$ and $\chi := 1/\Delta t$, by virtue of the notation proposed above, we introduce the following bilinear form,

$$a_i^n(w, v) := \chi(w, v) + (D_i \operatorname{Grad} w, \operatorname{Grad} v) - (\mathbf{v}_i^n w, \operatorname{Grad} v) + (r_i w, v) + \int_{\Gamma_{N,i}} \mathbf{v}_i^n \cdot \mathbf{n}_i w v.$$

Then the weak counterpart of problem 2 considered at time step t^n, reads as follows.

Problem 4 *Find $[c_1, c_2] \in V_1 \times V_2$ such that,*

$$\sum_{i=1,2} \left(a_i(c_i, v_i) + \int_\Gamma \zeta_i c_i v_i \right) = \sum_{i=1,2} (\mathcal{F}_i, v_i) + \int_\Gamma \zeta_2 c_2 v_1 + \int_\Gamma \zeta_1 c_1 v_2 \quad \forall v_i \in V_i, \; i = 1, 2. \tag{6}$$

On the other hand, the weak counterpart of problem 3 for a fixed time step t^n, is reported below.

Problem 5 *Given an initial guess $\lambda^0 \in H^{1/2}(\Gamma)$, a relaxation parameter $\theta > 0$ and a tolerance ϵ, for $k = 1, 2, \ldots$ find a sequence $[c_1^k, c_2^k] \in V$ such that,*

$$\begin{cases} a_1\left(c_1^k, v_1\right) + \int_\Gamma \zeta_1 c_1^k v_1 = (\mathcal{F}_1, v_1) + \int_\Gamma \zeta_2 \lambda^{k-1} v_1 \quad \forall v_1 \in V_1, \\ a_2\left(c_2^k, v_2\right) + \int_\Gamma \zeta_2 c_2^k v_2 = (\mathcal{F}_2, v_2) + \int_\Gamma \zeta_1 c_1^k v_2 \quad \forall v_2 \in V_2, \\ \lambda^k = \theta c_2^k|_\Gamma + (1-\theta)\lambda^{k-1}. \end{cases} \tag{7}$$

2.1 Reinterpretation of the multidomain problem via Steklov-Poincaré operators

In this section we introduce the so called Steklov-Poincaré operators, in order to rewrite a multidomain problem, for instance problem 1, as a problem on the interface between the subdomains. The main interest of this approach comes from the fact that, in this framework, the Robin-Robin iterative method can be reinterpreted as a preconditioned Richardson method for a suitable interface unknown. This reinterpretation leads to a straightforward characterization of convergence properties of the iterative method at hand. Moreover, the latter result is the starting point to devise efficient algorithms for the solution of the multidomain problem.

First of all, let us introduce Λ, a subset of $H^{1/2}(\Gamma)$ formally defined as

$$\Lambda := \{\lambda \in H^{1/2}(\Gamma) | \lambda = 0 \text{ on } \overline{\Gamma}_{D,i} \cap \overline{\Gamma},\ i = 1, 2\}.$$

For a precise definition of Λ in the framework of Sobolev spaces, we refer to [5]. Let us introduce the following lifting operators:

$\mathcal{R}_1 : \Lambda \to V_1,\ \mathcal{R}_1\lambda = u_1,$ with

$$\begin{cases} \chi u_1 + \operatorname{Div}\left(-D_1 \operatorname{Grad} u_1 + \mathbf{v}_1 u_1\right) \\ \qquad + r_1 u_1 = 0 \text{ in } \Omega_1, \\ u_1 = 0 \text{ on } \Gamma_{D,1}, \\ \partial_{\mathbf{n}_1} u_1 = 0 \text{ on } \Gamma_{N,1}, \\ -\hat{\partial}_{\mathbf{n}_1} u_1 + \zeta_1 u_1 = \zeta_2 \lambda \text{ on } \Gamma. \end{cases} \tag{8}$$

$\mathcal{R}_2 : \Lambda \to V_2,\ \mathcal{R}_2\lambda = u_2,$ with

$$\begin{cases} \chi u_2 + \operatorname{Div}\left(-D_2 \operatorname{Grad} u_2 + \mathbf{v}_2 u_2\right) \\ \qquad + r_2 u_2 = 0 \text{ in } \Omega_2, \\ u_2 = 0 \text{ on } \Gamma_{D,2}, \\ \partial_{\mathbf{n}_2} u_2 = 0 \text{ on } \Gamma_{N,2}, \\ u_2 = \lambda \text{ on } \Gamma. \end{cases} \tag{9}$$

$\mathcal{G}_1 : V_1' \to V_1,\ \mathcal{G}_1(\mathcal{F}_1) = g_1,$ with

$$\begin{cases} \chi g_1 + \operatorname{Div}\left(-D_1 \operatorname{Grad} g_1 + \mathbf{v}_1 g_1\right) \\ \qquad + r_1 g_1 = \mathcal{F}_1 \text{ in } \Omega_1, \\ g_1 = 0 \text{ on } \Gamma_{D,1}, \\ \partial_{\mathbf{n}_1} g_1 = 0 \text{ on } \Gamma_{N,1}, \\ -\hat{\partial}_{\mathbf{n}_1} g_1 + \zeta_1 g_1 = 0 \text{ on } \Gamma. \end{cases} \tag{10}$$

$\mathcal{G}_2 : V_2' \to V_2,\ \mathcal{G}_2(\mathcal{F}_2) = g_2,$ with

$$\begin{cases} \chi g_2 + \operatorname{Div}\left(-D_2 \operatorname{Grad} g_2 + \mathbf{v}_2 g_2\right) \\ \qquad + r_2 g_2 = \mathcal{F}_2 \text{ in } \Omega_2, \\ g_2 = 0 \text{ on } \Gamma_{D,2}, \\ \partial_{\mathbf{n}_2} g_2 = 0 \text{ on } \Gamma_{N,2}, \\ g_2 = 0 \text{ on } \Gamma. \end{cases} \tag{11}$$

Finally, we denote with $\mathcal{L}_i$ a general linear and continuous map from Λ to V_i, such that $\mathcal{L}_i\rho|_\Gamma = \rho, \forall\rho \in \Lambda$ in the sense of traces. A possible choice of the operator $\mathcal{L}_i$ is the *harmonic* lifting defined as, $\mathcal{H}_i : \Lambda \to V_i$, $\mathcal{H}_i\rho = v_i$, with

$$\begin{cases} \Delta v_i = 0 \text{ in } \Omega_i, \\ v_i = 0 \text{ on } \Gamma_{D,i}, \\ \partial_{\mathbf{n}_i} v_i = 0 \text{ on } \Gamma_{N,i}, \\ v_i = \rho \text{ on } \Gamma. \end{cases} \tag{12}$$

We are now in position to introduce the Steklov-Poincaré operators. We set,

$$\mathcal{S}_1 : \Lambda \to \Lambda' : \mathcal{S}_1\lambda = -\zeta_1(\mathcal{R}_1\lambda)|_\Gamma, \tag{13}$$

$$\mathcal{S}_2 : \Lambda \to \Lambda' : \mathcal{S}_2\lambda = -\hat{\partial}_{\mathbf{n}_2}\mathcal{R}_2\lambda + \zeta_2\lambda, \tag{14}$$

where Λ' is the dual space of Λ. We point out that this definition of $\mathcal{S}_i$, unlike the classical definition according to which $\mathcal{S}_i\lambda$ is the trace on Γ of the normal derivative of the harmonic lifting (see for example [7], Ch 1.), is defined specifically for matching conditions of Robin type, as (5d,e). By virtue of these operators, we are able to reformulate problem 2 as a problem on the interface Γ. To this aim, let us consider (5e), that has not been used in the definition of operators $\mathcal{R}_i$ yet. Moreover, we choose $\lambda = c_2|_\Gamma$ as interface unknown. Then, making use of the definitions of $\mathcal{R}_i, \mathcal{G}_i$ and observing that $\mathcal{R}_i\lambda + \mathcal{G}_i(\mathcal{F}_i) = c_i$, equation (5e) can be rewritten as,

$$-\hat{\partial}_{\mathbf{n}_2}\mathcal{R}_2\lambda + \zeta_2(\mathcal{R}_2\lambda)|_\Gamma - \zeta_1(\mathcal{R}_1\lambda)|_\Gamma = \hat{\partial}_{\mathbf{n}_2}g_2 - \zeta_2 g_2|_\Gamma + \zeta_1 g_1|_\Gamma.$$

Consequently, applying definitions (13,14) recalling that $g_2|_\Gamma = 0$, $\zeta_1 g_1|_\Gamma = \hat{\partial}_{\mathbf{n}_1}g_1$ and $\mathcal{R}_2\lambda|_\Gamma = \lambda$, the unknown λ satisfies the following equation,

$$\mathcal{S}_1\lambda + \mathcal{S}_2\lambda = \hat{\partial}_{\mathbf{n}_1}g_1 + \hat{\partial}_{\mathbf{n}_2}g_2.$$

In other words, defining $\mathcal{S} : \Lambda \to \Lambda'$ as $\mathcal{S} := \mathcal{S}_1 + \mathcal{S}_2$ and denoting $\eta := \hat{\partial}_{\mathbf{n}_1}g_1 + \hat{\partial}_{\mathbf{n}_2}g_2 \in \Lambda'$ we assert that the interface unknown λ satisfies $\mathcal{S}\lambda = \eta$, that is the Steklov-Poincaré counterpart of equation (5). Additionally, denoting with $\langle\cdot,\cdot\rangle$ the duality pairing between Λ' and Λ, the interface problem can be reformulated in the weak form as follows.

Problem 6 *Find $\lambda \in \Lambda$ such that,*

$$\langle\mathcal{S}\lambda, \rho\rangle = \langle\eta, \rho\rangle, \quad \forall\rho \in \Lambda. \tag{15}$$

Equation (15) can be rewritten in a more useful form by means of the weak reinterpretation of the Steklov-Poincaré operators. The weak counterpart of definitions (13,14) is,

$$\langle\mathcal{S}_1\lambda, \rho\rangle = -\int_\Gamma \zeta_1(\mathcal{R}_1\lambda)\rho, \quad \forall\rho \in \Lambda \tag{16}$$

$$\langle\mathcal{S}_2\lambda, \rho\rangle = \langle -\hat{\partial}_{\mathbf{n}_2}\mathcal{R}_2\lambda + \zeta_2\lambda, \rho\rangle, \quad \forall\rho \in \Lambda. \tag{17}$$

Considering the weak formulation of problems (8, 9, 10, 11), after usual manipulations we rewrite (16,17) in the following form,

$$\langle \mathcal{S}_1\lambda, \rho\rangle = a_1\left(u_1, \mathcal{L}_1\rho\right) - \int_\Gamma \zeta_2\lambda\rho, \quad \forall\rho \in \Lambda, \tag{18}$$

$$\langle \mathcal{S}_2\lambda, \rho\rangle = a_2\left(u_2, \mathcal{L}_2\rho\right) + \int_\Gamma \zeta_2\lambda\rho, \quad \forall\rho \in \Lambda. \tag{19}$$

Moreover we observe that, multiplying (10,11) by $\mathcal{L}_1\rho, \mathcal{L}_2\rho$ respectively and integrating by parts, we have,

$$\langle \hat{\partial}_{\mathbf{n}_i} g_i, \rho\rangle = (\mathcal{F}_i, \mathcal{L}_i\rho) - a_i\left(g_i, \mathcal{L}_i\rho\right) \quad \forall\rho \in \Lambda, i = 1, 2. \tag{20}$$

Consequently, we notice that, applying (18), (19) and (20), equation (15) can be rewritten as,

$$\sum_{i=1,2} a_i\left(\mathcal{R}_i\lambda + \mathcal{G}_i(\mathcal{F}_i), \mathcal{L}_i\rho\right) = \sum_{i=1,2} (\mathcal{F}_i, \mathcal{L}_i\rho) \quad \forall\rho \in \Lambda. \tag{21}$$

Reminding that $\mathcal{R}_i\lambda + \mathcal{G}_i(\mathcal{F}_i) = c_i$, it follows that equation (21) corresponds to equation (6) with $v_i = \mathcal{L}_i\rho$, which confirms that problems 6 and 4 are equivalent in the weak sense.

2.2 Properties of the Steklov-Poincaré operators

In this paragraph we introduce some properties of the lifting and of the Steklov-Poincaré operators that will be necessary in the sequel, to analyze the convergence of the relaxed Robin-Robin scheme. In order to proceed, let us introduce an assumption on the coefficients which will be required during our analysis. In particular we assume that the diffusivity is large enough and the time step is small enough. More precisely, being $K_{t,i}$ the constant of the trace inequality form Γ to Ω_i (namely the constant such that the inequality $\int_\Sigma v^2 \leq \|v\| H^{1^2}$ holds true for all $v \in H^1(\Omega)$ with $\Sigma \subset \partial\Omega$), we assume that there exist two positive constants $\alpha_i, i = 1, 2$ such that,

$$\begin{aligned} \mu_i - 2K_{t,i}\|J_v\| L^\infty(\Gamma) &\geq \alpha_i \\ \chi - 2K_{t,i}\|J_v\| L^\infty(\Gamma) &\geq \alpha_i. \end{aligned} \tag{22}$$

Then we introduce the following lemma.

Lemma 1 *Under assumption (22),*

1) there exists a positive constant M_1 *such that*

$$\|\mathcal{R}_1\lambda\| H^1 \leq M_1 \|\lambda\| \Lambda,$$

2) there exist positive constants m_2, M_2 such that

$$m_2 \|\lambda\| \Lambda \leq \|\mathcal{R}_2\lambda\| H^1 \leq M_2 \|\lambda\| \Lambda.$$

Proof.

Point 1. Let us consider the weak formulation of problem (8), we have

$$a_1(u_1, v_1) + \int_\Gamma \zeta_1 u_1 v_1 = \int_\Gamma \zeta_2 \lambda v_1.$$

Choosing $v_1 = u_1$ in the latter equation, counter-integrating by parts the transport term, reminding that $\mathrm{Div}\, \mathbf{v}_i = 0$ we obtain,

$$\chi \|u_1\| L^{2^2} + \mu_1 \|\mathrm{Grad}\, u_1\| L^{2^2}$$
$$+\frac{1}{2} \int\limits_{\Gamma_{N,1}} \mathbf{v}_1 \cdot \mathbf{n}_1 u_1^2 - \frac{1}{2}\int_\Gamma \mathbf{v}_1 \cdot \mathbf{n}_1 u_1^2 + \int_\Gamma \zeta_1 u_1^2 = \int_\Gamma \zeta_2 \lambda u_1 \leq \frac{1}{2}\int_\Gamma \zeta_2 u_1^2 + \frac{1}{2}\int_\Gamma \zeta_2 \lambda^2.$$

Then, observing that $\mathbf{v}_1 \cdot \mathbf{n}_1 = J_v$ on Γ, $\mathbf{v}_1 \cdot \mathbf{n}_1 \geq 0$ on $\Gamma_{N,1}$, $\zeta_1 - \zeta_2 = J_v$ and applying the trace inequality from Γ to Ω_1 with constant $K_{t,1}$, under assumption (22) there exists a positive constant α_1 such that

$$\alpha_1 \|u_1\| H^{1^2} \leq \frac{1}{2} \|\zeta_2\| L^\infty(\Gamma) \|\lambda\| L^2(\Gamma)^2.$$

Moreover, we notice that $\Lambda \subset L^2(\Gamma)$ with continuity, thus there exists K such that $\|\cdot\| L^2(\Gamma) \leq K \|\cdot\| \Lambda$. Consequently, the result follows with $M_1 = (K\|\zeta_2\| L^\infty(\Gamma)/2\alpha_1)^{1/2}$.

Point 2. We observe that $\mathcal{R}_2$ enjoys the property $\mathcal{R}_2\lambda|_\Gamma = \lambda$. Consequently, the lower bound $m_2 \|\lambda\| \Lambda \leq \|\mathcal{R}_2\lambda\| H^1$ is a direct consequence of the trace inequality. As regards the upper bound, we observe that $\mathcal{R}_2$ is a linear and continuous lifting operator from Λ to $V_2 \subset H^1(\Omega)$. Then $\mathcal{R}_2$ is bounded, namely there exists a positive constant M_2 such that $\|\mathcal{R}_2\lambda\| H^1 \leq M_2 \|\lambda\| \Lambda$, which concludes the proof. □

We recall that the general lifting operators $\mathcal{L}_i, i = 1, 2$ enjoy the same properties as $\mathcal{R}_2$. In the following lemma, we introduce the basic properties of $\mathcal{S}_1, \mathcal{S}_2$.

Lemma 2 *Under assumption (22), the Steklov-Poincaré operators satisfy:*

1) $\mathcal{S}_i$ are linear and continuous operators: there exist two positive constants β_i such that,

$$\langle \mathcal{S}_i\lambda, \rho\rangle \leq \beta_i \|\lambda\| \Lambda \|\rho\| \Lambda, \ \forall \lambda, \rho \in \Lambda, i = 1, 2.$$

2) $\mathcal{S}_2$ is coercive: there exists a positive constant α_2 such that,

$$\langle \mathcal{S}_2\lambda, \lambda\rangle \geq \alpha_2 \|\lambda\|_{\Lambda}^2, \ \forall \lambda \in \Lambda.$$

3) When $\mathbf{v}_2 = 0$, we have $\zeta_1 = \zeta_2$ (implying $J_v = 0$), then $\mathcal{S}_2$ is symmetric:

$$\langle \mathcal{S}_2\lambda, \rho\rangle = \langle \mathcal{S}_2\rho, \lambda\rangle, \ \forall \lambda, \rho \in \Lambda.$$

Proof.

Point 1. By virtue of the continuity of the bilinear forms $a_i(\cdot,\cdot)$, from (18,19) we have

$$\langle \mathcal{S}_i\lambda, \rho\rangle = a_i(\mathcal{R}_i\lambda, \mathcal{L}_i\rho) + (-1)^i \int_\Gamma \zeta_2\lambda\rho \leq K_i \|\mathcal{R}_i\lambda\|_{H^1} \|\mathcal{L}_i\rho\|_{H^1},$$

Then, applying lemma 1 we have,

$$\langle \mathcal{S}_i\lambda, \rho\rangle \leq K_i M_i^2 \|\lambda\|_{\Lambda} \|\rho\|_{\Lambda},$$

from which the result follows with $\beta_i = K_i M_i^2$.

Point 2. From (19) we have,

$$\langle \mathcal{S}_2\lambda, \lambda\rangle = a_2(\mathcal{R}_2\lambda, \mathcal{R}_2\lambda) + \int_\Gamma \zeta_2\lambda^2.$$

Proceeding as in lemma 1, under assumption (22), we may conclude that

$$\begin{aligned}
\langle \mathcal{S}_2\lambda, \lambda\rangle &= \chi \|\mathcal{R}_2\lambda\|_{L^2}^2 \\
&\quad + \mu_2 \|\mathrm{Grad}\,\mathcal{R}_2\lambda\|_{L^2}^2 + \frac{1}{2}\int_{\Gamma_{N,2}} \mathbf{v}_2\cdot\mathbf{n}_2(\mathcal{R}_2\lambda)^2 - \frac{1}{2}\int_\Gamma \mathbf{v}_2\cdot\mathbf{n}_2\lambda^2 \\
&\quad + \int_\Gamma P\lambda^2 - \int_\Gamma w_2 J_v\lambda^2 \geq \left(\chi - \frac{1+2w_2}{2}K_{t,2}\|J_v\|_{L^\infty(\Gamma)}\right)\|\mathcal{R}_2\lambda\|_{L^2}^2 \\
&\quad + \left(\mu_2 - \frac{1+2w_2}{2}K_{t,2}\|J_v\|_{L^\infty(\Gamma)}\right)\|\mathrm{Grad}\,\mathcal{R}_2\lambda\|_{L^2}^2 \\
&\geq \alpha_2 \|\mathcal{R}_2\lambda\|_{H^1}^2 \geq \alpha_2 m_2 \|\lambda\|_{\Lambda}.
\end{aligned}$$

Point 3. From (19), by specifying the definition of $a_2(\cdot,\cdot)$, in the special case $\mathbf{v}_2 = 0$ and $J_v = 0$ we have,

$$\langle \mathcal{S}_2\lambda, \rho\rangle = \chi(\mathcal{R}_2\lambda, \mathcal{L}_2\rho) + (D_2\,\mathrm{Grad}\,\mathcal{R}_2\lambda, \mathrm{Grad}\,\mathcal{L}_2\rho).$$

By setting $\mathcal{L}_2 = \mathcal{R}_2$, the result follows from the right hand side of the equation above, by virtue of the symmetry of the inner product in L^2. □

2.3 Reinterpretation of the Robin-Robin method with Steklov-Poincaré operators

Our aim is to rewrite problem 5 by means of the Steklov-Poincaré operators $\mathcal{S}_i$. This will allow us to show the equivalence of problem 5 with a Richardson iterative scheme, applied to approximate the solution of problem 6. Let us start from the first of equation (7). Reminding that $c_i = u_i + g_i$ and choosing $v_i = \mathcal{L}_i\rho$, we have

$$\int_\Gamma \zeta_1 c_1^k \rho = (\mathcal{F}_1, \mathcal{L}_1\rho) - a_1(g_1, \mathcal{L}_1\rho) - a_1(u_1, \mathcal{L}_1\rho) \\ + \int_\Gamma \zeta_2 \lambda^{k-1}\rho = \langle \hat{\partial}_{\mathbf{n}_1} g_1, \rho\rangle - \langle \mathcal{S}_1\lambda^{k-1}, \rho\rangle \quad \forall \rho \in \Lambda. \quad (23)$$

Considering the second of (7) with $v_2 = \mathcal{L}_2\rho$ and splitting $c_2 = u_2 + g_2$ we have,

$$a_2(u_2^k, \mathcal{L}_2\rho) + \int_\Gamma \zeta_2 u_2^k \rho = \int_\Gamma \zeta_1 c_1^k \rho + (\mathcal{F}_2, \mathcal{L}_2\rho) - a_2(g_2, \mathcal{L}_2\rho) \quad \forall \rho \in \Lambda.$$

Then, applying equation (19) and observing that $c_2|_\Gamma = u_2|_\Gamma$ we have,

$$\langle \mathcal{S}_2 c_2^k, \rho\rangle = \langle \mathcal{S}_2 u_2^k, \rho\rangle = \int_\Gamma \zeta_1 c_1^k \rho + \langle \hat{\partial}_{\mathbf{n}_2} g_2, \rho\rangle \quad \forall \rho \in \Lambda. \quad (24)$$

Consequently, replacing (23) into (24), denoting $\eta = \hat{\partial}_{\mathbf{n}_1} g_1 + \hat{\partial}_{\mathbf{n}_2} g_2$ we obtain,

$$\langle \mathcal{S}_2 c_2^k, \rho\rangle = -\langle \mathcal{S}_1\lambda^{k-1}, \rho\rangle + \langle \eta, \rho\rangle. \quad (25)$$

By applying the operator $\mathcal{S}_2$ to the third of (7) we have,

$$\langle \mathcal{S}_2\lambda^k, \rho\rangle = (1-\theta)\langle \mathcal{S}_2\lambda^{k-1}, \rho\rangle + \theta\langle \mathcal{S}_2 c_2^k, \rho\rangle \quad \forall \rho \in \Lambda. \quad (26)$$

Replacing (25) in (26) we finally obtain,

$$\langle \mathcal{S}_2(\lambda^k - \lambda^{k-1}), \rho\rangle = \theta\langle \eta - \mathcal{S}\lambda^{k-1}, \rho\rangle \quad \forall \rho \in \Lambda. \quad (27)$$

Equation (27) shows that the Robin-Robin method can be regarded as a *preconditioned stationary Richardson method*, see for example [6] Ch. 4, applied to equation (15), the role of preconditioner being played by the operator $\mathcal{S}_2$. Let us invoke the following abstract result, proved in [7] (Theorem 4.2.2 and Remark 4.2.4). In what follows, we denote with η, μ a generic couple of functions.

Theorem 3 *Let X be a real Hilbert space, X' its dual space, and $\langle \cdot, \cdot \rangle$ the duality paring between X' and X. Let $Q : X \to X'$ be a linear invertible continuous operator which can be split as $Q = Q_1 + Q_2$, where Q_i $(i = 1, 2)$ are linear operators. Consider the problem $Q\lambda = \eta$ where $\eta \in X'$ is given and $\lambda \in X$ is to be determined. Suppose that,*

1) Q_1 is continuous, i.e. there exists $\beta_1 > 0$ such that,

$$\langle Q_1\rho, \mu\rangle \leq \beta_1 \|\rho\| X \|\mu\| X \quad \forall \rho, \mu \in X$$

2) Q_2 is continuous and coercive, i.e. there exists $\beta_2 > 0$ such that

$$\langle Q_2\rho, \mu\rangle \leq \beta_2 \|\rho\| X \|\mu\| X \quad \forall \rho, \mu \in X$$

and there exists $\alpha_2 > 0$ such that

$$\langle Q_2\rho, \rho\rangle \geq \alpha_2 \|\rho\| X^2 \quad \forall \rho \in X.$$

3) there exists a constant $k^* > 0$ such that

$$\langle Q_2\rho, Q_2^{-1}Q\rho\rangle + \langle Q\rho, \rho\rangle \geq k^* \|\rho\| X^2 \quad \forall \rho \in X.$$

Then, for any given λ^0 in X and for any θ satisfying $0 < \theta < \theta_{max}$ with $\theta_{max} = (k^*\alpha_2^2)/[\beta_2(\beta_1 + \beta_2)^2]$, the sequence

$$\lambda^k = \lambda^{k-1} + \theta Q_2^{-1}(\eta - Q\lambda^{k-1}) \tag{28}$$

converges in X to the solution of $Q\lambda = \rho$. Moreover, for each $k \geq 1$ one has, $\|e^k\| Q_2 \leq K_\theta \|e^{k-1}\| Q_2$, where $\|\rho\| Q_2 = \langle Q_2\rho, \rho\rangle^{1/2}$, $K_\theta = 1 + \theta^2(\beta_1 + \beta_2)^2/\alpha_2^2 - \theta k^*/\beta_2$, and e^k is the splitting error $e^k = \lambda - \lambda^k \in \Lambda$.

In our case, equation (28) corresponds to (27) where operators $\mathcal{S}_1$ and $\mathcal{S}_2$ play the role of Q_1, Q_2. Then, if we can prove that the operators $\mathcal{S}_1, \mathcal{S}_2$ satisfy the assumptions of theorem 3 it follows that the Richardson method (27), as well as the corresponding relaxed Robin-Robin one, namely problem 5, are convergent iterative schemes for a suitable range of values of θ. To this purpose, we introduce the following lemma.

Lemma 4 *Under assumptions (22) and if $\mathbf{v}_2$ is small enough (in a sense that will be specified later) the Steklov-Poincaré operators $\mathcal{S}_1, \mathcal{S}_2$ satisfy the assumptions of theorem 3.*

Proof. $\mathcal{S}_1, \mathcal{S}_2$ satisfy assumptions $1), 2)$ as a direct consequence of lemma 2. The key point is to prove that assumption 3) holds true. For the sake of clarity, we split this proof in three parts, $a), b), c)$.

Part a. Let us denote $\mu := \mathcal{S}_2^{-1}\mathcal{S}\rho$, then we notice that $\langle \mathcal{S}_2\mu, \rho\rangle = \langle \mathcal{S}_2\mathcal{S}_2^{-1}\mathcal{S}\rho, \rho\rangle = \langle \mathcal{S}\rho, \rho\rangle$, thus we have,

$$\langle \mathcal{S}_2\rho, \mathcal{S}_2^{-1}\mathcal{S}\rho\rangle + \langle \mathcal{S}\rho, \rho\rangle = \langle \mathcal{S}_2\rho, \mu\rangle - \langle \mathcal{S}_2\mu, \rho\rangle + 2\langle \mathcal{S}\rho, \rho\rangle. \tag{29}$$

We are going to prove that there exist two positive constants k and α such that,

$$\langle \mathcal{S}_2\rho, \mu\rangle - \langle \mathcal{S}_2\mu, \rho\rangle \geq -k \left\|\rho\right\| \Lambda^2 \text{ with } \mu = \mathcal{S}_2^{-1}\mathcal{S}\rho, \quad \forall \rho \in \Lambda, \tag{30}$$

$$\langle \mathcal{S}\rho, \rho\rangle \geq \alpha \left\|\rho\right\| \Lambda^2 \quad \forall \rho \in \Lambda. \tag{31}$$

Then, applying (29), property 3) of theorem 3 holds true provided that there exists a positive constant k^* such that $2\alpha - k \geq k^* > 0$.

Part b. We prove the following equivalent form of (30),

$$\langle \mathcal{S}_2\mu, \rho\rangle - \langle \mathcal{S}_2\rho, \mu\rangle \leq k \left\|\rho\right\| \Lambda^2 \text{ with } \mu = \mathcal{S}_2^{-1}\mathcal{S}\rho, \quad \forall \rho \in \Lambda.$$

If $\mathcal{S}_2$ is symmetric, this inequality is trivial, indeed the equality holds true with $k = 0$. In the general case, for all $\rho, \mu \in \Lambda$, we have

$$\begin{aligned}
&\langle \mathcal{S}_2\mu, \rho\rangle - \langle \mathcal{S}_2\rho, \mu\rangle \\
&\quad = a_2\left(\mathcal{R}_2\mu, \mathcal{R}_2\rho\right) + \int_\Gamma \zeta_2 \mathcal{R}_2\mu\mathcal{R}_2\rho - a_2\left(\mathcal{R}_2\rho, \mathcal{R}_2\mu\right) - \int_\Gamma \zeta_2\mathcal{R}_2\rho\mathcal{R}_2\mu \\
&\quad = 2\left(\mathbf{v}_2 \operatorname{Grad} \mathcal{R}_2\mu, \mathcal{R}_2\rho\right) - \int_{\Gamma_{N,2}} (\mathbf{v}_2 \cdot \mathbf{n}_2)\mathcal{R}_2\rho\mathcal{R}_2\mu - \int_\Gamma (\mathbf{v}_2 \cdot \mathbf{n}_2)\mu\rho \\
&\leq \left[K_{t,2} \left\|\mathbf{v}_2 \cdot \mathbf{n}_2\right\| L^\infty(\Gamma \cup \Gamma_{N,2}) + 2\Big(\sum_{i=1}^d \left\|v_{2,i}\right\| L^4(\Omega_2)^2\Big)^{1/2}\right] \left\|\mathcal{R}_2\mu\right\| H^1 \left\|\mathcal{R}_2\rho\right\| H^1 \\
&\leq \left[M_2^2 \left(K_{t,2} \left\|\mathbf{v}_2 \cdot \mathbf{n}_2\right\| L^\infty(\Gamma \cup \Gamma_{N,2}) + 2\Big(\sum_{i=1}^d \left\|v_{2,i}\right\| L^4(\Omega_2)^2\Big)^{1/2}\right)\right] \left\|\mu\right\| \Lambda \left\|\rho\right\| \Lambda.
\end{aligned} \tag{32}$$

Let us now consider the following problem.

Problem 7 *Find $\rho \in \Lambda$ such that*

$$\langle \mathcal{S}_2\rho, \mu\rangle = \langle \eta, \mu\rangle \quad \forall \mu \in \Lambda.$$

We observe that, by virtue of lemma 2 and the Lax-Milgram lemma, this problem admits a unique weak solution. Then, the application $T : \Lambda' \to \Lambda$ associating to each $\eta \in \Lambda'$ the element $\rho \in \Lambda$, is a bijection. Consequently, the operator $\mathcal{S}_2$ is linear, continuous and invertible. Its inverse $\mathcal{S}_2^{-1} : \Lambda' \to \Lambda$ is linear and continuous. Hence, there exists a positive constant β' such that,

$$\left\|\mathcal{S}_2^{-1}\eta\right\| \Lambda \leq \beta' \left\|\eta\right\| \Lambda' \quad \text{with} \quad \left\|\eta\right\| \Lambda' := \sup_{\rho\in\Lambda, \rho\neq 0} \frac{|\langle \eta, \rho\rangle|}{\left\|\rho\right\| \Lambda}.$$

Moreover, since $\mathcal{S}_1$ and $\mathcal{S}_2$ are continuous operators with constants β_1, β_2, $\mathcal{S} = \mathcal{S}_1 + \mathcal{S}_2$ is continuous too, with constant $\beta = \beta_1 + \beta_2$. By virtue of these properties, we have

$$\left\|\mu\right\| \Lambda = \left\|\mathcal{S}_2^{-1}\mathcal{S}\rho\right\| \Lambda \leq \beta' \sup_{\rho\in\Lambda, \rho\neq 0} \frac{|\langle \mathcal{S}\rho, \rho\rangle|}{\left\|\rho\right\| \Lambda} \leq \beta'\beta \left\|\rho\right\| \Lambda. \tag{33}$$

Consequently, replacing (33) in (32) we have

$$\langle \mathcal{S}_2\mu,\rho\rangle - \langle \mathcal{S}_2\rho,\mu\rangle$$
$$\leq \beta'\beta\left[M_2^2\left(K_{t,2}\,\|\mathbf{v}_2\cdot\mathbf{n}_2\|\,L^\infty(\Gamma\cup\Gamma_{N,2}) + 2(\sum_{i=1}^{d}\|v_{2,i}\|\,L^4(\Omega_2)^2)^{1/2}\right)\right]\|\rho\|\,\Lambda^2$$
$$\forall\rho\in\Lambda,$$

which is equivalent to (30) with

$$k = \beta'\beta\left[M_2^2\left(K_{t,2}\,\|\mathbf{v}_2\cdot\mathbf{n}_2\|\,L^\infty(\Gamma\cup\Gamma_{N,2}) + 2(\sum_{i=1}^{d}\|v_{2,i}\|\,L^4(\Omega_2)^2)^{1/2}\right)\right].$$

Part c. In this paragraph we prove inequality (31) that corresponds to the coercivity of the operator $\mathcal{S}$. Let us start observing that, from the variational formulation of equation (8), choosing the test function equal to $\mathcal{R}_1\rho$ we have,

$$a_1(\mathcal{R}_1\rho,\mathcal{R}_1\rho) + \int_\Gamma \zeta_1(\mathcal{R}_1\rho)^2 - \int_\Gamma \zeta_2\rho\mathcal{R}_1\rho = 0 \tag{34}$$

Then, summing (34), (16) and (19), recalling that $u_1 = \mathcal{R}_1\rho, u_2 = \mathcal{R}_2\rho$ and choosing $\mathcal{L}_2(\cdot)\equiv\mathcal{R}_2(\cdot)$ and $\eta=\rho$ we obtain,

$$\langle\mathcal{S}\rho,\rho\rangle = \langle\mathcal{S}_1\rho,\rho\rangle + \langle\mathcal{S}_2\rho,\rho\rangle = \sum_{i=1,2} a_i(\mathcal{R}_i\rho,\mathcal{R}_i\rho)$$
$$+\int_\Gamma\zeta_1(\mathcal{R}_1\rho)^2 - \int_\Gamma\zeta_1(\mathcal{R}_1\rho)\rho + \int_\Gamma\zeta_2\rho^2 - \int_\Gamma\zeta_2(\mathcal{R}_1\rho)\rho. \tag{35}$$

Then, we point out that,

$$\int_\Gamma\zeta_1(\mathcal{R}_1\rho)^2 - \int_\Gamma\zeta_1(\mathcal{R}_1\rho)\rho + \int_\Gamma\zeta_2\rho^2 - \int_\Gamma\zeta_2(\mathcal{R}_1\rho)\rho$$
$$= \int_\Gamma(P+w_1J_v)\mathcal{R}_1\rho(\mathcal{R}_1\rho-\rho) - \int_\Gamma(P-w_2J_v)\rho(\mathcal{R}_1\rho-\rho)$$
$$= \int_\Gamma P(\mathcal{R}_1\rho-\rho)^2 + \int_\Gamma J_v[w_1(\mathcal{R}_1\rho)^2 - w_1(\mathcal{R}_1\rho)\rho + w_2(\mathcal{R}_1\rho)\rho - w_2\rho^2]$$
$$\geq -\|J_v\|\,L^\infty(\Gamma)\int_\Gamma(w_1(\mathcal{R}_1\rho)^2 + w_1|(\mathcal{R}_1\rho)\rho| + w_2|(\mathcal{R}_1\rho)\rho| + w_2\rho^2)$$
$$\geq -(3/2)\,\|J_v\|\,L^\infty(\Gamma)\int_\Gamma(\mathcal{R}_1\rho)^2 - (3/2)\,\|J_v\|\,L^\infty(\Gamma)\int_\Gamma\rho^2. \tag{36}$$

Consequently, replacing (36) in (35), observing that $\mathcal{R}_2\rho|_\Gamma = \rho$ and applying assumption (22) we have

$$\langle \mathcal{S}\rho, \rho\rangle \geq \sum_{i=1,2} \left(\chi \|\mathcal{R}_i\rho\| L^{2^2} + \mu_i \|\mathrm{Grad}\,\mathcal{R}_i\| L^{2^2}\right)$$

$$- \sum_{i=1,2} \left(\frac{1}{2}\int_\Gamma \mathbf{v}_i \cdot \mathbf{n}_i (\mathcal{R}_i\rho)^2 + \frac{3}{2}\|J_v\| L^\infty(\Gamma) \int_\Gamma (\mathcal{R}_i\rho)^2\right)$$

$$\geq \alpha_1 \|\mathcal{R}_1\rho\| H^{1^2} + \alpha_2 \|\mathcal{R}_2\rho\| H^{1^2} \geq \alpha_2 m_2 \|\rho\| \Lambda^2.$$

Then, inequality (31) holds true with $\alpha = \alpha_2 m_2$. Finally, if the velocity is such that

$$\beta\beta' M_2^2 \left(K_{t,2} \|\mathbf{v}_2 \cdot \mathbf{n}_2\| L^\infty(\Gamma \cup \Gamma_{N,2}) + 2(\sum_{i=1}^{d} \|v_{2,i}\| L^4(\Omega_2)^2)^{1/2} \right) < 2\alpha_2 m_2, \tag{37}$$

there exists a constant $k^* > 0$ such that $2\alpha - k \geq k^*$, and the proof is concluded.□

Remark 1 Since α_i is directly related to the diffusivity coefficient D_i on Ω_i, (37) is verified in the diffusion dominated case, where diffusivities are large and the modulus of the transport velocity $\mathbf{v}_2$ is low. Hence, the convergence analysis of problem 5 applies when the governing equation on Ω_2 is dominated by diffusion. This is indeed the case of the problem at hand. Actually, recalling that the domain Ω_1 represents a free fluid and Ω_2 represents a porous medium, we have that $\mathbf{v}_2$ and its normal component on the interface, $\mathbf{v}_2 \cdot \mathbf{n}_2|_\Gamma$, are small. Moreover, we point out that there is no restriction on the magnitude of $\mathbf{v}_1$.

2.4 Space discretization

In this paragraph we address the space discretization of the the coupled problem. First of all, for the sake of clarity, we introduce the space discrete counterpart of problem 5. To this aim, let $\mathcal{T}_{h,1}, \mathcal{T}_{h,2}$ be admissible and shape regular triangulations of $\overline{\Omega}_1, \overline{\Omega}_2$ respectively. Moreover, we assume that $\mathcal{T}_{h,1}, \mathcal{T}_{h,2}$ are conforming triangulations on Γ. In other words we require that $\mathcal{T}_h := \mathcal{T}_{h,l} \cup \mathcal{T}_{h,w}$ is an admissible triangulation for $\overline{\Omega}_1 \cup \overline{\Omega}_2$. Then, let $V_{h,i}, i = 1,2$ be the following linear finite element spaces,

$$V_{h,i} := \{v_{h,i} \in C^0(\overline{\Omega}_i) \mid v_{h,i} \in \mathbb{P}_1\ \forall K \in \mathcal{T}_{h,i},\ v_{h,i}|_{\Gamma_{D,i}} = 0, i = 1,2\}$$

and let us define $V_h := V_{h,1} \times V_{h,2}$. Finally let Λ_h be the space of the restrictions on Γ of functions in $V_{h,i}$. Then, the discrete counterpart of problem 4 reads as follows.

Problem 8 *Find* $[c_{h,1}, c_{h,2}] \in V_h$ *such that,*

$$\sum_{i=1,2}\left(a_i\left(c_{h,i},v_{h,i}\right)+\int_\Gamma \zeta_i c_{h,i}v_{h,i}\right)=$$
$$+\sum_{i=1,2}\left(\mathcal{F}_i,v_{h,i}\right)+\int_\Gamma \zeta_2 c_{h,2}v_{h,1}+\int_\Gamma \zeta_1 c_{h,1}v_{h,2} \quad \forall v_{h,i}\in V_{h,i},\ i=1,2.$$

The Robin-Robin iterative method applied to problem 8 reads as follows.

Problem 9 *Given an initial guess $\lambda_h^0 \in \Lambda_h$, a relaxation parameter $\theta > 0$ and a tolerance ϵ, for $k=1,2,\dots$ find a sequence $[c_{h,1}^k, c_{h,2}^k] \in V_h$ such that,*

$$\begin{cases} a_1\left(c_{h,1}^k, v_{h,1}\right)+\int_\Gamma \zeta_1 c_{h,1}^k v_{h,1} = \left(\mathcal{F}_1, v_{h,1}\right)+\int_\Gamma \zeta_2 \lambda_h^{k-1} v_{h,1} \quad \forall v_{h,1}\in V_{h,1}, \\ a_2\left(c_{h,2}^k, v_{h,2}\right)+\int_\Gamma \zeta_2 c_{h,2}^k v_{h,2} = \left(\mathcal{F}_{h,2}, v_{h,2}\right)+\int_\Gamma \zeta_1 c_{h,1}^k v_{h,2} \quad \forall v_{h,2}\in V_{h,2}, \\ \lambda_h^k = c_{h,2}^k|_\Gamma \ \textit{or}\ \lambda_h^k = \theta c_{h,2}^k|_\Gamma + (1-\theta)\lambda_h^{k-1}. \end{cases} \tag{38}$$

The convergence analysis of the Robin-Robin method can be carried out for the fully discrete case as well, since the discrete spaces of $V_{h,i}$ are subspaces of V_i. The key point that has to be investigated, however, is the dependence of the convergence rate on the characteristic mesh size h. Let us start introducing the definition and the properties of the discrete lifting and Steklov-Poincaré operators, that we call respectively $\mathcal{R}_{h,i}, \mathcal{G}_{h,i}, \mathcal{S}_{h,i}, i=1,2, \mathcal{S}_h=\mathcal{S}_{h,1}+\mathcal{S}_{h,2}$. These definitions follow in a straightforward manner from equations (8,9,10,11). For the sake of clarity, we introduce the definitions of $\mathcal{R}_{h,i}$ in the weak sense.

$\mathcal{R}_{h,1}:\Lambda_h\rightarrow V_{h,1},\ \mathcal{R}_{h,1}\lambda_h=u_{h,1}$, with

$$a_1\left(u_{h,1},v_{h,1}\right)+\int_\Gamma \zeta_1 u_{h,1}v_{h,1}=\int_\Gamma \zeta_2\lambda_h v_{h,1} \quad \forall v_{h,1}\in V_{h,1} \tag{39}$$

$\mathcal{R}_{h,2}:\Lambda_h\rightarrow V_{h,2},\ \mathcal{R}_{h,2}\lambda_h=u_{h,2}$, with

$$\begin{cases} a_2\left(u_{h,2},v_{h,2}\right)=0 \quad \forall v_{h,2}\in V_{h,2} \\ u_{h,2}=\lambda_h \text{ on } \Gamma \end{cases} \tag{40}$$

Lemma 5 *Operators $\mathcal{R}_{h,i}$ enjoy the properties of lemma 1 with constants M_1', M_2', m_2' independent of h.*

Proof. In the discrete case as well, the inequality $m_2'\|\lambda_h\|\Lambda \le \|\mathcal{R}_{h,2}\lambda_h\|H^1$ is a consequence of the trace theorem, consequently m_2' does not depend on h. As regards inequality $\|\mathcal{R}_{h,i}\lambda_h\|H^1 \le M_i'\|\lambda_h\|\Lambda$, we consider the case $i=1$ at first. Choosing $v_{h,1}=u_{h,1}-\mathcal{L}_{h,1}\lambda_h$ in (39), we have

$$a_1(u_{h,1}, u_{h,1}) + \int_\Gamma \zeta_1 u_{h,1}^2$$
$$= \int_\Gamma \zeta_2 \lambda_h u_{h,1} + a_1(u_{h,1}, \mathcal{L}_{h,1}\lambda_h) + \int_\Gamma \zeta_1 u_{h,1}\mathcal{L}_{h,1}\lambda_h - \int_\Gamma \zeta_2 \lambda_h^2.$$

Then, exploiting the definition of $a_1(\cdot,\cdot)$ and its continuity, under assumption (22), we have

$$\alpha_1 \|u_{h,1}\| H^{1^2} \leq \chi \|u_{h,1}\| L^{2^2} + \mu_1 \|\mathrm{Grad}\, u_{h,1}\| L^{2^2} - (3/2)\|J_v\| L^\infty(\Gamma) \int_\Gamma u_{h,1}^2$$
$$\leq a_1(u_{h,1}, u_{h,1}) + \int_\Gamma \zeta_1 u_{h,1}^2 \leq M_1 \|u_{h,1}\| H^1 \|\mathcal{L}_{h,1}\lambda_h\| H^1. \quad (41)$$

Let us now choose $\mathcal{L}_{h,1} \equiv \mathcal{H}_{h,1}$, the harmonic lifting defined in (12). Owing to the *finite element uniform extension theorem*, (see for example [7] Theorem 4.1.3 and references therein), $\|\lambda_h\| \Lambda$ is uniformly equivalent to $\|\mathcal{H}_{h,i}\lambda_h\| H^1$ with respect to h. This theorem, in the version proposed in [7] requires $\mathcal{H}_{h,i}$ to be the harmonic lifting operator. However, by repeating the proof reported there, it can be extended to more general lifting operators, as for instance $\mathcal{R}_2$, owing to the well posedness of second order elliptic problems. Then, by virtue of this theorem, there exists a positive constant k_1, independent of h such that $\|\mathcal{H}_{h,1}\lambda_h\| H^1 \leq k_1 \|\lambda_h\| \Lambda$. Consequently, the result follows from (41) with $M_1' = M_1 k_1$. Inequality $\|\mathcal{R}_{h,2}\lambda_h\| H^1 \leq M_2' \|\lambda_h\| \Lambda$ is proved choosing $v_{h,2} = u_{h,2} - \mathcal{H}_{h,2}\lambda_h$ in (40) and proceeding similarly. □

For the Steklov-Poincaré operators, we observe that equations (16,17) and (18,19) simply become,

$$\langle \mathcal{S}_{h,1}\lambda_h, \rho_h\rangle = -\int_\Gamma \zeta_1(\mathcal{R}_{h,1}\lambda_h)\rho_h = a_1(u_{h,1}, \mathcal{L}_{h,1}\rho_h) - \int_\Gamma \zeta_2\lambda_h\rho_h \quad \forall \rho_h \in \Lambda_h,$$

$$\langle \mathcal{S}_{h,2}\lambda_h, \rho_h\rangle = \langle -\hat{\partial}_{\mathbf{n}_2}\mathcal{R}_{h,2}\lambda_h + \zeta_2\lambda_h, \rho_h\rangle = a_2(u_{h,2}, \mathcal{L}_{h,2}\rho_h) + \int_\Gamma \zeta_2\lambda_h\rho_h$$
$$\forall \rho_h \in \Lambda_h.$$

Furthermore we introduce the following corollary of lemma 5.

Corollary 6 *Operators $\mathcal{S}_{h,i}$ enjoy the same properties of lemma 2, with constants $\alpha_2', \beta_1', \beta_2'$ independent of h.*

Proof. It follows from that of lemma 2, and the fact that the constants M_1', M_2', m_2' of lemma 5 are independent of h. □

Finally, from corollary 6 and the convergence theorem 3, we can conclude that the convergence rate of method 8 is independent of the mesh size h. We summarize this statement in the following final theorem. Let us remind that $\lambda_h = c_{h,2}|_\Gamma$ is defined in problem 8, while λ_h^k is the approximation with the Robin-Robin method, defined in problem 9.

Theorem 7 *The iterative scheme*

$$\langle \mathcal{S}_{h,2}(\lambda_h^k - \lambda_h^{k-1}), \rho_h \rangle = \theta \langle \rho - \mathcal{S}_h \lambda_h^{k-1}, \rho_h \rangle \quad \forall \rho_h \in \Lambda_h.$$

is convergent for any $\theta \in (0, \theta_{max})$, and its rate of convergence is independent of the spatial discretization. Precisely, for any $\theta \in (0, \theta_{max})$, there exists a constant $K_\theta < 1$, independent on h, such that

$$\left\| \lambda_h - \lambda_h^k \right\| \Lambda \le K_\theta \left\| \lambda_h - \lambda_h^{k-1} \right\| \Lambda.$$

An important consequence of this result is that the operator $\mathcal{S}_{h,2}$ is an *optimal* preconditioner of $\mathcal{S}_h$. Indeed, this will be the starting point for the discussion of efficient iterative methods to solve the global linear system arising from the discretization of problem 8.

Remark 2 In many applications the advection-diffusion phenomena are dominated by advection. This is the case, for example of the advection and diffusion of a chemical in a free flow, when the flow is fast. It is well known that the finite element methods and in general the Galerkin methods could be unappropriate when facing advection dominated problems, thus, resorting to stabilization techniques becomes mandatory. Different strategies can be pursued in this regard. When considering strongly consistent methods, and in particular the Streamline-Upwind/Petrov-Galerkin one (SUPG), the stabilization is obtained by adding suitable terms, $a_{h,i}(w,v)$ and $(\mathcal{F}_{h,i}, v)\,, i = 1,2$ to the bilinear forms and to the right hand sides of problem 8. Let us analyze the effect of the stabilizing bilinear form on the convergence of the Robin-Robin method. In the case of linear finite elements one has $\Delta w = 0$ on each element K of $\mathcal{T}_{h,i}, i = 1,2$ and for all $w \in V_{h,i}$, thus $a_{h,i}(w,v)$ is equal to

$$a_{h,i}(w,v) = \\ \frac{1}{2} \sum_{K \in \mathcal{T}_{h,i}} \delta_{K,i} \int_K \left[(\chi + r_i) w \frac{h_K}{|\mathbf{v}_i|} (\mathbf{v}_i \cdot \mathrm{Grad}\, v) + \frac{h_K}{|\mathbf{v}_i|} (\mathbf{v}_i \cdot \mathrm{Grad}\, w)(\mathbf{v}_i \cdot \mathrm{Grad}\, v) \right],$$

where we denote with h_K the diameter of K. Moreover, $\delta_{K,i}$ is a parameter weighting the stabilization term on each element K. Then, by applying the Young inequality to split the product $(\chi + r_i) w (\mathbf{v}_i \cdot \mathrm{Grad}\, v)$ in the two terms $(\chi + r_i)^2 w^2$ and $(\mathbf{v}_i \cdot \mathrm{Grad}\, v)^2$, the following estimate holds true,

$$a_{h,i}(v,v) \ge -\frac{1}{4} \sum_{K \in \mathcal{T}_{h,i}} \delta_{K,i} \int_K (\chi + r_i)^2 \frac{h_K}{|\mathbf{v}_i|} v^2.$$

Then, assuming that $\delta_{K,i} \le |\mathbf{v}_i| / [(\chi + r_i) h_K]$ for all K and considering for the sake of simplicity r_i to be constant we obtain,

$$a_{h,i}(v,v) \ge -\frac{1}{4} \sum_{K \in \mathcal{T}_{h,i}} \int_K (\chi + r_i) v^2 = -\frac{1}{4} (\chi + r_i) \left\| v \right\| L^{2^2}.$$

Thus theorem 7 holds in the stabilized case as well, under slightly more restrictive assumptions than (22). More precisely we require that there exist constants $\alpha_i > 0$ such that,

$$\mu_i - 2K_{t,i} \|J_v\| L^\infty(\Gamma) \geq \alpha_i,$$
$$(3/4)\chi_i - (1/4)r_i - 2K_{t,i} \|J_v\| L^\infty(\Gamma) \geq \alpha_i.$$

2.5 Algebraic reinterpretation of the Robin-Robin method

In this section, we consider the algebraic counterpart of problem 8. We observe that this is equivalent to solve, at each time step, the linear system of equations

$$A\mathbf{c} = \mathbf{b}. \tag{42}$$

Let us denote by N_1, N_2 the number of degrees of freedom associated to $\mathcal{T}_{h,1}, \mathcal{T}_{h,2}$ respectively, moreover let be N_Γ the degrees of freedom associated with the interface Γ, then the number of degrees of freedom associated with the inner nodes in Ω_2 is given by $N_{2,0} = N_2 - N_\Gamma$. Consequently in (42), $\mathbf{c} = [\mathbf{c}_1, \mathbf{c}_2, \mathbf{c}_\Gamma]^T \in \mathcal{R}^{N_1+N_2}$ is the vector of unknowns specifying the discrete solution, with $\mathbf{c}_1 \in \mathcal{R}^{N_1}$, $\mathbf{c}_2 \in \mathcal{R}^{N_{2,0}}$, $\mathbf{c}_\Gamma \in \mathcal{R}^{N_\Gamma}$. Moreover, $\mathbf{b} \in \mathcal{R}^{N_1+N_2}$ is function of the forcing terms, the boundary conditions and the solution at the previous time steps, and can be correspondingly split as: $\mathbf{b} = [\mathbf{b}_1, \mathbf{b}_2, \mathbf{b}_\Gamma]^T$, $\mathbf{b}_1 \in \mathcal{R}^{N_1}$, $\mathbf{b}_2 \in \mathcal{R}^{N_{2,0}}$, $\mathbf{b}_\Gamma \in \mathcal{R}^{N_\Gamma}$. Finally, $A \in \mathcal{R}^{(N_1+N_2)\times(N_1+N_2)}$ has the following block-wise pattern:

$$A = \begin{bmatrix} A_{11} & \mathbf{0} & A_{1\Gamma} \\ \mathbf{0} & A_{22} & A_{2\Gamma} \\ A_{\Gamma 1} & A_{\Gamma 2} & A_{\Gamma\Gamma} \end{bmatrix}, \tag{43}$$

where A_{11} is the $N_1 \times N_1$ matrix associated with the discretization of the bilinear form

$$a_1(u,v) + \int_\Gamma \zeta_1 uv \quad \forall u, v \in V_{h,1}.$$

Correspondingly, $A_{1\Gamma}$ arises from the discretization of the term $-\int_\Gamma \zeta_2 \lambda v$ with $\lambda \in \Lambda_h, v \in V_{h,1}$, while A_{22} is associated with the discretization of $a_2(u,v)$ for all $u, v \in V_{h,2,0}$, being $V_{h,2,0}$ the space of functions of in $V_{h,2}$ with null trace on Γ, then $A_{2\Gamma}$ refers to the discretization of $a_2(\mathcal{R}_{h,2}\lambda, v)$ for $\lambda \in \Lambda_h, v \in V_{h,2,0}$. Finally $A_{\Gamma\Gamma}$ is related to the discretization of

$$a_2(\mathcal{R}_{h,2}\lambda, \mathcal{R}_{h,2}\rho) + \int_\Gamma \zeta_2 \mathcal{R}_{h,2}\lambda \mathcal{R}_{h,2}\rho \quad \forall \lambda, \rho \in \Lambda_h,$$

and we notice that $A_{\Gamma 1} = A_{1\Gamma}^T$ and $A_{\Gamma 2} = A_{2\Gamma}^T$. The algebraic reinterpretation of the Robin-Robin iterative scheme readily follows. Indeed, problem 9 corresponds to

$$\begin{bmatrix} A_{11} & \mathbf{0} & \mathbf{0} \\ \mathbf{0} & A_{22} & A_{2\Gamma} \\ A_{\Gamma 1} & A_{\Gamma 2} & A_{\Gamma\Gamma} \end{bmatrix} \cdot \begin{bmatrix} \mathbf{c}_1^k \\ \mathbf{c}_2^k \\ \mathbf{c}_\Gamma^k \end{bmatrix} = \theta \left(\begin{bmatrix} \mathbf{b}_1 \\ \mathbf{b}_2 \\ \mathbf{b}_\Gamma \end{bmatrix} - \begin{bmatrix} A_{1\Gamma}\mathbf{c}_\Gamma^{k-1} \\ \mathbf{0} \\ \mathbf{0} \end{bmatrix} \right) + (1-\theta) \begin{bmatrix} A_{11} & \mathbf{0} & \mathbf{0} \\ \mathbf{0} & A_{22} & A_{2\Gamma} \\ A_{\Gamma 1} & A_{\Gamma 2} & A_{\Gamma\Gamma} \end{bmatrix} \cdot \begin{bmatrix} \mathbf{c}_1^{k-1} \\ \mathbf{c}_2^{k-1} \\ \mathbf{c}_\Gamma^{k-1} \end{bmatrix} \quad (44)$$

that can be rewritten in the preconditioned Richardson form as follows,

$$Q\left(\mathbf{c}_k - \mathbf{c}_{k-1}\right) = \theta(\mathbf{b} - A\mathbf{c}_{k-1}) = \theta \mathbf{r}_{k-1}, \quad k \geq 1, \quad (45)$$

where the preconditioner is the matrix

$$Q = \begin{bmatrix} A_{11} & \mathbf{0} & \mathbf{0} \\ \mathbf{0} & A_{22} & A_{2\Gamma} \\ A_{\Gamma 1} & A_{\Gamma 2} & A_{\Gamma\Gamma} \end{bmatrix}. \quad (46)$$

By means of some algebraic computations, we eliminate $\mathbf{c}_1$ and $\mathbf{c}_2$ in (43), obtaining the reduced system

$$\Sigma \mathbf{c}_\Gamma = \boldsymbol{\eta}, \quad (47)$$

with

$$\boldsymbol{\eta} = \mathbf{b}_\Gamma - A_{\Gamma 1}A_{11}^{-1}\mathbf{b}_1 - A_{\Gamma 2}A_{22}^{-1}\mathbf{b}_2,$$
$$\Sigma = A_{\Gamma\Gamma} - A_{\Gamma 1}A_{11}^{-1}A_{1\Gamma} - A_{\Gamma 2}A_{22}^{-1}A_{2\Gamma}.$$

Consequently, we obtain the splitting $\Sigma = \Sigma_2 + \Sigma_1$, where $\Sigma_2 = A_{\Gamma\Gamma} - A_{\Gamma 2}A_{22}^{-1}A_{2\Gamma}$, $\Sigma_1 = -A_{\Gamma 1}A_{11}^{-1}A_{1\Gamma}$. The matrix Σ is the *Schur complement* of matrix A. Equation (47) represents the finite dimensional counterpart of (15) and Σ is the algebraic counterpart of the Steklov-Poincaré operator $\mathcal{S}_h$, while Σ_1 and Σ_2 play the role of $\mathcal{S}_{h,1}$ and $\mathcal{S}_{h,2}$, respectively. By means of these definitions and the block *LU* factorization of A, we explicitly write the iteration matrix associated with (44), that is,

$$I - \theta Q^{-1}A = \begin{bmatrix} (1-\theta)I_1 & \mathbf{0} & -\theta A_{11}^{-1}A_{1\Gamma} \\ \mathbf{0} & (1-\theta)I_2 & -\theta A_{22}^{-1}A_{2\Gamma}\Sigma_2^{-1}A_{\Gamma 1}A_{11}^{-1}A_{1\Gamma} \\ \mathbf{0} & \mathbf{0} & I_\Gamma - \theta\Sigma_2^{-1}\Sigma \end{bmatrix}, \quad (48)$$

in particular, the third block of system (45) yields

$$\Sigma_2\left(\mathbf{c}_\Gamma^k - \mathbf{c}_\Gamma^{k-1}\right) = \theta\left(\boldsymbol{\eta} - \Sigma\mathbf{c}_\Gamma^{k-1}\right). \quad (49)$$

The latter relation puts into evidence the role of the matrix Σ_2 as a preconditioner for the Schur complement matrix in the interface problem (47). Indeed, as a consequence of theorem 7, we assert that if $\theta \in (0, \theta_{max})$ the sequence $\mathbf{c}_\Gamma^k$ defined by (49) converges to $\mathbf{c}_\Gamma$ at a rate independent of h. This is equivalent to assert that, if $\rho(I - \theta\Sigma_2^{-1}\Sigma)$ denotes the spectral radius of $I - \theta\Sigma_2^{-1}\Sigma$, we have $\rho(I - \theta\Sigma_2^{-1}\Sigma) \le K_\theta < 1$, given $\theta \in (0, \theta_{max})$ with K_θ independent of h. Moreover, from basic analysis, see for example [6] Ch. 4 Theorem 4.8, the Richardson method is convergent if and only if $2\mathbb{R}(\lambda_i)/(\theta|\lambda_i|^2) > 1$, where we have denoted with $[\lambda_i], i = 1, \ldots, N_\Gamma$ the eigenvalues of the preconditioned matrix $\Sigma_2^{-1}\Sigma$. Thus, we assert that conditions $2\mathbb{R}(\lambda_i)/(\theta|\lambda_i|^2) > 1$ and $\theta \in (0, \theta_{max})$ are equivalent. Then, we introduce the following corollary of theorem 7.

Corollary 8 *The Robin-Robin scheme or its algebraic counterpart*

$$Q\left(\mathbf{c}_{k+1} - \mathbf{c}_k\right) = \theta(\mathbf{b} - A\mathbf{c}_k), \quad k \ge 0,$$

is convergent for any $\theta \in (0, \min[\theta_{max}, 2])$ *and its rate of convergence is asymptotically independent of the spatial discretization.*

Proof. Let us observe from (48) that the eigenvalues of $Q^{-1}A$, denoted with $[\mu_i], j = 1, \ldots, N_1 + N_2$ are $[\mu_j] = [\lambda_j]_{j=1,\ldots,N_\Gamma} \cup [1]_{j=N_\Gamma,\ldots,N_1+N_2}$. Then, the scheme (45) converges if and only if,

$$\frac{2\mathbb{R}(\mu_j)}{\theta|\mu_j|^2} > 1 \Leftrightarrow \left\{ \begin{array}{l} \frac{2\mathbb{R}(\lambda_j)}{\theta|\lambda_j|^2} > 1, \; j = 1, \ldots, N_\Gamma \;\Leftrightarrow\; 0 < \theta < \theta_{max} \\ \\ \frac{2}{\theta} > 1, \; j = N_\Gamma, \ldots, N_1 + N_2 \;\Leftrightarrow\; 0 < \theta < 2 \end{array} \right\}$$
$$\Leftrightarrow \; 0 < \theta < \min[\theta_{max}, 2].$$

Moreover, its rate of convergence is independent of the spatial discretization. Indeed, we observe that,

$$\begin{aligned} \rho(I - \theta Q^{-1}A) &= \max_{j=1,N_1+N_2} |1 - \theta\mu_i| \\ &= \left\{ \begin{array}{l} \max\limits_{j=1,N_\Gamma} |1 - \theta\lambda_j| < 1 \text{ for } \theta \in (0, \theta_{max}) \\ \\ \max\limits_{j=N_\Gamma,N_1+N_2} |1 - \theta| < 1 \text{ for } \theta \in (0, 2) \end{array} \right\} \end{aligned} \tag{50}$$

Furthermore, we have that $\rho(I - \theta Q^{-1}A) = \inf_{\|\cdot\|} \left\|I - \theta Q^{-1}A\right\|$, thus (50) implies that there exists a norm, denoted with $\|\cdot\| *$ and a constant K'_θ independent of h such that $\left\|I - \theta Q^{-1}A\right\| * \le K'_\theta < 1$ for $\theta \in (0, \min[\theta_{max}, 2])$. The result $\left\|\mathbf{c} - \mathbf{c}^k\right\| * \le K'_\theta \left\|\mathbf{c} - \mathbf{c}^{k-1}\right\| *$ is straightforwardly proven. □

Indeed, corollary 8 states that Q is an optimal preconditioner for system (42). Finally, in order to apply efficiently the scheme (45) to the solution of system (42), we need to find a convenient relaxation parameter to accelerate the convergence. To this aim we observe that, the behavior of the iteration matrix $I - \theta Q^{-1}A$ is governed by the third diagonal block $I - \theta\Sigma_2^{-1}\Sigma$. In our case, Σ_2 and Σ are positive definite but not symmetric matrices since the convection terms are not symmetric. Therefore we cannot a priori assert that $\Sigma_2^{-1}\Sigma$ has real positive eigenvalues. Thus, an optimal static choice of θ is not straightforward. However, as a heuristic choice, parameter θ is selected as if $\Sigma_2^{-1}\Sigma$ was symmetric, setting $\theta = 2/(\lambda_{min} + \lambda_{max})$, where λ_{min} and λ_{max} denote the minimum and maximum eigenvalues of $\Sigma_2^{-1}\Sigma$. In that case, by observing that $I - \Sigma_2^{-1}\Sigma = -\Sigma_2^{-1}\Sigma_1$ is the iteration matrix associated with the un-relaxed case, denoting with $[\gamma_i], i = 1, \dots, N_\Gamma$ its spectrum, the following inequalities hold:

$$\lambda_i = 1 - \gamma_i \text{ and } \gamma_i \leq \left\| -\Sigma_2^{-1}\Sigma_1 \right\| p \quad \forall i = 1, \dots, N_\Gamma \quad \forall p > 0 \tag{51}$$

$$M_p := \frac{\left\| \mathbf{c}_\Gamma^k - \mathbf{c}_\Gamma^{k-1} \right\| p}{\left\| \mathbf{c}_\Gamma^{k-1} - \mathbf{c}_\Gamma^{k-2} \right\| p} \leq \left\| -\Sigma_2^{-1}\Sigma_1 \right\| p, \tag{52}$$

where $\|\cdot\| p$ denotes the vector norm $\|\mathbf{v}\| p = (\sum_{i=1}^n |v_i|^p)^{1/p}$ for all $\mathbf{v} \in \mathcal{R}^n$, as well as the matrix norm induced by this vector norm. Consequently, making the approximations, $M_p \simeq \left\| -\Sigma_2^{-1}\Sigma_1 \right\| p$, $\gamma_{max} \simeq M_p$, $\gamma_{min} \simeq 0$ and noticing that $\lambda_{min} + \lambda_{max} = 2 - \gamma_{max} - \gamma_{min}$ we make the choice, $\theta \simeq 2/(2 - M_p)$. This choice have been successfully applied in the computations, as shown in table 1.

2.6 The Robin-Robin method as a preconditioner

In the previous section we proved that Q is an optimal preconditioner for the global matrix A, and we discussed the behavior of the preconditioned Richardson scheme applied to the solution of system $A\mathbf{c} = \mathbf{b}$. However, the numerical solution of a large linear system of equations can be approached with several methods that can be more efficient than the Richardson one. Indeed, observing that A is positive definite but not symmetric, we focus our attention on the *Preconditioned Generalized Minimal Residual* method (P-GMRes). Our aim is to use the P-GMRes algorithm to solve system $A\mathbf{c} = \mathbf{b}$ using Q as a preconditioner. First of all we observe that the P-GMRes method applied to system $A\mathbf{c} = \mathbf{b}$ and preconditioned with the matrix Q, converges at a rate ρ independent of h. Indeed, the convergence rate of the P-GMRes method in a suitable norm $\|\cdot\| *$ is the constant ρ such that, $\left\|\mathbf{c} - \mathbf{c}^k\right\| * \leq \rho \left\|\mathbf{c} - \mathbf{c}^{k-1}\right\| *$ for $k = 1, 2, 3, \dots$, being $\mathbf{c}$ the exact solution of system $A\mathbf{c} = \mathbf{b}$ and $\mathbf{c}^k$ the approximations produced by the P-GMRes. As shown in Remark 4.2.6 [7], the constant ρ can be estimated as $\rho \leq \sqrt{1 - (1 - K^2)^2/16}$, being K a constant such that $\left\|I - \theta Q^{-1}A\right\| * \leq K < 1$ for a suitable value θ. Indeed, corollary 8 ensures that in our case there exist a norm $\|\cdot\| *$, an interval $(0, \theta_{max})$ and a constant K independent on h, such that $\left\|I - \theta Q^{-1}A\right\| * \leq K < 1$ for all $\theta \in (0, \theta_{max})$. Consequently the P-GMRes method, applied to system $A\mathbf{c} = \mathbf{b}$ and preconditioned with Q, converges at a rate independent on h. The theoretical results presented so far are confirmed by numerical experiences shown in table 1. In all tests we have set $\Omega_1 = (0,4) \times (0,1)$, $\Omega_1 = (0,4)\times(-1,0)$, $\mathbf{v}_1 = [y, -10^{-1}]^T$, $\mathbf{v}_2 = [0, -10^{-1}]^T$. Moreover, we set the physical parameters to $D_1 = D_2 = 1.0$ and $P = 1.0$ and the time discretization step to $\Delta t = 1$. The finer grids are obtained by means of a uniform refinement plus regularization, thus $N_h = \mathcal{O}(h^{-2})$. In this table we compare method 9 in the un-relaxed ($\theta = 1$) and in the relaxed case. First of all, we notice that the application of the relaxation technique is effective, indeed the relaxed Richardson method needs fewer iterations to converge than the unrelaxed one. Furthermore, we observe that the P-GMRes method performs definitely better than the Richardson one. Finally, we observe that in accordance with the conclusion of theorem 7, the number of iterations is strictly independent of h for all the considered methods.

Table 1. Comparison of the number of iterations to reach an incremental error $E_i^k = \|c_i^k - c_i^{k-1}\| L^1 / \|c_i^k\| L^1$ less than 10^{-10}.

h	N	Unrelax. Rich.	Relax. Rich.	P-GMRes
0.1	4000	11	10	6
0.05	16000	12	10	6
0.025	60000	12	10	6
0.01875	106000	12	10	6
0.012	260000	12	10	6

In practical applications, a very large number of degrees of freedom is required. Even if the P-GMRes algorithm converges in few iterations, the solution of the subsystems $Q\mathbf{z} = \mathbf{r}$ arising from the preconditioning step has a large computational cost. Consequently, the choice of an efficient iterative method must take into account not only the number of iterations, but also the computational cost of each iteration. Reduction of the time necessary to pursue each iteration is obtained by resorting to *flexible preconditioned iterative solvers* (see [10]). Among the iterative methods considered above, the only one that allows a flexible variant is the preconditioned GMRes. Let us briefly introduce it. From definition (46), we see that a preconditioning step $Q\mathbf{z} = \mathbf{r}$, being $\mathbf{r} = \mathbf{b} - A\mathbf{c}$ the residual, consists in solving the two subsystems,

$$A_{11}\mathbf{z}_1 = \mathbf{r}_1, \tag{53}$$

$$\begin{bmatrix} A_{22} & A_{2\Gamma} \\ A_{\Gamma 2} & A_{\Gamma\Gamma} \end{bmatrix} \begin{bmatrix} \mathbf{z}_2 \\ \mathbf{z}_\Gamma \end{bmatrix} = \begin{bmatrix} \mathbf{r}_2 \\ \mathbf{r}_\Gamma \end{bmatrix} - \begin{bmatrix} 0 \\ A_{\Gamma 1}\mathbf{z}_1 \end{bmatrix}. \tag{54}$$

This task can be carried out by means of iterative methods such as BiCGStab or GMRes. The aim of the *flexible* strategy is to pursue the preconditioning step by solving (53) and (54) with a large tolerance, reducing henceforth the computational cost of the solution process. This can be done through the algorithm described below. Let us denote with $\hat{Q}_j^{-1}$ $j = 1, 2, 3, \ldots$ an approximation of Q^{-1}, obtained in our case by a low cost solution of (53,54). Then the F-P-GMRes (*Flexible P-GMRes*) algorithm reads as follows:

F-P-GMRes:
Compute $\mathbf{z}_0 = \hat{Q}_0^{-1}(\mathbf{b} - A\mathbf{x}_0)$, $\beta = \|\mathbf{z}_0\|_2$, $\mathbf{v}_1 = \mathbf{r}_0/\beta$.
For $j = 1, \ldots, m$ Do:
  compute $\mathbf{w} = A\mathbf{v}_j$,
  compute $\mathbf{z}_j = \hat{Q}_j^{-1}\mathbf{w}$.
  For $i = 1, \ldots, j$ Do:
    $h_{i,j} = (\mathbf{w}, \mathbf{v}_i)$,
    $\mathbf{w} = \mathbf{w} - h_{i,j}\mathbf{v}_i$,
  End Do.
  Compute $h_{j+1,j} = \|\mathbf{w}\|_2$ and $\mathbf{v}_{j+1} = \mathbf{w}/h_{j+1,j}$.
  Define $Z_m = [\mathbf{z}_1, \ldots, \mathbf{z}_m]$, $H_m = \{h_{i,j}\}_{1 \leq i \leq j+1; 1 \leq j \leq m}$.
End Do.
Compute $\mathbf{y}_m = \min_{\mathbf{y} \in \mathcal{R}^m} \|\beta \mathbf{e}_1 - H_m \mathbf{y}\|_2$, and $\mathbf{x}_m = \mathbf{x}_0 + Z_m \mathbf{y}_m$.
If $\|\hat{Q}_m^{-1}(\mathbf{b} - A\mathbf{x}_m)\|_2 < tol = 10^{-p}$ $p \in N$ then end, else $\mathbf{x}_0 \leftarrow \mathbf{x}_m$.

Table 2. Quantitative comparison between P-GMRes and F-P-GMRes for the computation of the stationary solution of problem 1.

Stenotic Artery	P-GMRes	F-P-GMRes
Iterations	2	2
Time/iter. (s)	872	312

A possible strategy to choose $\hat{Q}_j^{-1}$ at each iteration is the following. Should 10^{-p} be the tolerance fixed in F-P-GMRes, then the tolerance of the iterative method applied to solve systems (53,54) is set to 10^{-q}, with $q = [p/2]$. For the application at hand, by adopting this strategy one obtains that F-P-GMRes requires less CPU time to converge than P-GMRes. Indeed, as shown in Table 2, the application of the F-P-GMRes method is very efficient. Comparing it with P-GMRes we see in fact that, although we approximate Q^{-1} with $\hat{Q}_j^{-1}$, the global number of iterations does not increase. Thus the reduction in the computational cost at each iteration directly provides a reduction in the time needed to solve $A\mathbf{c} = \mathbf{b}$. In table 2 we have considered a test case that is relevant for haemodynamics, precisely a stenotic artery. The considered domain has been discretized with 68734 nodes. The tolerance of the outer iterative procedure is always the same and equal to 10^{-8}, while the tolerance for the approximate solution of the preconditioning step is set to 10^{-4}.

In figure 2 we show the result of the numerical simulation of the oxygen concentration in the lumen and in the wall of the aforementioned stenotic artery.

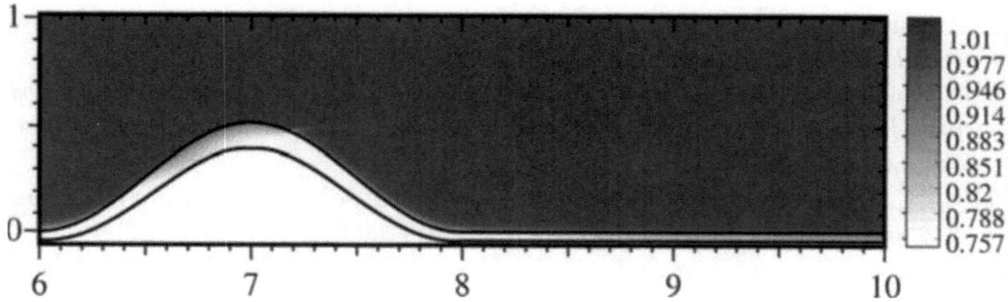

Fig. 2. oxygen concentration in a 2D geometry representing the lumen and the wall of a stenotic artery. More precisely, the lumen is the dark colored region where the concentration is close to the unity, while the wall is the region below, where the concentration is reasonably lower. Furthermore, the concentration is plotted in adimensional scale where the value of unity refers to the usual oxygen concentration in blood. The scales on the x,y axes represent the dimension of the vessel in centimeters, corresponding in this case to a large size artery.

We apply the multidomain model, namely problem 1, where Ω_1 represents the lumen and Ω_2 the wall of the artery. For the numerical approximation of the model we apply the Robin-Robin method, i.e. problem 3. In this case, the velocity of the blood in the artery, $\mathbf{v}_1$, is computed solving the incompressible Navier-Stokes equations numerically. On the other hand, the velocity into the wall has been approximated by the Darcy's law and its normal component across the interface turns out to be $J_v \simeq 10^{-7}$ cm/s. This value is small with respect to the diffusivity $D_1 = D_2 = 2.5\ 10^{-5}$ cm^2/s. As a result of that, this specific application should satisfy assumptions (22) and (37).

References

1. Yves Achdou, Patrick Le Tallec, Frédéric Nataf, and Marina Vidrascu, *A domain decomposition preconditioner for an advection-diffusion problem*, Comput. Methods Appl. Mech. Engrg. **184** (2000), no. 2-4, 145–170.
2. A. Alonso, R.T. Trotta, and A. Valli, *Coercive domain decomposition algorithms for advection-diffusion equations and systems*, J. Comput. Appl. Math. **96** (1998), 51–76.
3. M. Discacciati, E. Miglio, and A. Quarteroni, *Mathematical and numerical models for coupling surface and groundwater flows*, Applied Numerical Mathematics **43** (2002), 57–74.
4. G. Karner, K. Perktold, and H.P. Zehentner, *Intra and Extracorporeal Cardiovascular Fluid Dynamics*, ch. 7: Mass transport in large arteries and through the arterial wall, pp. 209–247, P. Verdonck and K. Perktold eds., WIT-Press - Computational Mechanics Publications, 2000.
5. J. L. Lions and E. Magenes, *Problèmes aux Limites non Homogènes et Applications*, vol. 1,2,3, Dunod, Paris, 1968.
6. A. Quarteroni, R. Sacco, and F. Saleri, *Numerical Mathematics*, Springer, New York, 2000.

7. A. Quarteroni and A. Valli, *Domain Decomposition Methods for Partial Differential Equations*, Oxford University Press, 1999.
8. A. Quarteroni, A. Veneziani, and P. Zunino, *A domain decomposition method for advection-diffusion processes with application to blood solutes*, SIAM J. Sci. Comput. **23** (2002), no. 6, 1959–1980.
9. ———, *Mathematical and numerical modelling of solute dynamics in blood flow and arterial walls*, SIAM J. Numer. Anal. **39** (2002), no. 5, 1488–1511.
10. Y. Saad, *Iterative Methods for Sparse Linear Systems*, PWS Publishing Company, 1996.
11. P. Zunino, *Mathematical and numerical modeling of mass transfer in the vascular system*, Ph.D. thesis, Ecole Polythéchnique Fédérale de Lausanne, 2002.

Efficient Preconditioning of Linear Systems Arising from the Discretization of Radiative Transfer Equation

Mohammed Seaïd and Axel Klar

Fachbereich Mathematik, TU Darmstadt, 64289 Darmstadt, Germany

Summary. Preconditioning techniques for iterative solvers of discretized radiative transfer equation are presented. Discrete ordinates collocation and Diamond differencing are used for angle and space discretizations respectively. To solve the resulting linear system we formulate source iteration, diffusion synthetic acceleration and Krylov subspace methods. We also introduce a fast multilevel algorithm. All these methods can be viewed as preconditioned iterative methods with different preconditioners. Numerical results along with comparisons of effectiveness and efficiency of these solvers are carried out on several test problems with both continuous and discontinuous variables.

1 Introduction

The radiative transfer problems are important in many applications such as astrophysics models, nuclear reactor design, glass manufacturing, and they are a fundamental part of many numerical methods used to solve complicated applications such as coupled radiative heat transfer and radiation hydrodynamics. Further physical and mathematical descriptions can be found in [15, 7, 14]. The starting point of radiation transfer is the general Boltzmann equation

$$\frac{1}{c}\frac{\partial \psi}{\partial t} + \omega \cdot \nabla \psi + \Sigma_t \psi = \int_{\omega'} \int_{\mathbf{E}'} \Sigma_s(\mathbf{E}' \to \mathbf{E}, \omega' \to \omega) \psi(t, \mathbf{x}, \nu, \omega', \mathbf{E}') d\omega' d\mathbf{E}' + \mathbf{q}(t, \mathbf{x}, \nu, \omega, \mathbf{E}), \quad (1)$$

where $\omega \cdot \nabla \psi$ models the loss of particles (photons or neutrons) due to leakage, Σ_t the loss due to collisions, the integral the gain due to the scattering and $\mathbf{q}(t, \mathbf{x}, \nu, \omega, \mathbf{E})$ models the gain from other external source terms. Here Σ_t is the total macroscopic cross section, Σ_s is a cross section for scattering from energy and direction $(\mathbf{E}', \omega')$ into $(\mathbf{E}, \omega)$ at time t with fixed speed c.

The main difficulties in solving numerically the integro-differential equation (1) are the large set of depend unknowns and the coupling between the transport and the integral operators through the scattering function Σ_s. For instance, ψ is a function of time variable t, space variable $\mathbf{x}$, frequency variable ν, direction variable ω and energy variable $\mathbf{E}$. Solving the large linear system of algebraic equations induced by discretizing these variables is computationally very demanding. On the other hand and due to the high speed c in (1) explicit time integrators tend to be inefficient for long term computations because of the CFL condition that restricts the size of time step to be extremely small.

Under the assumptions that the kernel Σ_s is often mildly anisotropic, becoming more isotropic as energy decreases, and the radiation time is small compared with all other time scales of the physical situation considered, we focus on the solution of steady-state, mono-energetic, frequency decoupled, isotropic radiative transfer problems in three space dimensions. However, all the methods presented in this paper can be straightforwardly extended to the more general problem (1). Hence the radiative transfer equation we consider in this paper reads

$$\begin{aligned} \omega\cdot\nabla\psi + (\sigma+\kappa)\psi &= \frac{\sigma}{4\pi}\int_{S^2}\psi(\mathbf{x},\omega')d\omega' + \kappa q(\mathbf{x}), \quad (\mathbf{x},\omega)\in\Omega\times S^2, \\ \psi(\hat{\mathbf{x}},\omega) &= g(\hat{\mathbf{x}}), \qquad (\hat{\mathbf{x}},\omega)\in\Gamma^-\times S^2, \end{aligned} \tag{2}$$

where Ω is a bounded subdomain of $\mathbb{R}^3$ with smooth boundary Γ, S^2 is the unit sphere, and $\Gamma^- = \{\hat{\mathbf{x}}\in\Gamma;\ \ \mathbf{n}(\hat{\mathbf{x}})\cdot\Omega < 0\}$ with $\mathbf{n}(\hat{\mathbf{x}})$ denotes the outward normal in $\hat{\mathbf{x}}$ with respect to Γ. The equation (2) models the changes of an intensity $\psi(\mathbf{x},\omega)$ as particles are passing through the domain Ω at position point $\mathbf{x} = (x,y,z)^T$ in the direction $\omega = (\mu,\eta,\xi)^T$ and are subject to loses due to absorption $\kappa = \kappa(\nu,\mathbf{x})$ and scattering $\sigma = \sigma(\nu,\mathbf{x})$, while their number grows due to the source $q(\mathbf{x})$ inside the domain Ω. Here $g(\mathbf{x})$ is known boundary function on Γ. We assume that σ and κ are nonnegative functions and we introduce the mean intensity φ as

$$\varphi(\mathbf{x}) = \frac{1}{4\pi}\int_{S^2}\psi(\mathbf{x},\omega')d\omega'. \tag{3}$$

We also define the scattering ratio γ and the opacity coefficient ϑ as

$$\gamma = \max_{\mathbf{x}\in\Omega}\left(\frac{\sigma(\mathbf{x})}{\sigma(\mathbf{x})+\kappa(\mathbf{x})}\right) \quad \text{and} \quad \vartheta = \min_{\mathbf{x}\in\Omega}\big(\sigma(\mathbf{x})+\kappa(\mathbf{x})\big)\mathrm{diam}(\Omega). \tag{4}$$

Here $\mathrm{diam}(\Omega)$ is the diameter of the space domain Ω. In applications, γ and ϑ are used to characterize the convergence rates of the iterative methods and the diffusion limits in the optically thick medium.

Note that using implicit time stepping schemes to integrate the original equation (1) and piecewise-constant difference methods to treat the energy variable, the equation (1) can be written in the same form as (2) where the time step is incorporated into σ and κ whereas the new variable $\psi = \psi_{\nu,k}$ approximates the intensity with the discrete energy E_k and the frequency ν.

Many numerical methods have been used to solve the equation (2). For a review on some of these methods see [14, 17]. It is well known [1] that the standard Source Iteration (SI) becomes extremely costly when the scattering ratio $\gamma \approx 1$. The standard Diffusion Synthetic Acceleration (DSA) has been widely used to accelerate the source iteration [4, 6]. The SI and DSA methods can be seen respectively, as Richardson iteration and preconditioned Richardson iteration with the diffusion approach as preconditioner.

In this paper we implement SI, DSA and a Krylov subspace method to solve the equation (2). We also propose a fast multilevel algorithm [12] which uses the approximate inverse operator as a preconditioner and solves the linear system only in the coarse meshes. Numerical results show this algorithm to be faster than DSA in many regimes. The robustness, efficiency and convergence rates of these methods are illustrated by several numerical test examples in both one and two space dimensions. Comparison of the results obtained by different methods is also included in this paper.

In section 2, we use discrete ordinates collocation for the angle discretization and Diamond differencing for the space discretization. In section 3, we show that the space and ordinates discretizations yield to a linear system of sparse matrices. Detailed matrix formulation is given on uniform space dicretization of the domain Ω. A brief overview of SI, DSA and Krylov subspace methods is given in section 4. In section 5, we formulate a fast multilevel algorithm using nested grids. The algorithm inherit the restriction and prolongation operators from the multigrid methodology for the intergrid transfer while Atkinson-Brakhage type smoother is applied at each level. Section 6 is devoted to numerical results for various test examples. We present computations for radiative transfer problems with thermal source term, subject to continuous, discontinuous variables and frequency dependence. Finally, in section 7, we set some concluding remarks.

2 Ordinates and Space Discretizations

We start with a discrete ordinates discretization in angle. This corresponds to expanding the integrals on the unit sphere S^2 in terms of N weighted quadrature rules,

$$\int_{S^2} \psi(\mathbf{x},\omega) d\omega \simeq \sum_{l=1}^{N} w_l \psi(\mathbf{x},\omega_l), \tag{5}$$

where $\omega_l = (\mu_l, \eta_l, \xi_l)^T$, for all $l = 1, \dots, N$, with $N = n(n+2)$, and n is the number of direction cosines. Since $\omega_l \in S^2$, we have

$$\mu_l^2 + \eta_l^2 + \xi_l^2 = 1, \qquad \text{for all} \quad l = 1, 2, \ldots, N.$$

We assume n an even number of quadrature points so that the points (μ_l, η_l, ξ_l) are nonzero, symmetric with respect to the x-, y- and z-axis and they are invariant under 90° rotations. Furthermore they satisfy the relation

$$\xi_i^2 = \xi_1^2 + 2\frac{i-1}{n-2}(1 - 3\xi_1^2),$$

for $i = 1, 2, \ldots, n/2$ and $0 < \xi_1 < 1/3$.

In (5) w_l are the corresponding weights chosen to be positive and satisfy

$$\sum_{l=1}^{N} w_l = 4\pi, \quad \sum_{l=1}^{N} w_l \mu_l = 0, \quad \sum_{l=1}^{N} w_l \eta_l = 0 \quad \text{and} \quad \sum_{l=1}^{N} w_l \xi_l = 0.$$

In practice we choose $w_l = 4\pi/N$ and (μ_l, η_l, ξ_l) are set in such a way the above conditions are guaranteed. Let S_N be a given set of N discrete directions in S^2, then a semi-discrete formulation of (2) is

$$\begin{aligned} \mu_l \frac{\partial \psi_l}{\partial x} + \eta_l \frac{\partial \psi_l}{\partial y} + \xi_l \frac{\partial \psi_l}{\partial z} + (\sigma + \kappa)\psi_l = \sigma\varphi(\mathbf{x}) + \kappa q(\mathbf{x}), \quad (\mathbf{x}, \omega_l) \in \Omega \times S_N, \\ \psi_l(\hat{x}, \hat{y}) = g(\hat{x}, \hat{y}), \qquad (\hat{\mathbf{x}}, \omega_l) \in \Gamma^- \times S_N. \end{aligned} \tag{6}$$

$\psi_l(\mathbf{x})$ denotes approximation to $\psi(\mathbf{x}, \mu_l, \eta_l, \xi_l)$ and φ is given by

$$\varphi(\mathbf{x}) = \frac{1}{4\pi} \sum_{l=1}^{N} w_l \psi_l(\mathbf{x}).$$

To discretize the equation (6) in space we suppose for simplicity, that the spatial domain is a box, $\Omega = [a_x, b_x] \times [a_y, b_y] \times [a_z, b_z]$. Then we cover the domain Ω with a uniform numerical mesh defined by

$$\begin{aligned} \Omega_h = \Big\{ \mathbf{x}_{ijk} = (x_i, y_j, z_k)^T, \quad x_i = a_x + i\Delta x, \quad y_j = a_y + j\Delta y, \\ z_k = a_z + k\Delta z, \quad i = 0, 1 \ldots, I, \quad j = 0, 1 \ldots, J, \quad k = 0, 1 \ldots, K \Big\}, \end{aligned}$$

where $x_0 = a_x$, $x_I = b_x$; $y_0 = a_y$, $y_J = b_y$; $z_0 = a_z$, $z_K = b_z$; and h denotes the maximum cell size. We define the averaged grid points

$$\Delta x = x_i - x_{i-1}, \quad \Delta y = y_j - y_{j-1}, \quad \Delta z = z_k - z_{k-1},$$

$$x_{i-\frac{1}{2}} = \frac{x_{i-1} + x_i}{2}, \quad y_{j-\frac{1}{2}} = \frac{y_{j-1} + y_j}{2}, \quad z_{k-\frac{1}{2}} = \frac{z_{k-1} + z_k}{2},$$

for $i = 1 \ldots, I$, $j = 1 \ldots, J$ and $k = 1 \ldots, K$. By using the notation f_{ijk} to denote the approximation value of the function f at the grid point (x_i, y_j, z_k), the fully discrete approximation for the equation (2) can be written as

$$\mu_l \frac{\psi_{l,ijk} - \psi_{l,i-1jk}}{\Delta x} + \eta_l \frac{\psi_{l,ijk} - \psi_{l,ij-1k}}{\Delta y} + \xi_l \frac{\psi_{l,ijk} - \psi_{l,ijk-1}}{\Delta z} +$$
$$(\sigma_{i-\frac{1}{2}j-\frac{1}{2}k-\frac{1}{2}} + \kappa_{i-\frac{1}{2}j-\frac{1}{2}k-\frac{1}{2}})\psi_{l,i-\frac{1}{2}j-\frac{1}{2}k-\frac{1}{2}} = \sigma_{i-\frac{1}{2}j-\frac{1}{2}k-\frac{1}{2}}\varphi_{i-\frac{1}{2}j-\frac{1}{2}k-\frac{1}{2}} +$$
$$\kappa_{i-\frac{1}{2}j-\frac{1}{2}k-\frac{1}{2}} q_{i-\frac{1}{2}j-\frac{1}{2}k-\frac{1}{2}}, \tag{7}$$

where the cell averages values of ψ are given by

$$\begin{aligned}
\psi_{l,i-1jk} &= \frac{1}{\Delta x}\int_{y_{j-1}}^{y_j}\int_{z_{k-1}}^{z_k} \psi_l(x_i, y, z)dydz,\\
\psi_{l,ij-1k} &= \frac{1}{\Delta y}\int_{x_{i-1}}^{x_i}\int_{z_{k-1}}^{z_k} \psi_l(x, y_j, z)dxdz,\\
\psi_{l,ijk-1} &= \frac{1}{\Delta z}\int_{x_{i-1}}^{x_i}\int_{y_{j-1}}^{y_j} \psi_l(x, y, z_k)dxdy,\\
\psi_{l,ijk} &= \frac{1}{\Delta x \Delta y \Delta z}\int_{x_{i-1}}^{x_i}\int_{y_{j-1}}^{y_j}\int_{z_{k-1}}^{z_k} \psi_l(x, y, z)dxdydz,
\end{aligned} \tag{8}$$

In this paper we use the Diamond difference method to approximate the fluxes in (8). The method consists on centred differences and approximating the function values at the cell centres $f_{l,i-\frac{1}{2}j-\frac{1}{2}k-\frac{1}{2}}$ by the average of their values at the eight neighbouring nodes as

$$f_{l,i-\frac{1}{2}j-\frac{1}{2}k-\frac{1}{2}} = \frac{1}{8}\Big[f_{l,i-1j-1k-1} + f_{l,i-1jk-1} + f_{l,i-1j-1k} + f_{l,i-1jk} +$$
$$f_{l,ij-1k-1} + f_{l,ijk-1} + f_{l,ij-1k} + f_{l,ijk}\Big]. \tag{9}$$

Hence the discrete mean intensity $\varphi_{i-\frac{1}{2}j-\frac{1}{2}k-\frac{1}{2}}$ in (7) is given by

$$\varphi_{i-\frac{1}{2}j-\frac{1}{2}k-\frac{1}{2}} = \sum_{l=1}^{N} w_l \psi_{l,i-\frac{1}{2}j-\frac{1}{2}k-\frac{1}{2}}.$$

Other discretizations using Legendre polynomial collocation in ordinates and finite element or Petrov-Galerkin methods in space can be used in the same manner, we refer to [14, 6, 18, 19] for details. For the discretization of the boundary conditions in (6) we can proceed as follows:

when $\hat{x} = a_x$, the normal $\mathbf{n}(\hat{\mathbf{x}}_{0jk}) = (-1, 0, 0)^T$, then $\mathbf{n}(\hat{\mathbf{x}}_{0jk}) \cdot \omega_l = -\mu_l$, and for $\mu_l > 0$ we have $\psi_{l,0jk} = g_{0jk}$

when $\hat{y} = a_y$, the normal $\mathbf{n}(\hat{\mathbf{x}}_{i0k}) = (0, -1, 0)^T$, then $\mathbf{n}(\hat{\mathbf{x}}_{i0k}) \cdot \omega_l = -\eta_l$, and for $\eta_l > 0$ we have $\psi_{l,i0k} = g_{i0k}$

when $\hat{z} = a_z$, the normal $\mathbf{n}(\hat{\mathbf{x}}_{ij0}) = (0, 0, -1)^T$, then $\mathbf{n}(\hat{\mathbf{x}}_{ij0}) \cdot \omega_l = -\xi_l$, and for $\xi_l > 0$ we have $\psi_{l,ij0} = g_{ij0}$

The other three cases can be discretized in a similar way. Needless to say that for a given $l = 1, 2, \ldots, N$ no component of ω_l is ever zero and only three of the above six cases can hold. Furthermore, in the discretization (7) there are $(I+1)(J+1)(K+1)$ unknowns $\psi_{l,ijk}$ and $IK + JI + JK + I + J + K + 1$ boundary equations.

3 Linear System Formulation

In order to simplify the notations and to get closer to a compact linear algebra formulation of (7), we first define the matrix entries

$$
\begin{aligned}
d_{l,i-\frac{1}{2}j-\frac{1}{2}k-\frac{1}{2}} &= \frac{|\mu_l|}{\Delta x} + \frac{|\eta_l|}{\Delta y} + \frac{|\xi_l|}{\Delta z} + \frac{\sigma_{i-\frac{1}{2}j-\frac{1}{2}k-\frac{1}{2}} + \kappa_{i-\frac{1}{2}j-\frac{1}{2}k-\frac{1}{2}}}{8},\\
u_{l,i-\frac{1}{2}j-\frac{1}{2}k-\frac{1}{2}} &= \frac{|\mu_l|}{\Delta x} + \frac{|\eta_l|}{\Delta y} + \frac{-|\xi_l|}{\Delta z} + \frac{\sigma_{i-\frac{1}{2}j-\frac{1}{2}k-\frac{1}{2}} + \kappa_{i-\frac{1}{2}j-\frac{1}{2}k-\frac{1}{2}}}{8},\\
\bar{u}_{l,i-\frac{1}{2}j-\frac{1}{2}k-\frac{1}{2}} &= \frac{|\mu_l|}{\Delta x} + \frac{-|\eta_l|}{\Delta y} + \frac{|\xi_l|}{\Delta z} + \frac{\sigma_{i-\frac{1}{2}j-\frac{1}{2}k-\frac{1}{2}} + \kappa_{i-\frac{1}{2}j-\frac{1}{2}k-\frac{1}{2}}}{8},\\
\underline{u}_{l,i-\frac{1}{2}j-\frac{1}{2}k-\frac{1}{2}} &= \frac{|\mu_l|}{\Delta x} + \frac{-|\eta_l|}{\Delta y} + \frac{-|\xi_l|}{\Delta z} + \frac{\sigma_{i-\frac{1}{2}j-\frac{1}{2}k-\frac{1}{2}} + \kappa_{i-\frac{1}{2}j-\frac{1}{2}k-\frac{1}{2}}}{8},\\
v_{l,i-\frac{1}{2}j-\frac{1}{2}k-\frac{1}{2}} &= \frac{-|\mu_l|}{\Delta x} + \frac{|\eta_l|}{\Delta y} + \frac{|\xi_l|}{\Delta z} + \frac{\sigma_{i-\frac{1}{2}j-\frac{1}{2}k-\frac{1}{2}} + \kappa_{i-\frac{1}{2}j-\frac{1}{2}k-\frac{1}{2}}}{8},\\
\bar{v}_{l,i-\frac{1}{2}j-\frac{1}{2}k-\frac{1}{2}} &= \frac{-|\mu_l|}{\Delta x} + \frac{|\eta_l|}{\Delta y} + \frac{-|\xi_l|}{\Delta z} + \frac{\sigma_{i-\frac{1}{2}j-\frac{1}{2}k-\frac{1}{2}} + \kappa_{i-\frac{1}{2}j-\frac{1}{2}k-\frac{1}{2}}}{8},\\
\underline{v}_{l,i-\frac{1}{2}j-\frac{1}{2}k-\frac{1}{2}} &= \frac{-|\mu_l|}{\Delta x} + \frac{-|\eta_l|}{\Delta y} + \frac{|\xi_l|}{\Delta z} + \frac{\sigma_{i-\frac{1}{2}j-\frac{1}{2}k-\frac{1}{2}} + \kappa_{i-\frac{1}{2}j-\frac{1}{2}k-\frac{1}{2}}}{8},\\
e_{l,i-\frac{1}{2}j-\frac{1}{2}k-\frac{1}{2}} &= \frac{-|\mu_l|}{\Delta x} + \frac{-|\eta_l|}{\Delta y} + \frac{-|\xi_l|}{\Delta z} + \frac{\sigma_{i-\frac{1}{2}j-\frac{1}{2}k-\frac{1}{2}} + \kappa_{i-\frac{1}{2}j-\frac{1}{2}k-\frac{1}{2}}}{8}.
\end{aligned}
$$

Next, we define the vectors

$$
\mathbf{\Psi}_l \equiv \begin{pmatrix} \mathbf{\Psi}_{l,0} \\ \vdots \\ \mathbf{\Psi}_{l,K} \end{pmatrix} \in I\!R^{(I+1)(J+1)(K+1)}, \quad \text{with}
$$

$$
\mathbf{\Psi}_{l,k} \equiv \begin{pmatrix} \Psi_{l,0k} \\ \vdots \\ \Psi_{l,Jk} \end{pmatrix} \in I\!R^{(I+1)(J+1)}, \qquad \Psi_{l,jk} \equiv \begin{pmatrix} \psi_{l,0jk} \\ \vdots \\ \psi_{l,Ijk} \end{pmatrix} \in I\!R^{(I+1)};
$$

$$
\mathbf{\Phi} \equiv \begin{pmatrix} \mathbf{\Phi}_{1-\frac{1}{2}} \\ \vdots \\ \mathbf{\Phi}_{K-\frac{1}{2}} \end{pmatrix} \in I\!R^{IJK}, \quad \text{with}
$$

$$\mathbf{\Phi}_{k-\frac{1}{2}} \equiv \begin{pmatrix} \mathbf{\Phi}_{1-\frac{1}{2}k-\frac{1}{2}} \\ \vdots \\ \mathbf{\Phi}_{J-\frac{1}{2}k-\frac{1}{2}} \end{pmatrix} \in I\!R^{IJ}, \qquad \mathbf{\Phi}_{j-\frac{1}{2}k-\frac{1}{2}} \equiv \begin{pmatrix} \varphi_{1-\frac{1}{2}j-\frac{1}{2}k-\frac{1}{2}} \\ \vdots \\ \varphi_{I-\frac{1}{2}j-\frac{1}{2}k-\frac{1}{2}} \end{pmatrix} \in I\!R^{I};$$

$$\mathbf{Q} \equiv \begin{pmatrix} \mathbf{Q}_{1-\frac{1}{2}} \\ \vdots \\ \mathbf{Q}_{K-\frac{1}{2}} \end{pmatrix} \in I\!R^{IJK}, \quad \text{with}$$

$$\mathbf{Q}_{k-\frac{1}{2}} \equiv \begin{pmatrix} Q_{1-\frac{1}{2}k-\frac{1}{2}} \\ \vdots \\ Q_{J-\frac{1}{2}k-\frac{1}{2}} \end{pmatrix} \in I\!R^{IJ}, \qquad Q_{j-\frac{1}{2}k-\frac{1}{2}} \equiv \begin{pmatrix} q_{1-\frac{1}{2}j-\frac{1}{2}k-\frac{1}{2}} \\ \vdots \\ q_{I-\frac{1}{2}j-\frac{1}{2}k-\frac{1}{2}} \end{pmatrix} \in I\!R^{I};$$

In what follows we define the matrix $\mathbf{H}_l$ (known as sweep matrix) for the first sweep case $\mu_l < 0$, $\eta_l < 0$, $\xi_l < 0$ and the other seven sweep cases can be derived similarly. In order to simplify the notation, we drop hereafter the space grids subscripts from the matrix entries unless otherwise stated. Thus,

$$\mathbf{H}_l \equiv \begin{pmatrix} \mathbf{D}_l & \mathbf{E}_l & & & \\ & \ddots & \ddots & & \\ & & \mathbf{D}_l & \mathbf{E}_l & \\ & & & \mathbf{D}_l & \mathbf{S} \\ & & & & \mathbf{S} \end{pmatrix} \in I\!R^{(I+1)(J+1)(K+1)\times(I+1)(J+1)(K+1)}, \qquad \text{with}$$

$$\mathbf{D}_l \equiv \begin{pmatrix} D_l & U_l & & \\ & \ddots & \ddots & \\ & & D_l & U_l \\ & & & S \end{pmatrix} \in I\!R^{(I+1)(J+1)\times(I+1)(J+1)}, \qquad \text{with}$$

$$D_l \equiv \begin{pmatrix} d & u & & \\ & \ddots & \ddots & \\ & & d & u \\ & & & 1 \end{pmatrix} \in I\!R^{(I+1)\times(I+1)}, \; U_l \equiv \begin{pmatrix} \underline{u} & \bar{u} & & \\ & \ddots & \ddots & \\ & & \underline{u} & \bar{u} \\ & & & 1 \end{pmatrix} \in I\!R^{(I+1)\times(I+1)};$$

$$\mathbf{E}_l \equiv \begin{pmatrix} V_l & W_l & & \\ & \ddots & \ddots & \\ & & V_l & W_l \\ & & & S \end{pmatrix} \in I\!R^{(I+1)(J+1)\times(I+1)(J+1)}, \qquad \text{with}$$

$$V_l \equiv \begin{pmatrix} \underline{v} & \bar{v} & & \\ & \ddots & \ddots & \\ & & \underline{v} & \bar{v} \\ & & & 1 \end{pmatrix} \in I\!R^{(I+1)\times(I+1)}, \; W_l \equiv \begin{pmatrix} v & \bar{e} & & \\ & \ddots & \ddots & \\ & & v & \bar{e} \\ & & & 1 \end{pmatrix} \in I\!R^{(I+1)\times(I+1)}.$$

$$\mathbf{S} \equiv \begin{pmatrix} S & S & & \\ & \ddots & \ddots & \\ & & S & S \\ & & & S \end{pmatrix} \in \mathbb{R}^{(I+1)(J+1)\times(I+1)(J+1)}, \quad \text{with}$$

$$S \equiv \begin{pmatrix} 1 & 1 & & \\ & \ddots & \ddots & \\ & & 1 & 1 \\ & & & 1 \end{pmatrix} \in \mathbb{R}^{(I+1)\times(J+1)}.$$

$$\mathbf{\Sigma}_l \equiv \begin{pmatrix} \mathbf{\Sigma}_{l,1-\frac{1}{2}} & & \\ & \ddots & \\ & & \mathbf{\Sigma}_{l,K-\frac{1}{2}} \\ & & \mathbf{0} \end{pmatrix} \in \mathbb{R}^{(I+1)(J+1)(K+1)\times IJK}, \qquad \text{with}$$

$$\mathbf{\Sigma}_{l,k-\frac{1}{2}} \equiv \begin{pmatrix} \Sigma_{l,1-\frac{1}{2}k-\frac{1}{2}} & & \\ & \ddots & \\ & & \Sigma_{l,J-\frac{1}{2}k-\frac{1}{2}} \\ & & \mathbf{0} \end{pmatrix} \in \mathbb{R}^{(I+1)(J+1)\times IJ}, \quad \text{and}$$

$$\Sigma_{l,j-\frac{1}{2}k-\frac{1}{2}} \equiv$$

$$\begin{pmatrix} \frac{\sigma_{1-\frac{1}{2}j-\frac{1}{2}k-\frac{1}{2}}+\kappa_{i-\frac{1}{2}j-\frac{1}{2}k-\frac{1}{2}}}{8} & & \\ & \ddots & \\ & & \frac{\sigma_{i-\frac{1}{2}j-\frac{1}{2}k-\frac{1}{2}}+\kappa_{i-\frac{1}{2}j-\frac{1}{2}k-\frac{1}{2}}}{8} \\ & & 0 \end{pmatrix} \in \mathbb{R}^{(I+1)\times I}.$$

Using these definitions with $\mathbf{\Psi}$ and $\mathbf{\Phi}$ being the unknowns, the fully discrete equation (7) can be written in matrix form as

$$\left(\begin{array}{ccc|c} \mathbf{H}_1 & & & -\mathbf{\Sigma}_1 \\ & \ddots & & \vdots \\ & & \mathbf{H}_N & -\mathbf{\Sigma}_N \\ \hline -\frac{w_1}{4\pi}\mathbf{S} & \dots & -\frac{w_N}{4\pi}\mathbf{S} & \mathbf{I} \end{array}\right) \begin{pmatrix} \mathbf{\Psi}_1 \\ \vdots \\ \mathbf{\Psi}_N \\ \hline \mathbf{\Phi} \end{pmatrix} = \begin{pmatrix} \mathbf{Q}_1 \\ \vdots \\ \mathbf{Q}_N \\ \hline \mathbf{0} \end{pmatrix}, \tag{10}$$

where $\mathbf{I}$ is the $IJK \times IJK$ identity matrix and $\mathbf{0}$ is the IJK null vector. The linear system (10) can be rewritten in common linear algebra notation as

$$\mathbf{A}\mathbf{x} = \mathbf{b} \tag{11}$$

with

$$\mathbf{A} \equiv \left(\begin{array}{ccc|c} \mathbf{H}_1 & & & -\mathbf{\Sigma}_1 \\ & \ddots & & \vdots \\ & & \mathbf{H}_N & -\mathbf{\Sigma}_N \\ \hline -\frac{w_1}{4\pi}\mathbf{S} & \dots & -\frac{w_N}{4\pi}\mathbf{S} & \mathbf{I} \end{array}\right), \quad \mathbf{x} \equiv \begin{pmatrix} \mathbf{\Psi}_1 \\ \vdots \\ \mathbf{\Psi}_N \\ \hline \mathbf{\Phi} \end{pmatrix}, \quad \text{and} \quad \mathbf{b} \equiv \begin{pmatrix} \mathbf{Q}_1 \\ \vdots \\ \mathbf{Q}_N \\ \hline \mathbf{0} \end{pmatrix}.$$

4 Preconditioning Techniques

In computational radiative transfer the desired quantity is usually the mean intensity $\mathbf{\Phi}$ which is a function only of position $\mathbf{x}$. Therefore we use the Gaussian elimination to eliminate the intensity $\mathbf{\Psi}_1, \dots, \mathbf{\Psi}_N$ from (10) and the reduced equation

$$\left(\mathbf{I} - \frac{1}{4\pi}\sum_{l=1}^{N} w_l \mathbf{S}\mathbf{H}_l^{-1}\mathbf{\Sigma}_l\right)\mathbf{\Phi} = \frac{1}{4\pi}\sum_{l=1}^{N} w_l \mathbf{S}\mathbf{H}_l^{-1}\mathbf{Q}_l, \tag{12}$$

is solved for $\mathbf{\Phi}$. We rewrite (12) in compact form as

$$\left(\mathbf{I} - \mathcal{A}\right)\mathbf{\Phi} = \mathbf{f}, \tag{13}$$

where the Schur matrix $\mathcal{A}$ and the right hand side $\mathbf{f}$ are given by

$$\mathcal{A} = \frac{1}{4\pi}\sum_{l=1}^{N} w_l \mathbf{S}\mathbf{H}_l^{-1}\mathbf{\Sigma}_l \quad \text{and} \quad \mathbf{f} = \frac{1}{4\pi}\sum_{l=1}^{N} w_l \mathbf{S}\mathbf{H}_l^{-1}\mathbf{Q}_l.$$

In this section we briefly discuss some numerical methods used in the literature to solve the linear system (13).

4.1 Source Iteration

The most popular iterative method to solve (12) is the Richardson iteration known in the radiative transfer community as Source Iteration (SI) method. Given an initial guess $\mathbf{\Phi}^{(0)}$, the $(m+1)$-iterate solution is obtained by

$$\mathbf{\Phi}^{(m+1)} = \frac{1}{4\pi}\sum_{l=1}^{N} w_l \mathbf{S}\mathbf{H}_l^{-1}\left(\mathbf{Q}_l + \mathbf{\Sigma}_l\mathbf{\Phi}^{(m)}\right), \quad m = 0, 1, \dots. \tag{14}$$

It is easy to see that iteration (14) is equivalent to preconditioned block Gauss-Seidel method applied to (11), where the preconditioner is the block lower triangle of the matrix $\mathbf{A}$. Thus, if $\mathbf{M}$ is the block lower triangle of $\mathbf{A}$,

$$\mathbf{M} \equiv \begin{pmatrix} \mathbf{H}_1 & & & \\ & \ddots & & \\ & & \mathbf{H}_N & \\ -\frac{w_1}{4\pi}\mathbf{S} & \dots & -\frac{w_N}{4\pi}\mathbf{S} & \mathbf{I} \end{pmatrix},$$

$\mathbf{x}$ and $\mathbf{b}$ are as given in (11), then

$$\mathbf{M}\mathbf{x}^{(m+1)} = (\mathbf{M} - \mathbf{A})\mathbf{x}^{(m)} + \mathbf{b},$$

and

$$\mathbf{x}^{(m+1)} = (\mathbf{I} - \mathbf{M}^{-1}\mathbf{A})\mathbf{x}^{(m)} + \mathbf{M}^{-1}\mathbf{b}. \tag{15}$$

Therefore the $(m+1)$-iterate mean intensity satisfies

$$\mathbf{\Phi}^{(m+1)} = \frac{1}{4\pi}\sum_{l=1}^{N} w_l \mathbf{S}\mathbf{H}_l^{-1}\mathbf{\Psi}_l^{(m+1)} = \frac{1}{4\pi}\sum_{l=1}^{N} w_l \mathbf{S}\mathbf{H}_l^{-1}\Big(\mathbf{Q}_l + \mathbf{\Sigma}_l\mathbf{\Phi}^{(m)}\Big),$$

which is identical to (14).

Formal results from linear algebra [11, 9] demonstrate that the preconditioned Richardson iteration (15) converges rapidly as long as the norm of the matrix $(\mathbf{I}-\mathbf{M}^{-1}\mathbf{A})$ is small. This condition is ensured by taking the scattering ratio γ small, compare [1] for analysis. For $\gamma \ll 1$ the SI method converges rapidly, but for $\gamma \approx 1$ (large optical opacity) convergence becomes slow and may restrict the efficiency of the SI algorithm. The SI algorithm can be implemented as follows

`Algorithm 1:` SI algorithm

given the initial guess $\mathbf{\Phi}^{(0)}$

do $m = 0, 1, \dots, itmax$

do $l = 1, 2, \dots, N$

a) set $\mathbf{w}_l \longleftarrow \mathbf{Q}_l + \mathbf{\Sigma}_l\mathbf{\Phi}^{(m)}$

b) solve for $\mathbf{y}_l$: $\quad \mathbf{H}_l\mathbf{y}_l = \mathbf{w}_l$

c) set $\mathbf{w}_l \longleftarrow \mathbf{S}\mathbf{y}_l$

end do

d) compute $\mathbf{\Phi}^{(m+1)} = \dfrac{1}{4\pi}\displaystyle\sum_{l=1}^{N} w_l\mathbf{w}_l$

e) compute $\mathbf{r}^{(m)} = \mathbf{\Phi}^{(m+1)} - \mathbf{\Phi}^{(m)}$

if $\left(\dfrac{\|\mathbf{r}^{(m)}\|_{L^2}}{\|\mathbf{r}^{(0)}\|_{L^2}} \leq \tau\right)$ stop

end do

Here $itmax$ is the maximum number of the iterations m, τ is a given tolerance and $\|.\|_{L^2}$ is the discrete L^2-norm. The step (b) can be solved directly using Gaussian elimination known in computational radiative transfer as sweeping procedure. Additionally, for each direction ω_l in S_N only one of the eight possible sweeps is needed.

4.2 Diffusion Synthetic Acceleration

Among the methods used to accelerate the SI algorithm are the synthetic acceleration procedures [1, 4]. The procedures consist on splitting the SI in to two-step iterations. Thus, we denote $\psi^{(m+\frac{1}{2})}$ as first iteration for the SI in the continuous form of the problem (2),

$$\begin{aligned} \omega \cdot \nabla \psi^{(m+\frac{1}{2})} + (\sigma+\kappa)\psi^{(m+\frac{1}{2})} &= \frac{\sigma}{4\pi}\int_{S^2} \psi^{(m)}(\mathbf{x},\omega')d\omega' + \kappa q(\mathbf{x}), \\ \psi^{(m+\frac{1}{2})}(\hat{\mathbf{x}},\omega) &= g(\hat{\mathbf{x}}), \end{aligned} \tag{16}$$

and an equation for $\psi^{(m+1)}$ is required in such a way to be more accurate approximation to ψ than $\psi^{(m+\frac{1}{2})}$. To perform this step with synthetic acceleration method, we subtract (16) from (2),

$$\begin{aligned} \omega \cdot \nabla \big(\psi - \psi^{(m+\frac{1}{2})}\big) + (\sigma+\kappa)\big(\psi - \psi^{(m+\frac{1}{2})}\big) &= \frac{\sigma}{4\pi}\int_{S^2} \big(\psi - \psi^{(m)}\big)(\mathbf{x},\omega')d\omega', \\ \big(\psi - \psi^{(m+\frac{1}{2})}\big)(\hat{\mathbf{x}},\omega) &= 0, \end{aligned} \tag{17}$$

then equations (17) are replaced by an approximate problem. The Diffusion Synthetic Acceleration (DSA) method [4] approximates the equation (17) by the diffusion problem

$$\begin{aligned} -\nabla \cdot \Big(\frac{1}{3(\sigma+\kappa)}\nabla\varphi\Big) + \kappa\varphi &= \frac{\sigma}{4\pi}\int_{S^2} \big(\psi - \psi^{(m)}\big)(\mathbf{x},\omega')d\omega', \quad \mathbf{x} \in \Omega, \\ \varphi + \frac{2}{3(\sigma+\kappa)}\mathbf{n}(\hat{\mathbf{x}})\cdot\nabla\varphi &= 0, \qquad \hat{\mathbf{x}} \in \Gamma. \end{aligned} \tag{18}$$

Here $\varphi(\mathbf{x})$ is an approximation to the mean intensity

$$\varphi(\mathbf{x}) \approx \frac{1}{4\pi}\int_{S^2} \big(\psi^{(m+1)} - \psi^{(m+\frac{1}{2})}\big)(\mathbf{x},\omega')d\omega'.$$

Thus the $(m+1)$-iterate mean intensity is given by

$$\varphi^{(m+1)} = \varphi^{(m+\frac{1}{2})} + \varphi.$$

Note that equation (18) does not depend on the angle variable ω, is linear elliptic equation and simple to solve numerically with less computational cost and memory requirement.

In order to build a discretization for the diffusion problem (18) which is consistent to the one used for the radiative transfer equation (2), we consider the same grid structure and the same notations as those used in section 2. Hence a space discretization for the equation (18) reads as

$$-\mathcal{D}_h^2\left(\frac{1}{3(\sigma+\kappa)}\varphi\right)_{ijk}+\kappa_{i-\frac{1}{2}j-\frac{1}{2}k-\frac{1}{2}}\varphi_{i-\frac{1}{2}j-\frac{1}{2}k-\frac{1}{2}}=p_{i-\frac{1}{2}j-\frac{1}{2}k-\frac{1}{2}}, \quad (19)$$

where $p_{i-\frac{1}{2}j-\frac{1}{2}k-\frac{1}{2}}=\sigma_{i-\frac{1}{2}j-\frac{1}{2}k-\frac{1}{2}}\big(\varphi^{(m+\frac{1}{2})}_{i-\frac{1}{2}j-\frac{1}{2}k-\frac{1}{2}}-\varphi^{(m)}_{i-\frac{1}{2}j-\frac{1}{2}k-\frac{1}{2}}\big)$ and the difference operator $\mathcal{D}_h^2$ is given by $\mathcal{D}_h^2=\mathcal{D}_x^2+\mathcal{D}_y^2+\mathcal{D}_z^2$, with

$$\mathcal{D}_x^2(\beta\varphi)_{ijk}=\frac{\beta_{ijk}+\beta_{i+1jk}}{2}\frac{\varphi_{i+1jk}-\varphi_{ijk}}{(\Delta x)^2}-\frac{\beta_{i-1jk}+\beta_{ijk}}{2}\frac{\varphi_{ijk}-\varphi_{i-1jk}}{(\Delta x)^2},$$
$$\mathcal{D}_y^2(\beta\varphi)_{ijk}=\frac{\beta_{ijk}+\beta_{ij+1k}}{2}\frac{\varphi_{ij+1k}-\varphi_{ijk}}{(\Delta y)^2}-\frac{\beta_{ij-1k}+\beta_{ijk}}{2}\frac{\varphi_{ijk}-\varphi_{ij-1k}}{(\Delta y)^2},$$
$$\mathcal{D}_z^2(\beta\varphi)_{ijk}=\frac{\beta_{ijk}+\beta_{ijk+1}}{2}\frac{\varphi_{ijk+1}-\varphi_{ijk}}{(\Delta z)^2}-\frac{\beta_{ijk-1}+\beta_{ijk}}{2}\frac{\varphi_{ijk}-\varphi_{ijk-1}}{(\Delta z)^2}.$$

The functions $\kappa_{i-\frac{1}{2}j-\frac{1}{2}k-\frac{1}{2}}$, $\varphi_{i-\frac{1}{2}j-\frac{1}{2}k-\frac{1}{2}}$ and $p_{i-\frac{1}{2}j-\frac{1}{2}k-\frac{1}{2}}$ appeared in (19) are given as in formula (9). The gradient in the boundary conditions is approximated by upwinding without using ghost points. For example, on the boundary surface $x=a_x$ of the domain Ω, the boundary discretization is

$$\varphi_{\frac{1}{2}j+\frac{1}{2}k+\frac{1}{2}}-\frac{2}{3(\sigma_{\frac{1}{2}j-\frac{1}{2}k-\frac{1}{2}}+\kappa_{\frac{1}{2}j-\frac{1}{2}k-\frac{1}{2}})}\frac{\varphi_{\frac{3}{2}j-\frac{1}{2}k-\frac{1}{2}}-\varphi_{\frac{1}{2}j-\frac{1}{2}k-\frac{1}{2}}}{\Delta x}=0,$$

and similar work has to be done for the other boundaries. All together, the above discretization leads to a linear system of form

$$\mathbf{D}\varphi=\mathbf{p}, \quad (20)$$

where $\mathbf{D}$ is $IJK\times IJK$ nonsymmetric positive definite matrix obtained from the difference diffusion operator (19) with boundary conditions included, and $\mathbf{p}$ is IJK vector containing the right hand side term.

Once again the DSA method can be viewed as preconditioned Richardson iteration for the linear system (11) with the diffusion matrix $\mathbf{D}$ like preconditioner,

$$\mathbf{x}^{(m+1)}=\big(\mathbf{I}-\mathbf{D}^{-1}\mathbf{A}\big)\mathbf{x}^{(m)}+\mathbf{D}^{-1}\mathbf{b},$$

and $\mathbf{D}^{-1}$ is obtained by solving the diffusion linear system (20). In terms of $\mathbf{\Phi}$ this is equivalent to

$$\mathbf{\Phi}^{(k+1)}=\Big(\mathbf{I}-(\mathbf{I}-\mathbf{D}^{-1})\mathcal{A}\Big)\mathbf{\Phi}^{(k)}+(\mathbf{I}-\mathbf{D}^{-1})\mathbf{b}.$$

The implementation of DSA method to approximate the solution of the radiative transfer equation (2) is carried out in the following algorithm

`Algorithm 2:` DSA algorithm

given the initial guess $\mathbf{\Phi}^{(0)}$

do $m=0,1,\ldots,itmax$

(a) – (d) are similar to `Algorithm 1` for the intermediate solution $\mathbf{\Phi}^{(m+\frac{1}{2})}$

(e) compute $\mathbf{p} = \sigma\big(\mathbf{\Phi}^{(m+\frac{1}{2})} - \mathbf{\Phi}^{(m)}\big)$

(f) solve for φ: $\mathbf{D}\varphi = \mathbf{p}$

(g) set $\mathbf{\Phi}^{(m+1)} = \mathbf{\Phi}^{(m+\frac{1}{2})} + \varphi$

(h) compute $\mathbf{r}^{(m)} = \mathbf{\Phi}^{(m+1)} - \mathbf{\Phi}^{(m)}$

if $\left(\dfrac{\|\mathbf{r}^{(m)}\|_{L^2}}{\|\mathbf{r}^{(0)}\|_{L^2}} \leq \tau\right)$ stop

end do

Note that the first lines in `Algorithm 2` are similar to the `Algorithm 1`. However, the source iteration algorithm gives only the intermediate solution $\mathbf{\Phi}^{(m+\frac{1}{2})}$ which has to be corrected by adding the solution φ obtained by the diffusion approach. Furthermore, if iterative methods are used for the diffusion approach, then an inner iteration loop has to be added to the iteration used by the SI algorithm and an outer SI iteration may require less accuracy from the inner iterations.

4.3 Krylov Subspace Methods

In general the matrices $\mathbf{A}$ and $\mathcal{A}$ in (11) and (13) respectively are nonsymmetric and not diagonally dominant. Furthermore, since σ and κ are nonnegative functions and S_N has nonzero directions, the matrix $\mathbf{A}$ has positive diagonal elements and nonpositive off-diagonal elements. In addition, if $e_{l,i-\frac{1}{2}j-\frac{1}{2}k-\frac{1}{2}} \leq 0$, for all l, i, j, k, then the matrix $\mathbf{A}$ is weakly diagonally dominant. This condition is equivalent to

$$h = \max\big(\Delta x, \Delta y, \Delta z\big) \leq \max_{ijk}\Bigg(\frac{8|\mu_l|}{\sigma_{i-\frac{1}{2}j-\frac{1}{2}k-\frac{1}{2}} + \kappa_{i-\frac{1}{2}j-\frac{1}{2}k-\frac{1}{2}}}, \frac{8|\eta_l|}{\sigma_{i-\frac{1}{2}j-\frac{1}{2}k-\frac{1}{2}} + \kappa_{i-\frac{1}{2}j-\frac{1}{2}k-\frac{1}{2}}}, \frac{8|\xi_l|}{\sigma_{i-\frac{1}{2}j-\frac{1}{2}k-\frac{1}{2}} + \kappa_{i-\frac{1}{2}j-\frac{1}{2}k-\frac{1}{2}}}\Bigg), \quad \forall\, l, \qquad (21)$$

which means physically that the cell size is no more than eight mean free paths of the particles being simulated. Needless to say that the condition (21) gives the bound of the coarser mesh should be used in the computations.

In this paper we propose two Krylov subspace based methods, namely the BI-Conjugate Gradient Stabilized (Bicgstab) [20] and the Generalized Minimal Residual (Gmres(m)) [16], where m stands for the number of restarts for Krylov subspace used in the orthogonalization. Bicgstab method has been applied early in [19] to solve (13). The main idea behind these approaches is that the Krylov subspace methods can be interpreted as weighted Richardson iteration

$$\mathbf{x}^{(m)} = \left(\mathbf{I} - \alpha\mathbf{P}^{-1}\mathbf{A}\right)\mathbf{x}^{(m-1)} + \alpha\mathbf{P}^{-1}\mathbf{b}, \quad 0 < \alpha < 2, \quad m = 1, \dots, \tag{22}$$

applied to the linear system (11), where the relaxation parameters α and the preconditioner $\mathbf{P}$ are variables within each iteration step. Note that when $\alpha = 1$ and $\mathbf{P} = \mathbf{M}$ the iteration (22) is reduced to the SI method.

The Bicgstab and Gmres(m) algorithms to solve the linear system (13) can be implemented in the conventional way as in [20, 16, 12], with the only difference that the dense matrix $\mathcal{A}$ can not be explicitly stored. All what is needed, however, is a subroutine that performs a matrix-vector multiplication as shown in the following algorithm

`Algorithm 3:` Matrix-vector multiplication

given a vector $\mathbf{u}$, to apply the matrix $\mathcal{A}$ to $\mathbf{u}$ we proceed as

do $l = 1, \dots, N$

a) set $\mathbf{v}_l \longleftarrow \mathbf{\Sigma}_l \mathbf{u}$

b) solve for $\mathbf{w}_l$: $\quad \mathbf{H}_l \mathbf{w}_l = \mathbf{v}_l$

c) set $\mathbf{v}_l \longleftarrow \mathbf{S}\mathbf{w}_l$

end do

d) set $\mathbf{u} \longleftarrow \mathbf{u} - \frac{1}{4\pi}\sum_{l=1}^{N} w_l \mathbf{v}_l$

Note that only three vectors ($\mathbf{u}$, $\mathbf{v}_l$ and $\mathbf{w}_l$) are needed to perform the multiplication of the matrix $\mathcal{A}$ to the vector $\mathbf{u}$. Moreover, only three calls for the algorithm 3 are required from the Bicgstab or Gmres(m) subroutines.

Preconditioned Bicgstab or Gmres(m) methods can be also used. For instance, in the case when the matrix $\mathcal{A}$ is diagonally dominant, the Bicgstab or Gmres(m) methods can be accelerated by using the diagonal as a preconditioner. This approach which requires additional computational work can be easily implemented. It is worth to say that incomplete Cholesky or ILU type preconditioners can not be used to solve (13) because the matrix $\mathcal{A}$ is never formed explicitly.

5 A Fast Multilevel Preconditioner

We describe in this section multilevel solvers for the linear system (13) using an approximate inverse operator as preconditioner on each level of the multigrid hierarchy. Early application of multilevel methods to radiative transfer problems goes to [12]. The author proposes two different type of smoothings to approximate solutions for the one dimensional version of (2) in slab geometry.

To formulate multilevel solvers we first modify our notation slightly. Using the discretizations introduced in section 2 we assume for simplicity, a given sequence of uniform, equidistant nested grids

$$\Omega_1 \subset \Omega_2 \subset \cdots \subset \Omega_{L-1} \subset \Omega_L = \Omega_h,$$

on Ω with respective mesh sizes $\Delta x = \Delta y = \Delta z = 2^{-l}$, $l = 1, \ldots, L$. We use the subscripts l and L to refer to the coarse and fine level respectively. Therefore the problem statement (13) becomes

$$\left(\mathbf{I} - \mathcal{A}_L\right)\mathbf{\Phi}_L = \mathbf{f}_L. \tag{23}$$

With $\mathbf{M}$ being the iteration matrix, multilevel procedure to solve (23) can be written as follows

$$\mathbf{M}\mathbf{\Phi}_L^{(m+1)} + \left(\mathbf{I} - \mathcal{A}_L - \mathbf{M}\right)\mathbf{\Phi}_L^{(m)} = \mathbf{f}_L.$$

This formulation is equivalent to,

$$\mathbf{\Phi}_L^{(m+1)} = \mathbf{\Phi}_L^{(m)} + \mathbf{M}^{-1}\left(\mathbf{f}_L - (\mathbf{I} - \mathcal{A}_L)\mathbf{\Phi}_L^{(m)}\right) = \mathbf{\Phi}_L^{(m)} + \mathbf{M}^{-1}\mathbf{r}^{(m)},$$

where $\mathbf{r}$ denotes the residual associated to (23) and is defined by

$$\mathbf{r} = \mathbf{f}_L - (\mathbf{I} - \mathcal{A}_L)\mathbf{\Phi}_L.$$

The preconditioner we consider in this section is the Atkinson-Brakhage approximate inverse [5] given as

$$\mathbf{M}^{-1} = \mathbf{B}_l^L = \mathbf{I} + (\mathbf{I} - \mathcal{A}_l)^{-1}\mathcal{A}_L. \tag{24}$$

Then the $(m+1)$-iterate solution for (23) is simply

$$\mathbf{\Phi}_L^{(m+1)} = \mathbf{\Phi}_L^{(m)} + \mathbf{B}_l^L\mathbf{r}^{(m)}.$$

Analysis of convergence for this kind of multilevel methods has been done in [12]. The central ideas in this analysis are the strong convergence and collective compactness of the operators generated by $\mathcal{A}_l$.

Let us first define the two-level (Grid2) algorithm. Applying the multilevel preconditioner (24) to the problem (23) we need the fine-to-coarse grid transfer operator $\mathcal{R}_L^l$ defined by

$$\mathcal{R}_L^l\mathbf{\Phi}_L = \mathbf{\Phi}_l,$$

and the coarse-to-fine grid transfer operator $\mathcal{P}_l^L$ defined by

$$\mathcal{P}_l^L\mathbf{\Phi}_l = \mathbf{\Phi}_L.$$

A natural way to choose these operators is, bilinear interpolation for $\mathcal{P}_l^L$ and simple injection for $\mathcal{R}_L^l$ as in the standard multigrid literature [10]. However, for radiative transfer equation with discontinuous variables these operators have to be changed to those given in [3] which are specially designed for problems with jumping coefficients. Note that to use these operators we require that any discontinuity of κ, σ or $\mathbf{q}$ in (2) is a spatial mesh point.

The Grid2 algorithm for solving (23) is detailed in the following steps

`Algorithm 4:` Grid2 algorithm

given the fine level $\{L, \mathcal{A}_L, \mathbf{f}_L\}$, the coarse level $\{l, \mathcal{A}_l, \mathbf{f}_l\}$, the initial guess $\mathbf{\Phi}_L^{(0)}$ and the tolerance τ

do $m = 0, 1, \ldots, itmax$

a) compute the residual $\mathbf{r}^{(m)} = \mathbf{f}_L - (\mathbf{I} - \mathcal{A}_L)\mathbf{\Phi}_L^{(m)}$

b) set $\mathbf{u}_L \longleftarrow \mathcal{A}_L \mathbf{r}^{(m)}$

c) restriction $\mathbf{u}_l \longleftarrow \mathcal{R}_L^l \mathbf{u}_L$

d) solve on the coarse level for $\mathbf{v}_l$: $(\mathbf{I} - \mathcal{A}_l)\mathbf{v}_l = \mathbf{u}_l$

Nyström interpolation

e) compute $\mathbf{w}_l \longleftarrow \mathcal{A}_l \mathbf{v}_l$

f) prolongation $\mathbf{w}_L \longleftarrow \mathcal{P}_l^L \mathbf{w}_l$

g) set $\mathbf{z}_L \longleftarrow \mathbf{u}_L + \mathbf{w}_L$

h) compute the preconditioner $\mathbf{p} \longleftarrow \mathbf{r} + \mathbf{z}_L$

i) update the solution $\mathbf{\Phi}_L^{(m+1)} = \mathbf{\Phi}_L^{(m)} + \mathbf{p}$

j) compute the residual $\mathbf{r}^{(m+1)} = \mathbf{f}_L - (\mathbf{I} - \mathcal{A}_L)\mathbf{\Phi}_L^{(m+1)}$

if $\left(\dfrac{\|\mathbf{r}^{(m+1)}\|_{L^2}}{\|\mathbf{r}^{(0)}\|_{L^2}} \leq \tau \right)$ stop

end do

The step (d) usually solves the coarse problem exaclty using direct methods. However, since $\mathcal{A}_l$ is dense matrix which is never explicitly computed nor stored, iterative solvers are required to perfom the step (d). In our numerical examples we used Gmres(m) method which has been discussed in section 4. Note that in our context the notation Grid2 for two-level algorithm does not necessarily mean that we consider two levels of mesh refinements. Thus, `Algorithm 4` is also applicable in cases where we have two different space discretizations on the same mesh ($L \neq l + 1$).

The fully multilevel algorithm (Grinest) or nested iteration as refered to in [10] can be implemented recursively as follows

`Algorithm 5:` Gridnest algorithm

given the finest level $\{l_{\max}, \mathcal{A}_{l_{\max}}, \mathbf{f}_{l_{\max}}\}$, the coarsest level $\{l_{\min}, \mathcal{A}_{l_{\min}}, \mathbf{f}_{l_{\min}}\}$ and the tolerances $\{\tau_l\}$, $l = l_{\min}, \dots, l_{\max}$

a) Solve on the coarsest level for $\mathbf{\Phi}_{l_{\min}}$: $\quad (\mathbf{I} - \mathcal{A}_{l_{\min}})\mathbf{\Phi}_{l_{\min}} = \mathbf{f}_{l_{\min}}$

do $k = l_{\min} + 1, \dots, l_{\max}$

b) set $\quad l \longleftarrow k - 1$

c) set $\quad L \longleftarrow k$

d) compute the right hand side $\mathbf{f}_L$

e) set $\quad \mathbf{\Phi}_L^{(0)} \longleftarrow \mathcal{P}_l^L \mathbf{\Phi}_l$

f) set $\quad \tau \longleftarrow \tau_l$

g) call Grid2 to solve for $\mathbf{\Phi}_L$: $\quad (\mathbf{I} - \mathcal{A}_L)\mathbf{\Phi}_L = \mathbf{f}_L$

end do

Some comments are in order. The steps (b)-(f) are needed only to set the inputs, fine level $\{L, \mathcal{A}_L, \mathbf{f}_L\}$, coarse level $\{l, \mathcal{A}_l, \mathbf{f}_l\}$, initial guess $\mathbf{\Phi}_L^{(0)}$ and tolerance τ to the algorithm Grid2. Recall that Gridnest uses coarse levels to obtain improved initial guesses for fine level problems. The tolerance parameters $\{\tau_l\}$, which determine how many iterations of the multilevel algorithm to do on each level, can be considered either fixed or adaptively chosen during the course of computation.

6 Numerical Results

The methods discussed in the above sections are tested on a PC with AMD-K6 200 processors using Fortran compiler. In all these methods the iterations are terminated when

$$\frac{\|\mathbf{r}^{(m)}\|_{L^2}}{\|\mathbf{r}^{(0)}\|_{L^2}} \leq 10^{-5}. \tag{25}$$

To solve the diffusion problem in DSA we used a preconditioned Bicgstab whereas, Gmres(10) is used in Grid2 and Gridnest to solve the coarse problems. These inner iterations are stopped as in (25) but with 10^{-2} instead of 10^{-5}. In all our computations for the two space dimension cases we used a discrete S_N-direction set with 60 directions from [8].

6.1 Radiative Transfer Equation in 1D Slab Geometry

Our first example is the equation (2) in 1D slab geometry

$$\mu\frac{\partial\psi}{\partial x} + (\sigma+\kappa)\psi = \frac{\sigma}{2}\int_{-1}^{1}\psi(x,\mu')d\mu' + \kappa q(x), \quad (x,\mu)\in[0,1]\times[-1,1],$$
$$\psi(0,\mu)=1, \quad \mu>0, \quad \text{and} \quad \psi(1,\mu)=0, \quad \mu<0. \tag{26}$$

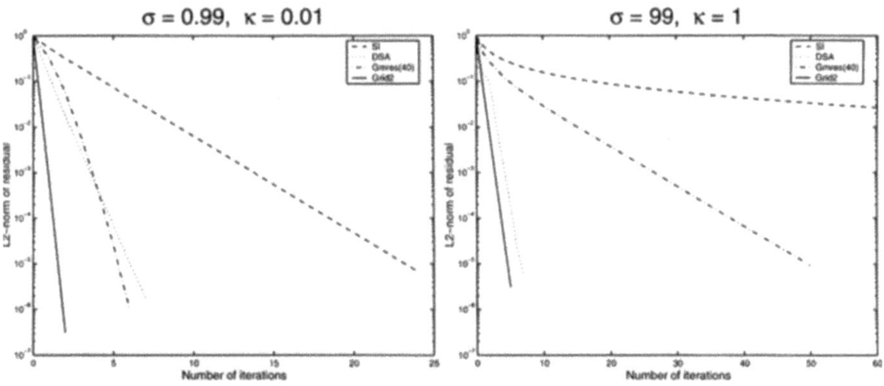

Fig. 1. Convergence plots for the radiative transfer equation in 1D slab geometry.

We set $q = 0$, we used 64 Gauss quadrature nodes in the discrete ordinates and a fine mesh of 512 gridpoints in the space dicrestization. The convergence results for two different values of σ and κ are showed in figure 1. The fast convergence of Grid2 is well demonstrated in both cases. Although the scattering ratio for the two cases is 0.99, Grid2 method shows strong reduction of number of iterations comparing to the other methods. Same observation is shown when the regime is optically thick ($\sigma = 99$, $\kappa = 1$).

Next we want to compare the efficiency of these methods in terms of CPU time and number of iterations when the scattering ratio γ runs in the range $(0, 1)$. To this end we set $\sigma = 10$ and we vary κ keeping the fine gridpoints fixed to 512. In figure 2 we plot the scattering ratio versus the number of iterations in the left and versus the CPU time in the right. Grid2 and Gmres(40) preserve roughly the same amount of computational work (referring to number of iterations and CPU time) in the whole interval, while the SI and DSA become costly for values of γ near 1.

6.2 Radiative Transfer Equation with Thermal Source

The second example is the equation (2) in the unit square with a thermal source $q = B(T)$, with B is the grey Blank function $B = a_R T^4$ with a_R is the radiation constant ($a_R = 1.8\times10^{-8}$). We fix the temperature to have the linear

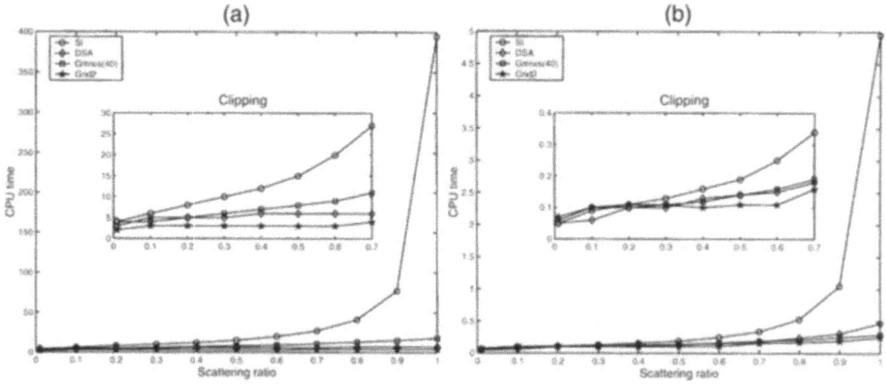

Fig. 2. Plots of scattering ratio γ versus the number of iterations in (a) and the CPU time (seconds) in (b). In both figures σ is fixed to 10.

profile $T(x, y) = 800x + 1000$ and the boundary function $g(\hat{x}, \hat{y}) = B(T(\hat{x}, \hat{y}))$. In figure 3 we show the convergence plots for different values of σ and κ using a fine mesh of 257×257 gridpoints. In all cases Grid2 algorithm presents faster convergence behaviour than DSA even in the diffusion regime ($\sigma = 100$ and $\kappa = 1$).

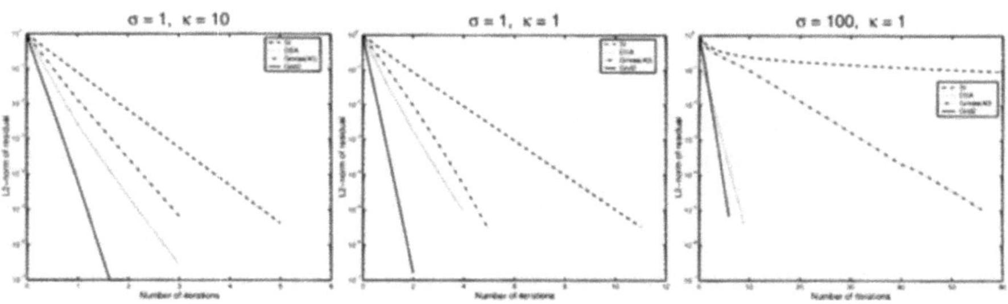

Fig. 3. Convergence plots for the radiative transfer equation with thermal source.

We summarize in table 1 the CPU time consumed for each method to perform the computations with the different values of σ and κ. In table 2 we report, the number of gridpoints $I \times J$ at each level, the iteration counter m for that level, the number of iterations in Gmres(10), i_{Gmres}, the L^∞-norm of the residual, $\|\mathbf{r}^{(m)}\|_{L^\infty}$, and the factor $\|\mathbf{r}^{(m)}\|_{L^\infty}/\|\mathbf{r}^{(m-1)}\|_{L^\infty}$.

The results presented in table 2 do not include those obtained for $\sigma = 1$, $\kappa = 1$ or $\sigma = 100$, $\kappa = 1$ because the number of iterations m in Grid2 is very large. For instance, for $\sigma = 100$, $\kappa = 1$ this number surpasses 15 on the coarsest mesh. Nevertheless, in all these test cases we have observed that there is very little variation in the number i_{Gmres} and the reduction factor $\|\mathbf{r}^{(m)}\|_{L^\infty}/\|\mathbf{r}^{(m-1)}\|_{L^\infty}$ as the meshes are refined.

Table 1. CPU time (in seconds)

	SI	DSA	Gmres(40)	Grid2
$\sigma = 1,\ \kappa = 10$	17.8	29.1	15.5	13.9
$\sigma = 1,\ \kappa = 1$	33.7	77.4	20.8	17.6
$\sigma = 100,\ \kappa = 1$	2389.8	82.3	174.4	119.3

Table 2. Results obtained by Gridnest for $\sigma = 1$ and $\kappa = 10$ at different levels

$I \times J$	m	i_{Gmres}	$\Vert\mathbf{r}^{(m)}\Vert_{L^\infty}$	$\Vert\mathbf{r}^{(m)}\Vert_{L^\infty}/\Vert\mathbf{r}^{(m-1)}\Vert_{L^\infty}$
33×33	0		0.73E-00	
	1	2	0.32E-01	0.43E-01
	2	2	0.12E-02	0.37E-01
	3	3	0.56E-05	0.46E-04
65×65	0		0.11E-01	
	1	2	0.33E-03	0.30E-01
	2	2	0.12E-05	0.36E-04
129×129	0		0.83E-02	
	1	2	0.67E-03	0.81E-01
	2	2	0.53E-05	0.79E-04
257×257	0		0.22E-02	
	1	2	0.73E-04	0.33E-01
	2	2	0.15E-06	0.20E-04

The results of these tables and figure 3 show various interesting features about the behaviours of the preconditioner used by each method. First, where the scattering ratio $\gamma = 0.99$ ($\sigma = 100$ and $\kappa = 1$), it is clear that the SI method is unacceptably slow to converge. The convergence rate is improved significantly by Gmres(40), and it is improved even more by DSA method but at the cost of extra work and storage. The most effective method for solving this example, however, is the multilevel Grid2 and Gridnest methods. Second, the number of iterations i_{Gmres} in Gridnset remain nearly the same in all levels and is bounded by the number in the coarsest level.

6.3 Radiative Transfer Equation with Discontinuous Variables

The aim of this example is to test the multilevel algorithm for radiative transfer problem with jumping coefficients. Thus, the problem statement is the equation (2) augmented by discontinuous scattering, absorption and source term [2]. The space domain geometry and the values of σ, κ and $\mathbf{s} = \kappa q$ for each subdomain are given in figure 4. We take in the first run of this example vacuum boundary conditions whereas, in the second run we use the nonhomogeneous boundary function g,

$$g(\hat{x},\hat{y}) = \begin{cases} 1 & \text{when} \quad \hat{x} = 0 \text{ and } 0 \leq \hat{y} \leq 1, \\ 1 & \text{when} \quad \hat{y} = 0 \text{ and } 0 < \hat{y} \leq 0.44, \\ 1 & \text{when} \quad \hat{y} = 1 \text{ and } 0 < \hat{y} \leq 0.44, \\ 0 & \text{otherwise.} \end{cases}$$

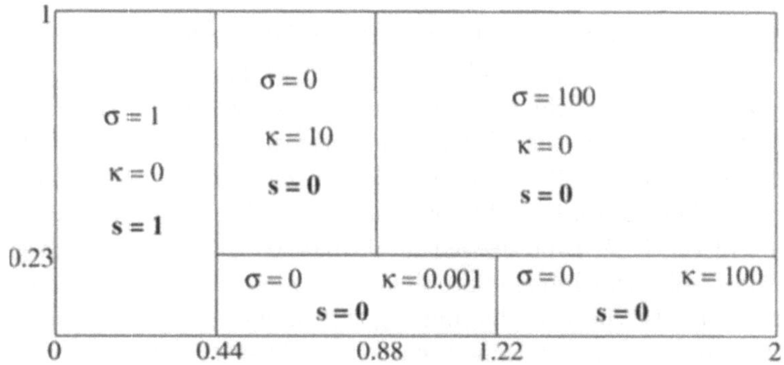

Fig. 4. Geometry and values of σ, κ, $\mathbf{s} = \kappa q$ for the discontinuous equation.

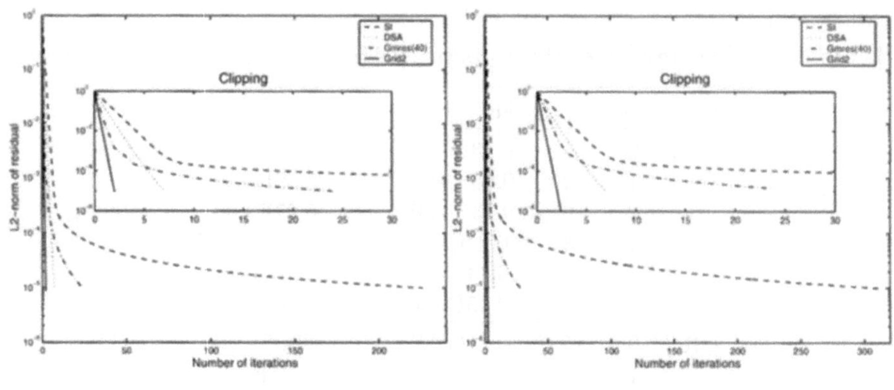

(a) (b)

Fig. 5. Convergence rates for the discontinuous equation subject to vacuum boundary conditions (a) and nonhomogeneous boundary conditions (b).

The space domain is discretized uniformly into 400×200 gridpoints at the finest level. We display in figure 5 the convergence rates for the two runs. Table 3 provides the running time used by each method for these computations. A simple inspection of figure 5 shows that Grid2 algorithm solves this problem more effectively than SI or Gmres(40) methods and with less iterations than DSA method.

Table 3. CPU time (in seconds)

	SI	DSA	Gmres(40)	Grid2
Vacuum boundary condition	794.04	304.42	192.25	171.04
Nonhomogeneous boundary condition	1100.18	397.53	110.07	105.31

Furthermore, we note that the Gmres(40) method performs poorly after the seventh iteration in both runs. This may be partly due to the fact that the discontinuous σ and κ coefficients change the matrix structure very badly. While, it is not surprising that the SI algorithm performs very poorly in this case. An examination of the CPU time in table 3 reveals that Gmres(40) consumes less computational work than the other methods. We have observed that the main part of the CPU time needed in DSA or Grid2 is used by Bicgstab or Gmres(10) to solve the diffusion problem in DSA or the coarse linear system in Grid2, respectively. However, by limiting the number of iterations in Bicgstab and Gmres(10) to 1, or increasing the tolerance from 10^{-2} to 10^{-1}, the results change favourably with significant advantage for Grid2.

Table 4 tabulates the results obtained by Gridnest using four levels. It can be clearly seen that there is very little variation in the number i_{Gmres} and the reduction factor $\|\mathbf{r}^{(m)}\|_{L^\infty}/\|\mathbf{r}^{(m-1)}\|_{L^\infty}$ as the meshes are refined.

Table 4. Gridnest results for the discontinuous problem with vacuum boundary

$I \times J$	m	i_{Gmres}	$\|\mathbf{r}^{(m)}\|_{L^\infty}$	$\|\mathbf{r}^{(m)}\|_{L^\infty}/\|\mathbf{r}^{(m-1)}\|_{L^\infty}$
100×50	0		0.13E+01	
	1	9	0.50E+00	0.39E+00
	2	9	0.13E+00	0.27E+00
	3	8	0.24E-01	0.19E+00
	4	7	0.14E-02	0.57E-01
	5	7	0.45E-05	0.32E-02
200×100	0		0.51E-01	
	1	8	0.10E-02	0.21E-01
	2	7	0.78E-05	0.73E-02
	3	7	0.32E-07	0.41E-02
400×200	0		0.79E-02	
	1	7	0.40E-04	0.51E-02
	2	7	0.13E-06	0.33E-02
800×400	0		0.38E-02	
	1	7	0.24E-04	0.64E-02
	2	7	0.94E-07	0.39E-02

6.4 Radiative Transfer Equation with Frequency Dependence

Our final example is the frequency-dependent problem

$$\begin{aligned}\omega \cdot \nabla \psi_\nu + (\sigma_\nu + \kappa_\nu)\psi_\nu &= \frac{\sigma_\nu}{4\pi}\int_{S^2} \psi_\nu(\mathbf{x}, \omega', \nu) d\omega' + \kappa_\nu B(T, \nu), \\ \psi_\nu(\hat{\mathbf{x}}) &= B(T(\hat{\mathbf{x}}), \nu), \qquad \hat{\mathbf{x}} \in \Gamma^-.\end{aligned} \tag{27}$$

Here $\psi_\nu = \psi(\mathbf{x}, \omega, \nu)$, $T = T(\mathbf{x})$, $\sigma_\nu = \sigma(\mathbf{x}, \nu)$ and $\kappa_\nu = \kappa(\mathbf{x}, \nu)$ denote respectively, the radiation intensity, the temperature, the scattering and the opacity within the frequency $\nu > 0$. B is the Planck function given by

$$B(T, \nu) = \frac{2\hbar\nu^3}{c^2}\left(e^{\hbar\nu/k_B T} - 1\right)^{-1},$$

where $\hbar$, k_B and c are Planck constant, Boltzmann constant and the speed of light, respectively. See [15] for the physical values of these constants.

In order to discretize the equations (27) respect to the frequency variable ν, we assume N_ν frequency bands $[\nu_\iota, \nu_{\iota+1}]$, $\iota = 1, \ldots, N_\nu$ with piecewise constant absorption,

$$\kappa_\nu = \kappa_\iota, \qquad \forall\, \nu \in [\nu_\iota, \nu_{\iota+1}] \quad \iota = 1, \ldots, N_\nu.$$

We define the frequency-averaged intensity in the band $[\nu_\iota, \nu_{\iota+1}]$ by

$$\psi_\iota = \int_{\nu_\iota}^{\nu_{\iota+1}} \psi_{\nu'}(\mathbf{x}, \omega, \nu') d\nu'.$$

Then, the equations (27) are transformed to a system of N_ν radiative transfer equations of the form

$$\begin{aligned}\omega \cdot \nabla \psi_\iota + (\sigma_\iota + \kappa_\iota)\psi_\iota &= \frac{\sigma_\iota}{4\pi}\int_{S^2} \psi_\iota(\mathbf{x}, \omega', \nu_\iota) d\omega' + \kappa_\iota \int_{\nu_\iota}^{\nu_{\iota+1}} B(T, \nu') d\nu', \\ \psi_\iota(\hat{\mathbf{x}}) &= \int_{\nu_\iota}^{\nu_{\iota+1}} B(T, \nu') d\nu', \qquad \hat{\mathbf{x}} \in \Gamma^-.\end{aligned} \tag{28}$$

To approximate the frequency integrals we used trapezoidal formula. In our numerical simulations we use eight frequency bands $[\nu_\iota, \nu_{\iota+1}]$, $\iota = 1, \ldots, 8$ from glass manufacturing [13]. These frequencies are reproduced in table 5.

We compute the solution of the equation (27) in a unit square, on the refined grid with 100×100 gridpoints, and 60 discrete ordinates. Hence the number of unknowns for each frequency band is 60×10^4, and these computations are done for the 8 frequencies such that the overall number of equations amounts 4.8×10^6. The scattering parameter σ is varying in the set $\{1, 10, 100\}$. For every frequency band we calculate it corresponding scattering ratio and, we store the number of iterations and the running time obtained by each method. The results are given in table 6.

Table 5. The eight frequency bands

Band ι	ν_ι	$\nu_{\iota+1}$	κ_ι
1	∞	5	0.40
2	5	0.3333	0.50
3	0.3333	0.2857	7.70
4	0.2857	0.2500	15.45
5	0.2500	0.2222	27.98
6	0.2222	0.1818	267.98
7	0.1818	0.1666	567.32
8	0.1666	0.1428	7136.06
	0.1428	0	opaque

All the algorithms iterate the solution in a large iteration numbers for the first frequency bands (with large scattering ratio), then these numbers go decreasing as the frequency bands grow until their reach the minimum for the last frequency band. In this case when the size of scattering ratio is changing dramatically over the frequency bands, only the DSA and Grid2 algorithms lead to satisfying results for all frequencies and also in diffusive limit ($\sigma = 100$). The superiority of Grid2 is clearly demonstrated in table 6.

7 Summary

In this paper we formulated and compared several numerical methods for the radiative transfer equation. We showed that all these methods can be interpreted as standard preconditioned iterations with different preconditioner operators. We have illustrated through a number of numerical examples the ability of these preconditioned algorithms to perform well in the presence of strong source terms and discontinuities. We have also compared the results of some radiative transfer problems obtained by SI and DSA methods, which are very often used in radiative transfer modelling, with that computed by multilevel Grid2 and Gridnest algorithms.

Parallel implementation of the methods presented in this paper is straight forward and only requires interprocessor communication to complete the matrix-vector and vector-vector products required at each iteration. The principal drawback of this implementation is that the number of sweeps needed to compute $\mathbf{H}^{-1}$ grows as the space dimension becomes larger.

Although we have restricted our numerical computations to the case of two-dimensional problems, the more important implications of our research concern the use of effective iterative methods for problems in three spatial dimensions coupled to convection-diffusion and hydrodynamic equations. We believe that for these problems, the multilevel preconditioners might be effec-

Table 6. Number of iterations and CPU time for the eight frequency-bands problem

	Band	Scattering ratio	SI	Gmres(40)	DSA	Grid2
	1	0.71428	13	6	6	3
	2	0.66666	13	6	6	3
	3	0.11494	7	5	4	3
	4	0.06077	6	4	4	3
$\sigma = 1$	5	0.03450	5	4	3	3
	6	0.00371	4	3	3	2
	7	0.00175	3	3	3	2
	8	0.00014	3	2	2	2
	CPU(sec)	——	30.45	26.36	39.38	17.89
	1	0.96153	117	17	8	5
	2	0.95238	114	17	8	5
	3	0.56497	21	8	5	4
	4	0.39292	14	7	5	3
$\sigma = 10$	5	0.26329	10	5	5	3
	6	0.03597	5	3	3	3
	7	0.01732	4	3	3	3
	8	0.00173	3	2	2	2
	CPU(sec)	——	108.30	36.61	35.13	21.33
	1	0.99601	1577	79	16	15
	2	0.99502	1637	79	17	14
	3	0.92850	151	22	15	10
	4	0.86617	80	16	14	9
$\sigma = 100$	5	0.78137	48	12	15	8
	6	0.27175	10	5	5	4
	7	0.14985	8	4	4	4
	8	0.01381	4	3	3	2
	CPU(sec)	——	1190.73	107.80	57.27	50.62

tive since they reduce the number of iterations needed for convergence and they are relatively inexpensive to implement. Our current effort is therefore to extend these methods for radiative heat transfer equations and radiation hydrodynamics in higher space dimensions.

Acknowledgements. The work presented here has been funded by the German research foundation, SFB 568.

References

1. Adams, M.L., Larsen, E.W.: Fast Iterative Methods for Deterministic Particle Transport Computations. Preprint
2. Adams, M.L.: Subcell Balance Formulations for Radiative Transfer on Arbitrary Grids. Transport Theory and Statistical Physics **26** (1997) 385–431

3. Alcouffe R., Brandt, A., Dendy, J., Painter, J.: The Multigrid Method for Diffusion Equations with Strongly Discontinuous Coefficients. SIAM J. Sci. Stat. Comp. **2** (1981) 430–454
4. Alcouffe R.: Diffusion Synthetic Acceleration Methods for the Diamond-Differenced Discrete-Ordinates Equations. Nucl, Sci, Eng. **64** (1977) 344–355
5. Atkinson K.E.: Iterative Variants of the Nyström Method for the Numerical Solution of Integral Equations. Numer. Math. **22** (1973) 17–31
6. Brown, P.N.: A Linear Algebraic Development of Diffusion Synthetic Acceleration for Three-Dimensional Transport Equations. SIAM. J. Numer. Anal. **32** (1995) 179–214
7. Chandrasekhar, S.: Radiative Transfer. Oxford Univ. Press, London (1950)
8. Fiveland, W.A.: The Selection of Discrete Ordinate Quadrature Sets for Anisotropic Scattering. ASME HTD. Fundam. Radiat. Heat Transfer **160**, (1991) 89–96
9. Greenbaum, A.: Iterative Methods for Solving Linear Systems. SIAM. Philadelphia, (1997)
10. Hackbusch, W.: Multi-Grid Methods and Applications. Springer Series in Computational Mathematics vol **4** Springer-Verlag, New York, (1985)
11. Kelly, C.T.: Iterative Methods for Linear and Nonlinear Equations. SIAM. Philadelphia, (1995)
12. Kelley, C.T.: Multilevel Source Iteration Accelerators for the Linear Transport Equation in Slab Geometry. Transp. Theory and Stat. Phy. **24**, (1995) 679–707
13. Larsen, E., Thömmes, G., Klar, A., Seaïd, M., Götz, Th.: Simplified P_N Approximations to the Equations of Radiative Heat Transfer and Applications. J. Comp. Phys., **183** (2002) 652–675
14. Lewis, E.E., Miller W.F.Jr.: Computational Methods of Neutron Transport. John Wiley & Sons, New York (1984)
15. Mihalas, D., Mihalas, B.W.: Foundations of Radiation Hydrodynamics. Oxford Univ. Press, New York (1984)
16. Saad Y., Schultz M.H.: GMRES: A Generalized Minimal Residual Algorithm for Solving Nonsymmetric Linear Systems. SIAM. J. Sci. Statist. Comput. **7** (1986) 856–869
17. Seaïd, M.: Notes on Numerical Methods for Two-Dimensional Neutron Transport Equation. Technical Report Nr. **2232**, TU Darmstadt (2002)
18. Turek, S.: An Efficient Solution Technique for the radiative Transfer Equation. IMPACT, Comput. Sci. Eng. **5** (1993) 201–214
19. Turek, S.: A Generalized Mean Intensity Approach for the Numerical Solution of the Radiative Transfer Equation. Computing **54** (1995) 27–38
20. Van der Vorst, H.A.: BI-CGSTAB: A Fast and Smoothly Converging Variant of BI-CG for the Solution of Nonsymmetric Linear Systems. SIAM. J. Sci. Statist. Comput. **13** (1992) 631–644

On Using Common Lisp for Scientific Computing

Nicolas Neuss

Interdisziplinäres Zentrum für wissenschaftliches Rechnen (IWR), Universität Heidelberg, Im Neuenheimer Feld 368, D-69120 Heidelberg, Germany

Summary. Lisp is a very flexible and powerful language, but up to now it has not been used intensively for applications in scientific computing. The main reason is the prejudice that Lisp is slow. While this prejudice may have been true in early stages of Lisp's history, it is not really true today. Furthermore, the virtues of Lisp are becoming more and more important. In this contribution, we support this point of view: first, by comparing the efficiency of BLAS routines written in C and Common Lisp and second, by discussing the recently developed toolbox FEMLISP for solving partial differential equations with finite element methods and multigrid.

1 Introduction

Lisp is the second-oldest high-level computer language after Fortran. It is a very flexible and powerful language, and many problems arising in the computational sciences were first solved using Lisp as a vehicle. Besides being well-known as *the* language for solving problems in artificial intelligence, Lisp was often used for prototyping, i.e. as the language in which an application was first implemented. For example, the world's first computer algebra system, MACSYMA, was implemented in Lisp. Lisp is also the language of the CAD system AUTOCAD and the extensible editor EMACS. In spite of these successes, Lisp has not yet become a mainstream language. The reasons are mostly historical: in the first decades of computer history, resources were scarce, and a language for which the first implementations were interpreters, which has automatic memory management as an essential component, and which is used best inside large development environments could not compete with lightweight languages that were more or less sufficient for solving the problems of small complexity which could be treated at that time.

Nowadays, the situation is drastically different. Computing power has increased tremendously, rendering it possible to have powerful Lisp environments on personal computers while still using only a fraction of the available memory and CPU power. Due to Java, automatic memory management has become a mainstream feature. In the meantime, Lisp itself has also grown up into *Common Lisp*, which is a powerful, object-oriented language, suited to real-world applications, with many implementations that support compilation to native code. Thus, using Lisp also for problems outside its original realm of artificial intelligence is a natural choice today. Indeed, this has been known for some time and was discussed at several points in literature, e.g. in [6], but, unfortunately, only with small impact outside the Lisp community. Yet today, there seems to be a growing interest in Lisp especially in the mathematical community, as the recent applications KENZO [13] or FEMLISP [12], as well as the flourishing of the free computer algebra system MAXIMA [9] show.

The structure of this article is as follows. In Section 2, we give an overview of Common Lisp. In Section 3, we study the floating-point efficiency for compiled Lisp code and, in Section 4, we briefly sketch the FEMLISP toolbox. We end with conclusions and acknowledgements.

2 Common Lisp

In this section, we want to give a brief overview of Common Lisp, which is one dialect of Lisp. Lisp itself was designed at the end of the 1950s and is thus the second-oldest high-level computer language after Fortran, which was formulated some years earlier in 1954. Thus, it had a long time for growing-up and changed considerably from its beginnings. During its first decades, it diversified into a large variety of dialects which were then unified during the 1980s. The result of this unification effort is Common Lisp, which became an ANSI Standard in 1994. Nowadays, it is the dominant dialect in the Lisp family of languages, though other dialects coexist, notably Scheme (which has a small language core and is adapted well to educational purposes), Emacs Lisp (the configuration language of the Emacs editor), and AUTOLISP (the configuration language of AUTOCAD).

In brief, Common Lisp has the following features:

1. It is an interactive language which is usually used in an integrated development environment. That is, compiler, debugger, and application all are parts of the same Lisp environment. The editor is tightly integrated as well (often being a part of the same environment). This allows expressions to be evaluated with one key stroke, as well as information and documentation of functions to be available when editing.
2. It has powerful commands for list handling. (This does not mean that lists are the only data structure. Instead, Common Lisp possesses a wealth of other built-in data structures, e.g. multi-dimensional arrays, hash-tables, or bit-vectors.)

3. It allows for *symbolic manipulation*: every token which is read is interned as an entity called "symbol" in name spaces called "packages". Thus, comparison of symbols is fast (equality of pointers), and composite data structures can be constructed efficiently.
4. Probably its most obvious characteristic is a consistent prefix syntax of the form `(operator argument1 ...)`, e.g. `(+ 3 5)`, `(sin x)` or `(if (< x 0) (- x) x)`.
5. Because of the three preceding points, Lisp programs have a tree representation which can be efficiently built and manipulated. Consequently, a very strong mechanism for syntactic transformations called "macros" becomes feasible. Most of Common Lisp basic operators are indeed macros. For example, the loop expressions from Table 2 are transformed into more elementary expressions involving conditionals and even jumps.
6. Lisp is dynamically typed: variables can contain data of arbitrary type, and the data themselves contain type information which is analysed at runtime.
7. Common Lisp possesses a complete numerical tower: there are short and arbitrarily long integers, rational numbers, real and complex floating-point numbers, together with generic arithmetic operating on them.
8. Lisp has automatic memory management, i.e. memory does not have to be freed manually, but is reclaimed automatically by garbage collection in case it is not referenced any more. Garbage collection has turned out to be a crucial ingredient for many advanced software techniques and is built-in in most computer languages designed in recent years.
9. Lisp supports "functional programming", i.e. every expression has a return value, and functions are first-class objects: they can be generated (and compiled) in a local environment and can appear as a value of a variable or return-value of a function.
10. Common Lisp is also a very strong object-oriented programming language. The *Common Lisp Object System* (CLOS) has multiple inheritance, multiple dispatch, class redefinition, introspection, and a meta-object protocol allowing for extension and modification.
11. Common Lisp supports low-level programming: C or Fortran code can usually be translated into the equivalent Common Lisp code.

3 Numerical efficiency of Common Lisp

In this section, we test Common Lisp for its numerical efficiency. We do this by considering two important operations from the *Basic Linear Algebra Subroutines* (BLAS), see [8], [5].

Most computations in numerical analysis and scientific computing have to perform at least one of the following operations:

Scalar product: Compute the *scalar product* or *dot product* between two vectors $x = (x_1, \ldots, x_n)^t$ and $y = (y_1, \ldots, y_n)^t$. The formula for this dot-operation is

$$\mathtt{dot(x,y)} = x \cdot y = \sum_{i=1}^{n} x_i y_i \,. \tag{1}$$

Matrix-vector multiplication: Compute $b = Ax$ for

$$A = \begin{pmatrix} a_{11} & \ldots & a_{1n} \\ \vdots & & \ldots \\ a_{m1} & \ldots & a_{mn} \end{pmatrix}, \; x = \begin{pmatrix} x_1 \\ \vdots \\ x_n \end{pmatrix}, \; b = \begin{pmatrix} b_1 \\ \vdots \\ b_m \end{pmatrix}. \tag{2}$$

The formula for this operation is

$$b_i = \sum_{i=1}^{n} A_{ij} x_j \,, \; i = 1, \ldots, m \,. \tag{3}$$

¿From this formula, we can see that the computation of b can be done in two different ways. The first possibility is to compute each component b_i as the dot product between x and the i-th row of A. Alternatively, one can sum up scalar multiples of columns of A. This latter operation is composed of elementary axpy-operations

$$y_i := y_i + a x_i \,, \; i = 1, \ldots, n \,. \tag{4}$$

Solving linear equations: Solve $Ax = b$ for x where

$$A = \begin{pmatrix} a_{11} & \ldots & a_{1n} \\ \vdots & & \ldots \\ a_{n1} & \ldots & a_{nn} \end{pmatrix}, \; x = \begin{pmatrix} x_1 \\ \vdots \\ x_n \end{pmatrix}, \; b = \begin{pmatrix} b_1 \\ \vdots \\ b_n \,. \end{pmatrix} \tag{5}$$

Often, this is done by transforming A into a product of a lower and an upper triangular matrix. Here, the elementary operation is again a call to axpy.

Both dot and axpy have an operation count of n additions and n multiplications or n *arithmetic operations* where an arithmetic operation consists of one addition and one multiplication.

An implementation of these routines in C for vectors of double-precision floating-point numbers is given in Table 1 while an equivalent implementation in Common Lisp is given in Table 2.[1] We see that both versions present the

[1] The C and Lisp code used in these tests can be downloaded from the author's web page [10]. The C code has been published in [11].
Note also that a translation of C or Fortran code to Common Lisp can be done automatically for a large part of the existing code base, see e.g. [6].

```
double ddot (double *x, double *y,
             int n)
{
  int j;
  double sum = 0.0;
  for (j=0; j<n; j++)
    sum += x[j]*y[j];
  return sum;
}
```

```
void daxpy (double a, double *x,
            double *y, int n)
{
  int j;
  for (j=0; j<n; j++)
     y[j] += a*x[j];
}
```

Table 1. C code for ddot and daxpy.

```
(defun ddot (x y n)
  (declare (type fixnum n)
           (type (simple-array double-float (*)) x y))
  (loop for j of-type fixnum below n
        summing (* (aref x j) (aref y j)) of-type double-float))

(defun daxpy (a x y n)
  (declare (type fixnum n) (type double-float a)
           (type (simple-array double-float (*)) x y))
  (loop for i of-type fixnum below n do
        (incf (aref y i) (* a (aref x i)))))
```

Table 2. Common Lisp code for ddot and daxpy.

same information to their compiler, such that both compilers should be able to produce optimal machine code. The Lisp version is more verbose, however. In fact, Lisp's syntax was not designed for putting type declarations everywhere, and Lisp programmers mostly write code without type declarations. Also in this case, this would make the above code shorter and more generally applicable, e.g. to complex-valued vectors. On the other hand, it would result in lower performance, because the compiler is not able to deduce this type information without help and has to produce code handling general numbers. Thus, the type declarations become necessary if such routines present a performance bottleneck for the respective application.

Now, we measure the speed with which those routines are performed either on a pair of short vectors ($n = 256$) or on a pair of long vectors ($n = 10^6$). The size of n makes a crucial difference because of the higher data locality in the first case. To see why, remember that the CPU can usually process data at a higher speed than it can be delivered from main memory by random access. As a remedy, the so-called *cache* consisting of fast memory chips is installed as a buffer between CPU and main memory. Now, in the case $n = 256$, both vectors fit in the so-called primary cache whereas, in the second case, they do not even fit in the so-called secondary cache (at least for the hardware which we used for our tests).

Table 3 shows the performance which we measured on two machines: a two-year-old laptop (Pentium 2, 400 MHz) and a new PC (Pentium 4, 2.4 GHz).

	Pentium 2 (400 MHz)				Pentium 4 (2.4 GHz)			
	$n = 256$		$n = 10^6$		$n = 256$		$n = 10^6$	
	`ddot`	`daxpy`	`ddot`	`daxpy`	`ddot`	`daxpy`	`ddot`	`daxpy`
C	172	117	45	24	910	1140	300	140
CMUCL	94	87	35	18	690	480	300	140

Table 3. Some BLAS performance results (MFLOPS).

We compiled the C code with the GNU C compiler (`gcc`) and `-O3` optimization. The Lisp code was compiled with the optimization settings

```
(declaim (optimize (debug 0) (safety 0) (space 0)
                   (compilation-speed 0) (speed 3))
```

using CMU Common Lisp (CMUCL) from [4].

As a result of these tests, we see that the machine code produced by `CMUCL` is upto 60% slower than the machine code produced by `gcc` if all data is inside the cache (but equally fast if memory bandwidth is the limiting factor). The reason is that the compiler back end of `gcc` is more elaborate, presumably because many more people have improved on it. Further, writing a good compiler for Common Lisp is a very demanding task, not because of the compiler back end, but because of obtaining good efficiency for advanced language features, for example CLOS. Thus, many commercial or free implementations put the emphasis there and accept a slightly reduced low-level performance.

Nevertheless, it is interesting that judicious use of the advanced language features of Common Lisp can also result in a *higher* performance than a pure C implementation can obtain. The key feature here is that Common Lisp source code can be generated and compiled to machine code at runtime, while conventional languages do not allow this in a portable fashion. This allows for the compilation of code adapted to some special runtime situation. This is the equivalent of doing a controlled just-in-time compilation of arbitrary complexity.

Examples where this can be used abound. For example, in interactive programs for solving partial differential equations (see Section 4), coefficient functions can be read and compiled to native code. This presents a large problem for other approaches, because interpreting those expressions is complicated and may be too slow at runtime. Another example is the application of stencils to functions defined on structured grids, which arises for filters applied in image-processing, wavelet transformations, and finite-difference discretizations. We elaborate this in the following.
Two versions of C code for applying a stencil to a function defined on a structured two-dimensional grid of dimension $n \times n$ are shown in Table 4. In the second version, the dimensions of the grid as well as the stencil are assumed to be known (the stencil arises from discretizing the Laplace equation with

```
/* dynamic version: n, stencil_size, stencil_values unknown
   at compile time */
for (pos1=n; pos1<(n-1)*n; pos1+=n)
  for (pos2=pos1+1; pos2<pos1+n-1; pos2++)
  {
      register double s = 0.0;
        for (k=0; k<stencil_size; k++)
          s += stencil_values[k]*u[pos2+stencil_shift[k]];
        u[pos2] = s;
  }

/* static version: n=1000, nine-point stencil
   for Laplace equation */
for (pos1=1000; pos1<(1000-1)*1000; pos1+=1000)
  for (pos2=pos1+1; pos2<pos1+999; pos2++)
        u[pos2] += 8.0/3.0 * u[pos2] - 1.0/3.0 *
          (u[pos2-1001]+u[pos2-1]+u[pos2+999]+u[pos2-1000]+
             u[pos2+1000]+u[pos2-999]+u[pos2+1]+u[pos2+1001]);
```

Table 4. Applying a stencil to a function on a structured grid.

	C (dynamic)	C (static)	CMUCL
P2 (400 MHz)	4.3 sec	1.1 sec	1.6 sec
P4 (2.4 GHz)	0.59 sec	0.30 sec	0.48 sec

Table 5. Timings of C and CL for 10 applications of a difference stencil.

bilinear finite elements). It is obvious that the C compiler can produce much more efficient code for the second version, but this is only possible if both n and the stencil are known at compilation time. In contrast, with Common Lisp, it is easy to generate and compile the code for the second version at runtime.[2] Comparing the efficiency of the dynamic and static C code compiled by `gcc` with the efficiency of the code generated by CMUCL we obtain the values shown in Table 5, and we see that the Lisp code performs somewhat worse than the static C version because of the reasons mentioned above, but better than the (dynamic) C code with the same functionality.

4 Femlisp

The theoretical and practical treatment of partial differential equations (PDEs) is a highly complex domain of mathematics with a multitude of possible phenomena arising. Consequently, the efficient numerical solution is also intricate and the implementation of such methods inherits this intricacy. Thus, originally, software in this domain could handle only special situations

[2] Of course, compilation introduces an overhead (for our example about 0.02 seconds on the P4), but these are fixed costs, independent of the size of the grid and independent of the number of stencil applications.

well and was not applicable to different problems. But this changed in the last decade, when more and more software packages appeared, claiming to be multi-purpose tools for solving PDEs. Before C++ became popular, this work was done in Fortran (e.g. [1]) or C (e.g. [2]), but nowadays, most developments in this application area are done in C++.

Unfortunately, C++ is a double-edged sword. C++ improves on its predecessor C, while still maintaining a very machine-oriented programming style. It does not have most of the advanced features described in Section 2. Additionally, it suffers from the compromises it has to make to ensure efficient code, see [7]. Thus, it appears to be a rather inadequate choice for implementing a complicated framework for solving partial differential equations.

This observation was the reason for the development of the toolbox FEMLISP which can solve partial differential equations using the finite element method (FEM), see e.g. [3]. It benefits from the choice of Common Lisp as the implementation language, especially because of the powerful interactive environment and due to the expressiveness of the language which results in a considerable reduction of source code.[3]

At the moment, FEMLISP has the following functionality. It handles unstructured meshes consisting of arbitrary tensor-product cells in arbitrary space dimensions. This functionality has not yet been implemented in any other software. At the moment, the discretization is done with Lagrange finite elements of arbitrary order, but the extension to other types of finite elements should pose no problems. Isoparametric and non-parametric cell mappings for obtaining a good resolution of boundaries and interfaces are available. For the solution of the arising linear systems, geometric and algebraic multigrid methods are available. The graphic display of meshes, coefficients and finite element functions is done by linking to Gnuplot and IBM's Data Explorer.

FEMLISP was first presented at the CISC 2002 in Berlin and the EMG 2002 in Hohenwart. It is described in more detail in [12]. Its first public release is scheduled for January, 2003.

5 Conclusions

Lisp is certainly one of the most expressive computer languages available today, if not the most expressive. Because of its long history, it is a very mature language suited very well to large-scale applications. More and more, such a language will become a necessity also for scientific computing, where

[3] The amount of source code reduction is difficult to quantify exactly, but is said to be in the order of 1.5–5. One hint for a large reduction factor is that the FEMLISP core comprises only about 12.000 lines of source code, which appears to be very little when compared with similar programs written in C or C++.

the complexity of real-world applications requires the *combination* of many intricate algorithms to obtain a solution.

In this article, we have shown that the efficiency of Lisp code is no longer a problem, whereas the strength of Common Lisp opens up new possibilities for interesting applications in scientific computing. One of these applications is the solution of partial differential equations, where this power is certainly required for handling the multitude of phenomena and algorithms from a software engineering standpoint.

Acknowledgements

This article was written while I was a research assistant at the IWR with the Technical Simulation group of Prof. G. Wittum, whom I want to thank for giving me the opportunity to work on this non-standard topic. Further, I am indebted to the interesting and entertaining *comp.lang.lisp* newsgroup as well as the CMUCL mailing list for much information from the year 2000 onwards.

References

1. R. E. Bank. *PLTMG: A Software Package for Solving Elliptic Partial Differential Equations – Users' Guide 7.0*, volume 15 of *Frontiers in Applied Mathematics.* SIAM Books, Philadelphia, 1994.
2. P. Bastian, K. Birken, K. Johannsen, S. Lang, N. Neuss, H. Rentz-Reichert, and C. Wieners. UG – a flexible software toolbox for solving partial differential equations. *Comput. Visual. Sci.*, 1:27–40, 1997.
3. S. C. Brenner and L. R. Scott. *The Mathematical Theory of Finite Element Methods.* Texts in Applied Mathematics. Springer–Verlag, New York, 1994.
4. CMUCL. Homepage. http://www.cons.org/cmucl.
5. J. J. Dongarra. Performance of various computers using standard linear equations software. Technical report, Computer Science Department, University of Tennessee, 1998.
6. R. Fateman, K. A. Broughan, D. K. Willcock, and D. Rettig. Fast floating-point processing with Common Lisp. *ACM Trans. on Math. Software*, 21:26–62, 1995.
7. I. Joyner. A critique of C++ and Programming and Language Trends of the 1990s. http://www.literateprogramming.com/c++critique.pdf, 1996.
8. C. Lawson, R. Hanson, D. Kincaid, and F. Krogh. Basic linear algbra subprograms for Fortran usage. *ACM Trans. Math. Softw.*, 5:308–323, 1979.
9. Maxima. Homepage. http://sourceforge.net/~maxima.
10. N. Neuss. http://www.iwr.uni-heidelberg.de/~Nicolas.Neuss.
11. N. Neuss. A new sparse matrix storage method for adaptive solving of large systems of reaction-diffusion-transport equations. *Computing*, 68,1:19–36, 2002.
12. N. Neuss. Femlisp — a multi-purpose tool for solving partial differential equations. *Comput. Vis. Sci.*, (submitted).
13. F. Sergeraert. Common Lisp, Typing, and Mathematics. Satellite talk at the EACA Congress in Ezcaray, 2001.

The Discontinuous Galerkin Finite Element Method for Singularly Perturbed Problems

Hans-Görg Roos[1] and Helena Zarin[2,*]

[1] Institut für Numerische Mathematik, Technische Universität Dresden, 01062 Dresden, Germany
[2] Institute of Mathematics, Faculty of Science, University of Novi Sad, 21000 Novi Sad, Yugoslavia

Summary. A nonsymmetric discontinuous Galerkin finite element method with interior penalties is considered for two–dimensional singularly perturbed problems. On an anisotropic Shishkin mesh with bilinear elements we prove error estimates (uniformly in the perturbation parameter) in an integral norm associated with this method. We perform separate analyses for the cases of reaction–diffusion and convection–diffusion problems. On different types of interelement edges we derive the values of discontinuity–penalization parameters. Numerical experiments support the theoretical results.

1 Introduction

Discontinuous Galerkin finite element methods (DGFEMs) were developed in the early 1970s for various classes of partial differential equations. So far, many variants have been introduced and studied; for instance, [1] presents a detailed survey for second–order elliptic problems. In this paper we are interested in the numerical treatment of diffusion–convection–reaction problems. The idea of applying the DGFEMs to this type of problems is not new; see for example [3], [4], [6]. But recently there has been a renewed interest in the use of these methods, basically motivated by the absence of additional stabilization such as streamline–diffusion.

Here we analyze a h–version of the nonsymmetric DGFEM with interior penalties. Our aim is to derive error estimates (uniform in the perturbation parameter), applying the technique from [6] on a specially chosen shape–irregular mesh. The analysis of a hp–version of the same method on axiparallel anisotropic meshes has so far been done only for the reaction–diffusion problems in [3]. Here we use the same type of mesh, constructed a priori in order

* The work of this author was supported by DFG grant Ro975/6-3.

to resolve the layer regions of the exact solution. Unlike the L_2–projection of the error onto a finite element space considered in [6], we use a bilinear interpolant to split the error. This allows us to invoke well–known interpolation error estimates for singularly perturbed problems in the final error bound. The use of meshes with hanging nodes is a task for further research.

The paper is organized as follows. In the next section we present the two–dimensional singularly perturbed problem with its assumptions. In Sect. 3 the discontinuous Galerkin finite element method is described, while in Sect. 4 we give a general analysis of the DGFEM on an anisotropic mesh. The last two sections are devoted to the reaction–diffusion and convection–diffusion problems respectively. There we derive error estimates in our integral norm and also present the results of numerical tests.

Notation.

Throughout the paper C represents a positive constant independent of the mesh and the perturbation parameter. The Sobolev space of order k on a set D is denoted by $H^k(D)$, while $\| \ \|_{H^k(D)}$ and $| \ |_{H^k(D)}$ denote the corresponding norm and seminorm, respectively.

2 The Singularly Perturbed Problem

Consider the problem

$$-\varepsilon\Delta u + \mathbf{b}\cdot\nabla u + cu = f \qquad \text{in} \quad \Omega = (0,1)^2\,, \tag{1}$$

$$u = 0 \qquad \text{on} \quad \Gamma = \partial\Omega\,, \tag{2}$$

where $0 < \varepsilon \ll 1$ is a perturbation parameter. It is well–known that for small values of ε, standard numerical methods on quasiuniform meshes for (1)–(2) are unstable and usually fail to give accurate results. Therefore a large number of different meshes and methods have been developed in order to obtain ε–uniform error estimates for our problem (see the book [9] for a survey). Depending on the convective part $\mathbf{b}$, we divide the analysis of the DGFEM in two cases: $\mathbf{b} = 0$ (reaction–diffusion problem) and $\mathbf{b} > 0$ (convection–diffusion problem with exponential layers). For both types of problem, the solution u of (1)–(2) typically exhibits boundary and corner layers.

Let $\mathbf{b}$, c and f be sufficiently smooth functions on $\overline{\Omega}$. The existence and uniqueness of the solution of (1)–(2) are guaranteed under the additional assumptions:

- reaction–diffusion problem

$$c \geq \gamma > 0\,, \qquad f \in L_2(\Omega)\,, \tag{3}$$

- convection–diffusion problem

$$\mathbf{b} = (b_1, b_2) \geq (\beta_1, \beta_2) > 0\,, \qquad c \geq 0\,, \qquad f \in L_2(\Omega)\,, \tag{4}$$

$$c - \frac{1}{2}\nabla \cdot \mathbf{b} \geq \gamma_0 > 0\,. \tag{5}$$

Here γ, β_1, β_2, γ_0 are positive constants. Let us mention that some of the previous inequalities can be relaxed. We also assume existence of a decomposition

$$u = S + E \tag{6}$$

of u into a smooth part S (derivatives of S up to a certain order are ε–uniformly bounded) and layer parts E. Because sufficient conditions under which such a decomposition exists are difficult to describe, we refer to [5] for the reaction–diffusion problem and [8] for the convection–diffusion problem with exponential layers. The existence of such a decomposition as described in [5] and [8] allows one to prove the required interpolation error estimates on Shishkin meshes as in [2], which are the basis of our error estimates in Sects. 5 and 6.

3 The Discontinuous Galerkin Finite Element Method

In this section we present the version of the DGFEM that we later use, but first let us give some notation.

Let $\mathcal{T}$ be a general partitioning of the domain $\Omega = (0,1)^2$ consisting of disjoint open axiparallel rectangles κ such that $\overline{\Omega} = \cup_{\kappa \in \mathcal{T}} \overline{\kappa}$. In contrast to [6] we allow anisotropic meshes but assume in this first study on uniform convergence that there are no hanging nodes. To each element $\kappa \in \mathcal{T}$ we assign a nonnegative integer s_κ and define the broken Sobolev space with

$$H^{\mathbf{s}}(\Omega, \mathcal{T}) = \{v \in L_2(\Omega) : v|_\kappa \in H^{s_\kappa}(\kappa),\, \forall \kappa \in \mathcal{T}\}\,, \quad \mathbf{s} = \{s_\kappa : \kappa \in \mathcal{T}\}\,.$$

The corresponding broken Sobolev norm and seminorm are

$$\|v\|_{\mathbf{s},\mathcal{T}} = \left(\sum_{\kappa \in \mathcal{T}} \|v\|^2_{H^{s_\kappa}(\kappa)}\right)^{1/2}, \qquad |v|_{\mathbf{s},\mathcal{T}} = \left(\sum_{\kappa \in \mathcal{T}} |v|^2_{H^{s_\kappa}(\kappa)}\right)^{1/2},$$

respectively. In the case where $s_\kappa = s$ for all $\kappa \in \mathcal{T}$, we use the notation $H^s(\Omega, \mathcal{T})$, $\|v\|_{s,\mathcal{T}}$ and $|v|_{s,\mathcal{T}}$. For $v \in H^1(\Omega, \mathcal{T})$, the broken gradient $\nabla_{\mathcal{T}} v$ of a function v is defined by $(\nabla_{\mathcal{T}} v)|_\kappa = \nabla(v|_\kappa)$, $\kappa \in \mathcal{T}$. Let us assume that each element $\kappa \in \mathcal{T}$ is an affine image of a fixed reference element $\hat{\kappa} = (-1,1)^2$, i.e., $\kappa = F_\kappa(\hat{\kappa})$. Then the finite element space is

$$S(\Omega, \mathcal{T}, \mathbf{F}) = \{v \in L_2(\Omega) : v|_\kappa \circ F_\kappa \in \mathbb{Q}_1(\hat{\kappa})\}\,,$$

where $\mathbf{F} = \{F_\kappa : \kappa \in \mathcal{T}\}$ and $\mathbb{Q}_1(\hat{\kappa})$ is the space of bilinear functions on $\hat{\kappa}$. Let $\mathcal{E}$ be the set of all open one–dimensional element interfaces associated with the partitioning $\mathcal{T}$, and $\mathcal{E}_{\text{int}} \subset \mathcal{E}$ be the set of all edges $e \in \mathcal{E}$ contained in Ω. Also, let $\Gamma_{\text{int}} = \{x \in \Omega : x \in e \text{ for some } e \in \mathcal{E}_{\text{int}}\}$. Then for each $e \in \mathcal{E}_{\text{int}}$ there exist indices i and j such that $i > j$ and $\kappa := \kappa_i$ and $\kappa' := \kappa_j$ share the interface e. The (element–numbering–dependent) jump of a function $v \in H^1(\Omega, \mathcal{T})$ across e and the mean value of v on e are defined by

$$[v]_e = v|_{\partial\kappa \cap e} - v|_{\partial\kappa' \cap e}\,, \qquad \langle v \rangle_e = \frac{1}{2}\left(v|_{\partial\kappa\cap e} + v|_{\partial\kappa'\cap e}\right)\,,$$

respectively, where $\partial\kappa$ denotes the union of all open edges of the element κ. With each interface $e \in \mathcal{E}_{\text{int}}$ we associate the unit normal vector ν pointing from κ to κ'; if $e \subset \Gamma$, we take ν to be the unit outward normal vector μ. We also define the inflow and outflow parts of $\partial\kappa$ by

$$\partial_-\kappa = \{x \in \partial\kappa : \mathbf{b}(x) \cdot \mu_\kappa(x) < 0\}\,, \qquad \partial_+\kappa = \{x \in \partial\kappa : \mathbf{b}(x) \cdot \mu_\kappa(x) \geq 0\}\,,$$

respectively, where $\mu_\kappa(x)$ represents the unit outward normal vector to $\partial\kappa$ at the point $x \in \partial\kappa$. For any element $\kappa \in \mathcal{T}$ and $v \in H^1(\kappa)$, we denote by v_κ^+ the interior trace of $v|_\kappa$ on $\partial\kappa$. In the case $\partial_-\kappa \setminus \Gamma \neq \emptyset$ for some $\kappa \in \mathcal{T}$, for each $x \in \partial_-\kappa \setminus \Gamma$ there exists a unique $\kappa' \in \mathcal{T}$ such that $x \in \partial_+\kappa'$. Now for a function $v \in H^1(\Omega, \mathcal{T})$ and for some $\kappa \in \mathcal{T}$ with the property $\partial_-\kappa \setminus \Gamma \neq \emptyset$, we define the outer trace v_κ^- of v on $\partial_-\kappa \setminus \Gamma$ relative to κ as the inner trace $v_{\kappa'}^+$ relative to the element κ' such that $\partial_+\kappa' \cap (\partial_-\kappa \setminus \Gamma) \neq \emptyset$. The jump of v across $\partial_-\kappa \setminus \Gamma$ is defined by $\lfloor v \rfloor_\kappa = v_\kappa^+ - v_\kappa^-$.

Now we can present the DGFEM from [6] for the problem (1)–(2). Consider the bilinear form $B(\cdot,\cdot)$ defined by

$$\begin{aligned} B(u,v) = \sum_{\kappa\in\mathcal{T}} \Bigg(&\int_\kappa \varepsilon \nabla u \cdot \nabla v \, dx + \int_\kappa (\mathbf{b}\cdot\nabla u + cu)v \, dx \\ &- \int_{\partial_-\kappa\cap\Gamma} (\mathbf{b}\cdot\mu) u^+ v^+ \, ds - \int_{\partial_-\kappa\setminus\Gamma} (\mathbf{b}\cdot\mu)\lfloor u \rfloor v^+ \, ds \Bigg) \\ &+ \int_\Gamma \varepsilon\left(u(\nabla v\cdot\mu) - (\nabla u \cdot \mu)v\right) \, ds + \int_\Gamma \sigma u v \, ds \\ &+ \int_{\Gamma_{int}} \varepsilon\left([u]\langle \nabla v \cdot \nu\rangle - \langle \nabla u \cdot \nu \rangle [v]\right) \, ds + \int_{\Gamma_{int}} \sigma [u][v] \, ds\,, \end{aligned}$$

where σ represents a discontinuity–penalization parameter defined for each $e \in \mathcal{E}$ by $\sigma|_e = \sigma_e > 0$. The choice of σ_e will be presented in the following sections. In the case $\mathbf{b} = 0$, we write the above bilinear form as $B_0(u,v)$.

The associated nonsymmetric DGFEM with interior penalties for the problem (1)–(2) is: find $u_h \in S(\Omega, \mathcal{T}, \mathbf{F})$ such that

$$B(u_h, v) = \sum_{\kappa \in \mathcal{T}} \int_\kappa fv \, dx \qquad \forall v \in S(\Omega, \mathcal{T}, \mathbf{F}) . \tag{7}$$

In [6] the authors proved existence and uniqueness of the solution u_h of the discrete problem (7). All error estimates in that paper are derived in the DG–norm which is given by

$$\|v\|_{\mathrm{DG}}^2 = B(v, v) , \qquad v \in H^2(\Omega, \mathcal{T}) . \tag{8}$$

Introducing the notation

$$(v, w)_\tau = \int_\tau |\mathbf{b} \cdot \mu_\kappa| vw \, ds , \qquad \tau \subset \partial\kappa ,$$

and $\|v\|_\tau^2 = (v, v)_\tau$, the DG–norm can be written as

$$\|v\|_{\mathrm{DG}}^2 = \sum_{\kappa \in \mathcal{T}} \left(\varepsilon |v|_{1,\kappa}^2 + \|c_0 v\|_{0,\kappa}^2 \right) + \int_\Gamma \sigma v^2 \, ds + \int_{\Gamma_{\mathrm{int}}} \sigma [v]^2 \, ds$$
$$+ \sum_{\kappa \in \mathcal{T}} \left(\frac{1}{2} \|v^+\|_{\partial_-\kappa \cap \Gamma}^2 + \frac{1}{2} \|v^+ - v^-\|_{\partial_-\kappa \setminus \Gamma}^2 + \frac{1}{2} \|v^+\|_{\partial_+\kappa \cap \Gamma}^2 \right) , \tag{9}$$

where $\|\cdot\|_{0,\kappa}$ and $|\cdot|_{1,\kappa}$ represent the L_2–norm and seminorm on κ. Recalling the assumptions (3)–(5), we define $c_0^2 = c - (\nabla \cdot \mathbf{b})/2$.

Let us now give a brief summary of the error analysis from [6]. A starting point is the error representation $u - u_h = (u - \Pi u) + (\Pi u - u_h) \equiv \eta + \xi$ for some projector Π onto the finite element space. Using the Galerkin orthogonality property it follows that $\|\xi\|_{\mathrm{DG}}^2 = B(\xi, \xi) = -B(\eta, \xi)$. Under the assumption

$$\mathbf{b} \cdot \nabla_{\mathcal{T}} v \in S(\Omega, \mathcal{T}, \mathbf{F}) , \qquad \forall v \in S(\Omega, \mathcal{T}, \mathbf{F}) , \tag{10}$$

[6] shows that $B(\eta, \xi)$, and therefore $\|\xi\|_{\mathrm{DG}}$, can be estimated using only η. The final error estimate is then obtained from the triangle inequality.

Specializing the result from [6] to the problem (1)–(2) on a *shape–regular* mesh of maximum element diameter h, we obtain

$$\|u - u_h\|_{\mathrm{DG}}^2 \le C \left(\varepsilon h^2 + h^3 + h^4 \right) \|u\|_{H^2(\Omega)}^2 , \tag{11}$$

under the assumptions (3)–(5), (10), and for $\sigma_e = \varepsilon h_e^{-1}$, where h_e represents the length of an edge $e \in \mathcal{E}$. In the case $\mathbf{b} = 0$, the term h^3 vanishes in (11). In general, the estimate (11) is useless because $\|u\|_{H^2(\Omega)} \to \infty$ for $\varepsilon \to 0$. Therefore, we use an a–priori–constructed layer–adapted mesh and show that on such a mesh robust convergence is guaranteed. For simplicity, we avoid hanging nodes because then we can use well–known properties of the bilinear interpolant u^I of u on Shishkin meshes.

4 Error Analysis

We begin the error analysis by first introducing the error decomposition

$$u - u_h = (u - u^I) + (u^I - u_h) \equiv \eta + \xi ,$$

where u^I is the continuous piecewise bilinear interpolant of u that vanishes on the boundary Γ. Further we separately analyze $\|\eta\|_{\mathrm{DG}}$ and $\|\xi\|_{\mathrm{DG}}$, since the final estimate can be obtained from the triangle inequality.

Let us begin with the case $\mathbf{b} = 0$. Then the DG–norm (9) of a function v reduces to

$$\|v\|_{\mathrm{DG}}^2 = \sum_{\kappa\in\mathcal{T}} \left(\varepsilon|v|_{1,\kappa}^2 + \|\sqrt{c}v\|_{0,\kappa}^2\right) + \int_\Gamma \sigma v^2 \,\mathrm{d}s + \int_{\Gamma_{\mathrm{int}}} \sigma[v]^2 \,\mathrm{d}s .$$

For the term $\xi = u^I - u_h$, we estimate $B_0(\eta,\xi)$ since from the Galerkin orthogonality property one has $\|\xi\|_{\mathrm{DG}}^2 = -B_0(\eta,\xi)$. First we have

$$\begin{aligned} B_0(\eta,\xi) = {} & \sum_{\kappa\in\mathcal{T}} \int_\kappa (\varepsilon\nabla\eta\cdot\nabla\xi + c\eta\xi) \,\mathrm{d}x - \int_\Gamma \varepsilon(\nabla\eta\cdot\mu)\xi \,\mathrm{d}s \\ & + \int_{\Gamma_{\mathrm{int}}} \varepsilon\left([\eta]\langle\nabla\xi\cdot\nu\rangle - \langle\nabla\eta\cdot\nu\rangle[\xi]\right) \mathrm{d}s + \int_{\Gamma_{\mathrm{int}}} \sigma[\eta][\xi] \,\mathrm{d}s . \end{aligned} \tag{12}$$

As in [6], it can be proved that

$$\left|\sum_{\kappa\in\mathcal{T}} \int_\kappa (\varepsilon\nabla\eta\cdot\nabla\xi + c\eta\xi) \,\mathrm{d}x\right| \le C\|\xi\|_{\mathrm{DG}} \left(\sum_{\kappa\in\mathcal{T}} \left(\varepsilon|\eta|_{1,\kappa}^2 + \|\sqrt{c}\eta\|_{0,\kappa}^2\right)\right)^{1/2} \tag{13}$$

$$\left|\int_\Gamma \varepsilon(\nabla\eta\cdot\mu)\xi \,\mathrm{d}s\right| \le C\|\xi\|_{\mathrm{DG}} \left(\int_\Gamma \frac{\varepsilon^2}{\sigma}|\nabla\eta\cdot\mu|^2 \,\mathrm{d}s\right)^{1/2} . \tag{14}$$

Since our mesh consists of axiparallel rectangles, let us by $\Gamma_{\mathrm{int}}^{\mathrm{H}}$ (Γ^{H}) and $\Gamma_{\mathrm{int}}^{\mathrm{V}}$ (Γ^{V}) denote the sets of all interior (boundary) horizontal and vertical edges, respectively. Then we can write

$$\int_\Gamma \frac{\varepsilon^2}{\sigma}|\nabla\eta\cdot\mu|^2 \,\mathrm{d}s = \sum_{e\subset\Gamma^{\mathrm{H}}} \int_e \frac{\varepsilon^2}{\sigma}(\partial_y\eta)^2 \,\mathrm{d}s + \sum_{e\subset\Gamma^{\mathrm{V}}} \int_e \frac{\varepsilon^2}{\sigma}(\partial_x\eta)^2 \,\mathrm{d}s . \tag{15}$$

To estimate the remaining terms in (12), let us first introduce an element numbering in the partitioning $\mathcal{T}$ in the following way: the first element in the mesh is on the left–bottom position and the numbering is from the left to the right going from the bottom to the top, as depicted in Fig. 1.

Let us return to estimating $\|\xi\|_{\mathrm{DG}}^2$. The next term in (12) we can bound by

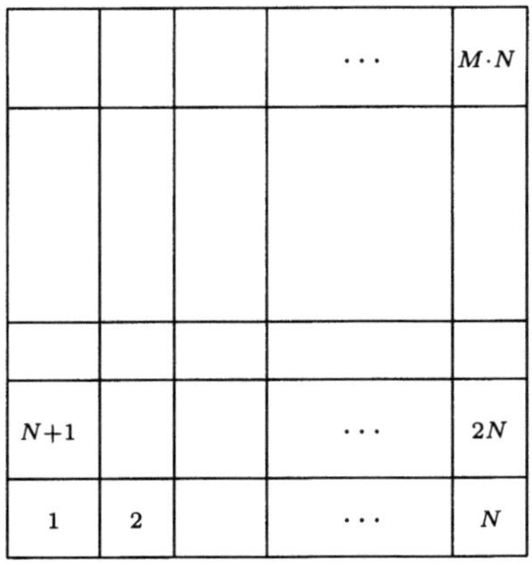

Fig. 1. The element numbering in a partitioning $\mathcal{T}$

$$\left|\int_{\Gamma_{\text{int}}} \varepsilon[\eta]\langle\nabla\xi\cdot\nu\rangle \,\mathrm{d}s\right| \le \sum_{e\subset\Gamma_{\text{int}}^{\text{H}}} \left|\int_e \varepsilon[\eta]_e\langle-\partial_y\xi\rangle_e \,\mathrm{d}s\right| + \sum_{e\subset\Gamma_{\text{int}}^{\text{V}}} \left|\int_e \varepsilon[\eta]_e\langle-\partial_x\xi\rangle_e \,\mathrm{d}s\right| .$$

Furthermore, for a fixed horizontal edge e, we have

$$\begin{aligned}\left|\int_e \varepsilon[\eta]_e\langle-\partial_y\xi\rangle_e \,\mathrm{d}s\right| \le \frac{1}{2}\Bigg(&\left|\int_e \varepsilon\eta|_{\partial\kappa\cap e}(\partial_y\xi)|_{\partial\kappa\cap e} \,\mathrm{d}s\right| + \left|\int_e \varepsilon\eta|_{\partial\kappa\cap e}(\partial_y\xi)|_{\partial\kappa'\cap e} \,\mathrm{d}s\right| \\ &+ \left|\int_e \varepsilon\eta|_{\partial\kappa'\cap e}(\partial_y\xi)|_{\partial\kappa\cap e} \,\mathrm{d}s\right| + \left|\int_e \varepsilon\eta|_{\partial\kappa'\cap e}(\partial_y\xi)|_{\partial\kappa'\cap e} \,\mathrm{d}s\right|\Bigg)\end{aligned}$$

where we assume e is a common edge for elements κ and κ'. An analogous bound can be derived for a vertical edge $e\subset\Gamma_{\text{int}}^{\text{V}}$. Now the Cauchy–Schwarz inequality and an inverse inequality yield

$$\begin{aligned}&\sum_{e\subset\Gamma_{\text{int}}^{\text{H}}} \left|\int_e \varepsilon\eta|_{\partial\kappa\cap e}(\partial_y\xi)|_{\partial\kappa\cap e} \,\mathrm{d}s\right| \\ &\quad\le C\sum_{e\subset\Gamma_{\text{int}}^{\text{H}}} \left(\int_e \tau_e\eta^2|_{\partial\kappa\cap e} \,\mathrm{d}s\right)^{1/2} \left(\int_e \frac{\varepsilon^2}{\tau_e}(\partial_y\xi)^2|_{\partial\kappa\cap e} \,\mathrm{d}s\right)^{1/2} \\ &\quad\le C\|\xi\|_{\mathrm{DG}} \left(\sum_{e\subset\Gamma_{\text{int}}^{\text{H}}} \int_e \frac{\varepsilon}{k_\kappa}\eta^2|_{\partial\kappa\cap e} \,\mathrm{d}s\right)^{1/2},\end{aligned}$$

for $\tau_e = \varepsilon/k_\kappa$, where $k_\kappa = y_j - y_{j-1}$ and $\kappa := (x_{i-1},x_i)\times(y_{j-1},y_j)$. Using similar arguments we finally obtain the bound

$$\left|\int_{\Gamma_{\rm int}} \varepsilon[\eta]\langle\nabla\xi\cdot\nu\rangle\ {\rm d}s\right|$$
$$\leq C\|\xi\|_{\rm DG}\left(\sum_{e\subset\Gamma_{\rm int}^{\rm H}}\int_e\left(\frac{\varepsilon}{k_\kappa}+\frac{\varepsilon}{k_{\kappa'}}\right)\left(\eta^2|_{\partial\kappa\cap e}+\eta^2|_{\partial\kappa'\cap e}\right)\ {\rm d}s\right.$$
$$\left.+\sum_{e\subset\Gamma_{\rm int}^{\rm V}}\int_e\left(\frac{\varepsilon}{h_\kappa}+\frac{\varepsilon}{h_{\kappa'}}\right)\left(\eta^2|_{\partial\kappa\cap e}+\eta^2|_{\partial\kappa'\cap e}\right)\ {\rm d}s\right)^{1/2}, \quad (16)$$

with $h_\kappa = x_i - x_{i-1}$ and $\kappa := (x_{i-1}, x_i)\times(y_{j-1}, y_j)$. The second integral in $B_0(\eta,\xi)$ over all interior edges from $\Gamma_{\rm int}$ can be treated similarly. Hence

$$\left|\int_{\Gamma_{\rm int}} \varepsilon\langle\nabla\eta\cdot\mu\rangle[\xi]\ {\rm d}s\right|$$
$$\leq C\|\xi\|_{\rm DG}\left(\sum_{e\subset\Gamma_{\rm int}^{\rm H}}\int_e\frac{\varepsilon^2}{\sigma_e}\left((\partial_y\eta)^2|_{\partial\kappa\cap e}+(\partial_y\eta)^2|_{\partial\kappa'\cap e}\right)\ {\rm d}s\right.$$
$$\left.+\sum_{e\subset\Gamma_{\rm int}^{\rm V}}\int_e\frac{\varepsilon^2}{\sigma_e}\left((\partial_x\eta)^2|_{\partial\kappa\cap e}+(\partial_x\eta)^2|_{\partial\kappa'\cap e}\right)\ {\rm d}s\right)^{1/2}. \quad (17)$$

The last term from (12) can be estimated as in [6] by

$$\left|\int_{\Gamma_{\rm int}}\sigma[\eta][\xi]\ {\rm d}s\right|\leq C\|\xi\|_{\rm DG}\left(\int_{\Gamma_{\rm int}}\sigma[\eta]^2\ {\rm d}s\right)^{1/2}. \quad (18)$$

Finally, collecting all the expressions from (13)–(18) into $B_0(\eta,\xi)$, from the triangle inequality we conclude that in the final error bound $\|u-u_h\|_{\rm DG}$ for the case $\mathbf{b}=0$, the following terms are to be estimated (and depend only on η):

$$I_1 \equiv \left(\sum_{\kappa\in\mathcal{T}} \varepsilon|\eta|^2_{1,\kappa}\right)^{1/2}, \qquad I_2 \equiv \left(\sum_{\kappa\in\mathcal{T}} \|\sqrt{c}\eta\|^2_{0,\kappa}\right)^{1/2}, \tag{19}$$

$$I_3 \equiv \left(\sum_{e\subset\Gamma^{\mathrm{H}}} \int_e \frac{\varepsilon^2}{\sigma_e}(\partial_y\eta)^2 \,\mathrm{d}s + \sum_{e\subset\Gamma^{\mathrm{V}}} \int_e \frac{\varepsilon^2}{\sigma_e}(\partial_x\eta)^2 \,\mathrm{d}s\right)^{1/2}, \tag{20}$$

$$I_4 \equiv \left(\sum_{e\subset\Gamma^{\mathrm{H}}_{\mathrm{int}}} \int_e \left(\frac{\varepsilon}{k_\kappa}+\frac{\varepsilon}{k_{\kappa'}}\right)\left(\eta^2|_{\partial\kappa\cap e}+\eta^2|_{\partial\kappa'\cap e}\right) \,\mathrm{d}s\right.$$

$$\left. + \sum_{e\subset\Gamma^{\mathrm{V}}_{\mathrm{int}}} \int_e \left(\frac{\varepsilon}{h_\kappa}+\frac{\varepsilon}{h_{\kappa'}}\right)\left(\eta^2|_{\partial\kappa\cap e}+\eta^2|_{\partial\kappa'\cap e}\right) \,\mathrm{d}s\right)^{1/2}, \tag{21}$$

$$I_5 \equiv \left(\sum_{e\subset\Gamma^{\mathrm{H}}_{\mathrm{int}}} \int_e \frac{\varepsilon^2}{\sigma_e}\left((\partial_y\eta)^2|_{\partial\kappa\cap e}+(\partial_y\eta)^2|_{\partial\kappa'\cap e}\right) \,\mathrm{d}s\right.$$

$$\left. + \sum_{e\subset\Gamma^{\mathrm{V}}_{\mathrm{int}}} \int_e \frac{\varepsilon^2}{\sigma_e}\left((\partial_x\eta)^2|_{\partial\kappa\cap e}+(\partial_x\eta)^2|_{\partial\kappa'\cap e}\right) \,\mathrm{d}s\right)^{1/2}, \tag{22}$$

$$I_6 \equiv \left(\int_\Gamma \sigma\eta^2 \,\mathrm{d}s\right)^{1/2}, \qquad I_7 \equiv \left(\int_{\Gamma_{\mathrm{int}}} \sigma[\eta]^2 \,\mathrm{d}s\right)^{1/2}. \tag{23}$$

In the rest of this section we present the error analysis for the case $\mathbf{b} > 0$. The corresponding bilinear form can be written as

$$B(u,v) = B_0(u,v) + \sum_{\kappa\in\mathcal{T}} \int_\kappa (\mathbf{b}\cdot\nabla u)v \,\mathrm{d}x$$

$$- \sum_{\kappa\in\mathcal{T}} \left(\int_{\partial_-\kappa\cap\Gamma} (\mathbf{b}\cdot\mu)u^+v^+ \,\mathrm{d}s + \int_{\partial_-\kappa\backslash\Gamma} (\mathbf{b}\cdot\mu)\lfloor u\rfloor v^+ \,\mathrm{d}s\right) \tag{24}$$

Again let us start from $\|\xi\|^2_{\mathrm{DG}} = -B(\eta,\xi)$. To obtain bounds on $B(\eta,\xi)$, we use all the existing estimates for $B_0(\eta,\xi)$ and analyze only additional integrals.

Performing an integration by parts, we get first

$$\sum_{\kappa\in\mathcal{T}} \left(\int_\kappa (\mathbf{b}\cdot\nabla\eta)\xi \,\mathrm{d}x - \int_{\partial_-\kappa\cap\Gamma} (\mathbf{b}\cdot\mu)\eta^+\xi^+ \,\mathrm{d}s - \int_{\partial_-\kappa\backslash\Gamma} (\mathbf{b}\cdot\mu)\lfloor\eta\rfloor\xi^+ \,\mathrm{d}s\right)$$

$$= \sum_{\kappa\in\mathcal{T}} \left(2\int_\kappa c_0^2\xi\eta \,\mathrm{d}x - 2\int_\kappa c\xi\eta \,\mathrm{d}x + \int_{\partial_+\kappa\cap\Gamma} (\mathbf{b}\cdot\mu)\xi^+\eta^+ \,\mathrm{d}s\right.$$

$$\left. + \int_{\partial_+\kappa\backslash\Gamma} (\mathbf{b}\cdot\mu)\xi^+\eta^+ \,\mathrm{d}s + \int_{\partial_-\kappa\backslash\Gamma} (\mathbf{b}\cdot\mu)\xi^+\eta^- \,\mathrm{d}s - \int_\kappa (\mathbf{b}\cdot\nabla\xi)\eta \,\mathrm{d}x\right) \tag{25}$$

Then from the Cauchy–Schwarz inequality it follows that

$$\left|\sum_{\kappa\in\mathcal{T}}\int_\kappa c\xi\eta\ \mathrm{d}x\right| \le \|\xi\|_{\mathrm{DG}}\left(\sum_{\kappa\in\mathcal{T}}\left\|\frac{c}{c_0}\eta\right\|^2_{0,\kappa}\right)^{1/2}, \tag{26}$$

$$\left|\sum_{\kappa\in\mathcal{T}}\int_\kappa c_0^2\xi\eta\ \mathrm{d}x\right| \le \|\xi\|_{\mathrm{DG}}\left(\sum_{\kappa\in\mathcal{T}}\|c_0\eta\|^2_{0,\kappa}\right)^{1/2}, \tag{27}$$

$$\left|\sum_{\kappa\in\mathcal{T}}\int_{\partial_+\kappa\cap\Gamma}(\mathbf{b}\cdot\mu)\xi^+\eta^+\ \mathrm{d}s\right| \le C\|\xi\|_{\mathrm{DG}}\left(\sum_{\kappa\in\mathcal{T}}\|\eta^+\|^2_{\partial_+\kappa\cap\Gamma}\right)^{1/2}. \tag{28}$$

The last two integrals in (25) can be estimated as in [6]; we obtain

$$\left|\sum_{\kappa\in\mathcal{T}}\int_{\partial_+\kappa\setminus\Gamma}(\mathbf{b}\cdot\mu)\xi^+\eta^+\ \mathrm{d}s + \int_{\partial_-\kappa\setminus\Gamma}(\mathbf{b}\cdot\mu)\xi^+\eta^-\ \mathrm{d}s\right|$$
$$\le C\|\xi\|_{\mathrm{DG}}\left(\sum_{\kappa\in\mathcal{T}}\|\eta^-\|^2_{\partial_-\kappa\setminus\Gamma}\right)^{1/2}. \tag{29}$$

In Sect. 6 we present a special analysis of the remaining term

$$\sum_{\kappa\in\mathcal{T}}\int_\kappa(\mathbf{b}\cdot\nabla\xi)\eta\ \mathrm{d}x\,.$$

At the end, we combine (26)–(29) into (24). In order to get a bound for the global error $u-u_h$ in the DG–norm for the case $\mathbf{b}>0$, we apply the triangle inequality to the error decomposition $u-u_h=\eta+\xi$. Finally, in addition to the terms $I_1, I_2, \ldots, I_7$ coming from the estimation of $|B_0(\eta,\xi)|$, we introduce new expressions that form part of the final error bound:

$$I_8 \equiv \left|\sum_{\kappa\in\mathcal{T}}\int_\kappa(\mathbf{b}\cdot\nabla\xi)\eta\ \mathrm{d}x\right|, \qquad I_9 \equiv \left(\sum_{\kappa\in\mathcal{T}}\left\|\frac{c}{c_0}\eta\right\|^2_{0,\kappa}\right)^{1/2}, \tag{30}$$

$$I_{10} \equiv \left(\sum_{\kappa\in\mathcal{T}}\|c_0\eta\|^2_{0,\kappa}\right)^{1/2}, \qquad I_{11} \equiv \left(\sum_{\kappa\in\mathcal{T}}\|\eta^+\|^2_{\partial_+\kappa\cap\Gamma}\right)^{1/2}, \tag{31}$$

$$I_{12} \equiv \left(\sum_{\kappa\in\mathcal{T}}\|\eta^-\|^2_{\partial_-\kappa\setminus\Gamma}\right)^{1/2}, \qquad I_{13} \equiv \left(\sum_{\kappa\in\mathcal{T}}\|\eta^+\|^2_{\partial_-\kappa\cap\Gamma}\right)^{1/2}, \tag{32}$$

$$I_{14} \equiv \left(\sum_{\kappa\in\mathcal{T}}\|\eta^+-\eta^-\|^2_{\partial_-\kappa\setminus\Gamma}\right)^{1/2}. \tag{33}$$

5 The Reaction–Diffusion Problem

In this section we derive error estimates for the problem

$$-\varepsilon\triangle u + cu = f \qquad \text{in} \quad \Omega = (0,1)^2\,, \tag{34}$$

$$u = 0 \qquad \text{on} \quad \Gamma\,. \tag{35}$$

As was already mentioned in Sect. 3, the essential ingredient of our analysis is the interpolation error and its estimates based on a decomposition (6) of the solution u. For the reaction–diffusion problem (34)–(35) under certain compatibility conditions, such a decomposition has been proved in [5].

As a triangulation we use a layer–adapted Shishkin mesh, which is piecewise uniform and constructed a priori in order to resolve the layer structure of the exact solution. Let λ be the mesh transition parameter defined by

$$\lambda = \min\left\{\frac{1}{4}, 2\sqrt{\frac{\varepsilon}{\gamma}}\ln N\right\}\,,$$

where γ is a positive constant defined in (3). For an even positive number $N \geq 4$, we specify the mesh points $\Omega_h = \{(x_i, y_j) \in \Omega : i,j = 0,1,\ldots,N\}$ by

$$x_i = y_i = \begin{cases} 4\lambda i/N, & i = 0,1,\ldots,N/4,\\ \lambda + 2(1-2\lambda)(i-N/4)/N, & i = N/4+1,\ldots,3N/4,\\ 1-4\lambda(1-i/N), & i = 3N/4+1,\ldots,N. \end{cases} \tag{36}$$

The mesh is constructed by drawing axiparallel lines through these mesh points. This divides Ω into a set $\mathcal{T}$ of mesh rectangles κ (see Fig. 2). Also we split the domain as $\overline{\Omega} = \Omega_{\mathrm{c}} \cup \Omega_{\mathrm{f}}$, $\Omega_{\mathrm{c}} = [\lambda, 1-\lambda] \times [\lambda, 1-\lambda]$, $\Omega_{\mathrm{f}} = \overline{\Omega \setminus \Omega_{\mathrm{c}}}$. We often refer to Ω_{f} as the layer region.

Assuming $\lambda = 2\sqrt{\varepsilon/\gamma}\ln N$ (otherwise N^{-1} is exponentially small compared to ε), let us by h and H denote the mesh widths (in both directions) inside and outside the boundary layers respectively:

$$h = 4\lambda N^{-1}\,, \qquad H = 2(1-2\lambda)N^{-1}\,. \tag{37}$$

In the following error analysis we successively use L_∞– and L_2–bounds for the interpolation error η. These estimates on the Shishkin mesh (36) can be proved as in [2] and [10] for the convection–diffusion problem. Therefore, with $\Omega_{\mathrm{c}}^* = [\lambda + H, 1-\lambda-H]^2$, we have

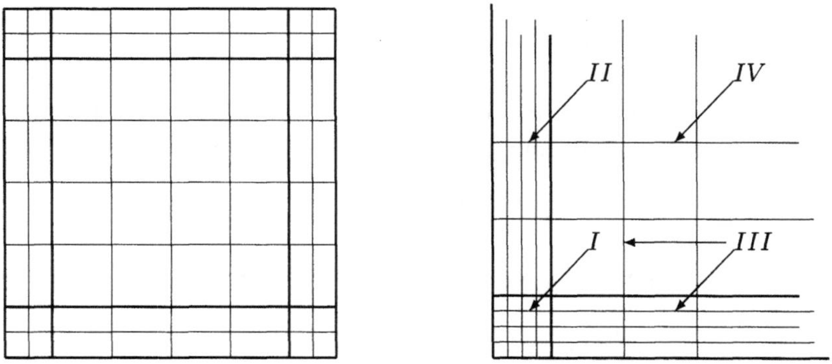

Fig. 2. Shishkin mesh for the reaction–diffusion problem, $N = 8$, (left) and four types of interelement edges (right)

$$\|\eta\|_{\infty,\Omega_\mathrm{f}} \le CN^{-2}\ln^2 N\,, \qquad \|\eta\|_{\infty,\Omega_\mathrm{c}} \le CN^{-2}\,, \tag{38}$$

$$\|\eta\|_0 \le CN^{-2}\,, \tag{39}$$

$$\varepsilon^{1/4}|\eta|_{1,\Omega_\mathrm{f}} \le CN^{-1}\ln N\,, \qquad \varepsilon^{1/4}|\eta|_{1,\Omega_\mathrm{c}} \le CN^{-1}\,, \tag{40}$$

$$\sqrt{\varepsilon}\,\|\partial_x\eta\|_{\infty,\Omega_\mathrm{f}} \le CN^{-1}\ln N\,, \qquad \|\partial_x\eta\|_{\infty,\Omega_\mathrm{c}^*} \le CN^{-1}\,, \tag{41}$$

$$\sqrt{\varepsilon}\,\|\partial_y\eta\|_{\infty,\Omega_\mathrm{f}} \le CN^{-1}\ln N\,, \qquad \|\partial_y\eta\|_{\infty,\Omega_\mathrm{c}^*} \le CN^{-1}\,. \tag{42}$$

Before starting to estimate $\|u - u_h\|_{\mathrm{DG}}$, we need to introduce several types of edges in the partitioning $\mathcal{T}$ (see Fig. 2, right). Depending on the type, we later determine the discontinuity–penalization parameter σ_e for each edge $e \in \mathcal{E}$. The edges of the first type belong to the set $([0,\lambda) \cup (1-\lambda,1])^2$. They are part of the layer region (more precisely, they lie in the corner layers) and their length is h. Since we assume $\lambda = 2\sqrt{\varepsilon/\gamma}\ln N$ and $\varepsilon \le CN^{-1}$, then $h \ll H$, so we describe these edges as short; other edges are called long. Edges of the second type are also short and belong to the rest of the layer region $\Omega_\mathrm{f} \setminus ([0,\lambda) \cup (1-\lambda,1])^2$. Since this region contains also long edges, we refer to them as being of the third type. Type III also contains long edges near the layers from the set $\Omega_\mathrm{c} \setminus \Omega_\mathrm{c}^*$. Finally, edges of type IV are long and lie in Ω_c^*.

Now we can begin to estimate the terms $I_1, I_2, \ldots, I_7$ from Sect. 4. First using (40) and (39) in (19), we obtain

$$I_1 \le C\varepsilon^{1/4}N^{-1}\ln N\,, \qquad I_2 \le CN^{-2}\,. \tag{43}$$

The treatment of the expressions $I_3, I_4, \ldots, I_7$ depends on the type of edge; more precisely, we look for the contribution from each edge to these terms. Let $I_{j,\mathrm{I}}$, $I_{j,\mathrm{II}}$, $I_{j,\mathrm{III}}$ and $I_{j,\mathrm{IV}}$ denote the contributions of edges of the type I, II, III and IV to the terms I_j, for $j = 3, 4, \ldots, 7$ respectively.

For type I boundary edges, from (42) we have

$$\int_e \frac{\varepsilon^2}{\sigma_e}(\partial_y \eta)^2 \, \mathrm{d}s \le C\frac{\varepsilon\sqrt{\varepsilon}}{\sigma_e} N^{-3} \ln^3 N \,, \qquad e \subset \Gamma^{\mathrm{H}} \,,$$

and the same estimate can be obtained for $e \subset \Gamma^{\mathrm{V}}$. In the rest of the analysis we assume that σ_e is the same for all edges of the same type. This yields

$$I_{3,\mathrm{I}} \le C\varepsilon^{3/4}\sigma_e^{-1/2} N^{-1} \ln^{3/2} N \,.$$

For the term $I_{6,\mathrm{I}}$ we use (38) and $h = C\sqrt{\varepsilon}N^{-1}\ln N$ in order to get the bound

$$I_{6,\mathrm{I}} \le C\varepsilon^{1/4}\sqrt{\sigma_e} N^{-2} \ln^{5/2} N \,.$$

Now choosing $\sigma_e = \sqrt{\varepsilon}N/\ln N$, we have $I_{3,\mathrm{I}} + I_{6,\mathrm{I}} \le C\sqrt{\varepsilon}N^{-3/2}\ln^2 N$. In the case of type I interior edges, since $k_\kappa = k_{\kappa'} = h_\kappa = h_{\kappa'} = h$, for the term $I_{4,\mathrm{I}}$ from (38) we obtain $I_{4,\mathrm{I}} \le C\sqrt{\varepsilon}N^{-1}\ln^2 N$. Using (41) and (42) in (22), we find that

$$I_{5,\mathrm{I}} \le C\varepsilon^{3/4}\sigma_e^{-1/2} N^{-1/2} \ln^{3/2} N \,.$$

Similarly as for $I_{6,\mathrm{I}}$, we can show that

$$I_{7,\mathrm{I}} \le C\varepsilon^{1/4}\sqrt{\sigma_e} N^{-3/2} \ln^{5/2} N$$

because $[\eta]_e^2 \le CN^{-4}\ln^4 N$. Taking the value

$$\sigma_e = \sqrt{\varepsilon}N/\ln N \tag{44}$$

for the parameter, we obtain the error bound

$$I_{4,\mathrm{I}} + I_{5,\mathrm{I}} + I_{7,\mathrm{I}} \le C\sqrt{\varepsilon}N^{-1}\ln^2 N \,, \tag{45}$$

which is the final contribution of type I edges to $\|u - u_h\|_{\mathrm{DG}}$.

Let us now analyze edges of type II. Since they can only be part of Γ_{int}, we consider $I_{4,\mathrm{II}}$, $I_{5,\mathrm{II}}$ and $I_{7,\mathrm{II}}$. For the term $I_{4,\mathrm{II}}$, since $h \le H$ we have

$$I_{4,\mathrm{II}} \le C\sqrt{\varepsilon h^{-1}\left(N^{-2}\ln^2 N\right)^2 hN^2} \le C\sqrt{\varepsilon}N^{-1}\ln^2 N \,.$$

From (41) and (42) we get

$$I_{5,\mathrm{II}} \le C\varepsilon^{3/4}\sigma_e^{-1/2} N^{-1/2} \ln^{3/2} N \,,$$

while from (38) it follows that $I_{7,\mathrm{II}} \le C\varepsilon^{1/4}\sqrt{\sigma_e}N^{-3/2}\ln^{5/2}$. Now with the choice (44) we conclude that

$$I_{4,\mathrm{II}} + I_{5,\mathrm{II}} + I_{7,\mathrm{II}} \le C\sqrt{\varepsilon}N^{-1}\ln^2 N \,. \tag{46}$$

For the type III edges we distinguish four different cases. The first case is dealing with boundary edges, for which we have estimates

$$I_{3,\mathrm{III}}+I_{6,\mathrm{III}} \le C\sqrt{\varepsilon}\sigma_e^{-1/2}N^{-1}\ln N + C\sqrt{\sigma_e}N^{-2}\ln^2 N \le C\varepsilon^{1/4}N^{-3/2}\ln^{3/2} N\,,$$

with $\sigma_e = \sqrt{\varepsilon}N/\ln N$. The second case involves edges from the boundary of Ω_{c}. Then

$$I_{4,\mathrm{III}} \le C\sqrt{\varepsilon h^{-1}\left(N^{-2}\ln^2 N\right)^2 HN^2} \le C\varepsilon^{1/4}N^{-1}\ln^{3/2} N\,,$$

$$I_{5,\mathrm{III}} \le C\sqrt{\varepsilon^2\sigma_e^{-1}\varepsilon^{-1}\left(N^{-1}\ln N\right)^2 HN^2} \le C\sqrt{\varepsilon}\sigma_e^{-1/2}N^{-1/2}\ln N\,,$$

$$I_{7,\mathrm{III}} \le C\sqrt{\sigma_e\left(N^{-2}\ln^2 N\right)^2 HN^2} \le C\sqrt{\sigma_e}N^{-3/2}\ln^2 N\,.$$

Again we can choose σ_e from (44) in order to obtain

$$I_{4,\mathrm{III}} + I_{5,\mathrm{III}} + I_{7,\mathrm{III}} \le C\varepsilon^{1/4}N^{-1}\ln^{3/2} N\,. \tag{47}$$

The third case comprises edges from the layer region Ω_{f} for which in the same manner we can prove the previous estimate. Finally, the last case involves type III edges from the set $\Omega_{\mathrm{c}} \setminus \Omega_{\mathrm{c}}^*$. For them we derive the bounds

$$I_{4,\mathrm{III}} \le C\sqrt{\varepsilon}N^{-3/2}\,, \quad I_{5,\mathrm{III}} \le C\sqrt{\varepsilon}\sigma_e^{-1/2}N^{-1}\ln N\,, \quad I_{7,\mathrm{III}} \le C\sqrt{\sigma_e}N^{-2}\,.$$

Thus, with

$$\sigma_e = \sqrt{\varepsilon}N\ln N \tag{48}$$

consequently we have

$$I_{4,\mathrm{III}} + I_{5,\mathrm{III}} + I_{7,\mathrm{III}} \le C\varepsilon^{1/4}N^{-3/2}\ln^{1/2} N\,. \tag{49}$$

To complete this analysis, we consider long edges from the interior of Ω_{c}^*. First we have

$$I_{4,\mathrm{IV}} \le C\sqrt{\varepsilon}N^{-1}\,, \qquad I_{5,\mathrm{IV}} \le C\varepsilon\sigma_e^{-1/2}N^{-1/2}\,, \qquad I_{7,\mathrm{IV}} \le C\sqrt{\sigma_e}N^{-3/2}\,,$$

where we have used interpolation error estimates (38), (41) and (42), as well as the property $H \le CN^{-1}$. Now, for

$$\sigma_e = \varepsilon N \tag{50}$$

we obtain

$$I_{4,\mathrm{IV}} + I_{5,\mathrm{IV}} + I_{7,\mathrm{IV}} \le C\sqrt{\varepsilon}N^{-1}\,. \tag{51}$$

In order to present a more readable version of how the different choices of the penalization parameter σ_e depend on the edge type, we summarize the results in Table 1.

Collecting the estimates (43), (45)–(47), (49) and (51), we can state the main result for the DGFEM and the reaction–diffusion problem.

Table 1. Different types of edges and corresponding parameters σ_e for the reaction–diffusion problem

edge type	parameter σ_e	
I	$\sqrt{\varepsilon}N/\ln N$	
II	$\sqrt{\varepsilon}N/\ln N$	
III	$\sqrt{\varepsilon}N\ln N$	if $e \subset \Omega_c \setminus \Omega_c^*$
	$\sqrt{\varepsilon}N/\ln N$	otherwise
IV	εN	

Theorem 1. *Let u be a solution of the reaction–diffusion problem (34)–(35), and let u_h be a solution of the discrete problem (7) with the bilinear form B_0 on the Shishkin mesh (36). Assuming $\varepsilon \le CN^{-1}$ and (3), and choosing the penalty parameter from (44), (48) and (50), we have*

$$\|u - u_h\|_{\mathrm{DG}} \le C\varepsilon^{1/4}N^{-1}\ln^{3/2} N\,.$$

In the rest of this section we experimentally verify the theoretical result of Theorem 1. We test the DGFEM (7) on the anisotropic Shishkin mesh (36) when applied to the problem

$$-\varepsilon\Delta u + u = f \qquad \text{in} \qquad \Omega = (0,1)^2\,, \tag{52}$$

$$u = 0 \qquad \text{on} \qquad \Gamma\,. \tag{53}$$

The function f is chosen such that

$$u(x,y) = \left(1 - e^{-x/\sqrt{\varepsilon}}\right)\left(1 - e^{-(1-x)/\sqrt{\varepsilon}}\right)\left(1 - e^{-y/\sqrt{\varepsilon}}\right)\left(1 - e^{-(1-y)\sqrt{\varepsilon}}\right)$$

is the exact solution. We take $\varepsilon = 10^{-8}$ which is small enough to bring out the singularly perturbed nature of the problem.

Table 2 presents the values of the DG–norm of the error $e_h \equiv u - u_h$ as well as an estimate of the maximum norm of the error, denoted by $\|e_h\|_{d,\infty}$. We compute this estimate using an auxiliary mesh that contains 25 uniformly distributed mesh points per element. We also compute the values of the L_2–norm of the error $\|e_h\|_0$, as well as the L_∞–norm of the jumps along the edges $\|[u]\|_\infty$. All integrals in these norms are approximated using the quadrature formulae

$$\int_\kappa f \;\mathrm{d}x \approx \frac{meas(\kappa)}{4} f(x_S, y_S)\,, \qquad \int_e f \;\mathrm{d}s \approx \frac{h_e}{2} f(z_S)\,,$$

where (x_S, y_S) is the midpoint of an element κ and z_S is the midpoint of an edge e. Rates of convergence are computed in the usual way.

Table 2. The DGFEM on the Shishkin mesh (36) for the problem (52)–(53)

	$\|e_h\|_{DG}$		$\|e_h\|_{d,\infty}$		$\|e_h\|_0$		$\|[u]\|_\infty$	
N	*error*	*rate*	*error*	*rate*	*error*	*rate*	*error*	*rate*
8	7.374(−3)	0.664	4.329(−1)	0.819	5.920(−3)	0.676	1.473(−1)	1.029
16	4.655(−3)	0.991	2.453(−1)	1.304	3.707(−3)	1.021	7.217(−2)	1.332
32	2.342(−3)	1.264	9.937(−2)	1.829	1.826(−3)	1.286	2.886(−2)	1.598
64	9.751(−4)	1.443	2.796(−2)	1.850	7.489(−4)	1.470	9.531(−3)	1.874
128	3.585(−4)	–	7.753(−3)	–	2.703(−4)	–	2.600(−3)	–

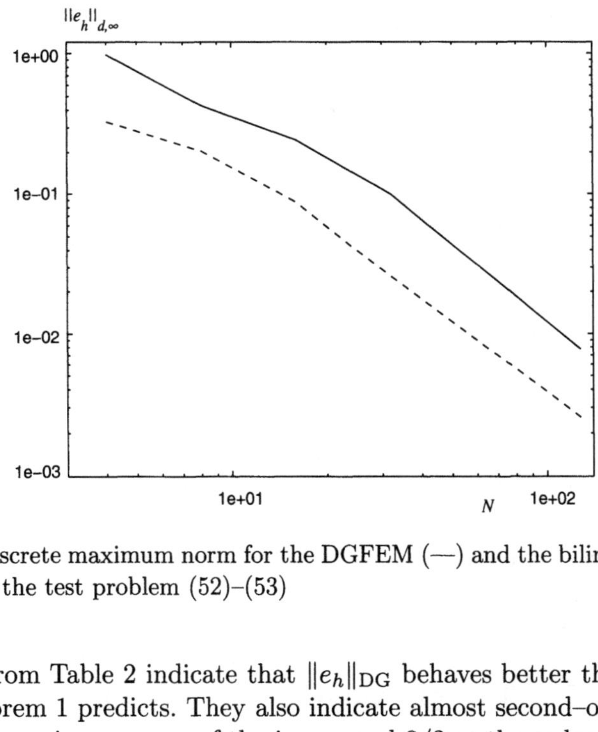

Fig. 3. The discrete maximum norm for the DGFEM (—) and the bilinear Galerkin FEM (– –) for the test problem (52)–(53)

The results from Table 2 indicate that $\|e_h\|_{DG}$ behaves better than the estimate of Theorem 1 predicts. They also indicate almost second–order convergence for the maximum norm of the jumps and 3/2 as the order for $\|e_h\|_0$.

It is important to mention that the linear system corresponding to the discrete problem (7) has dimension $4N^2$, which is four times larger than for the standard Galerkin method. Figure 3 shows that in the discrete maximum norm the bilinear Galerkin finite element method behaves better than the DGFEM for the test problem (52)–(53). As a common independent variable we take the same dimension of the corresponding linear systems.

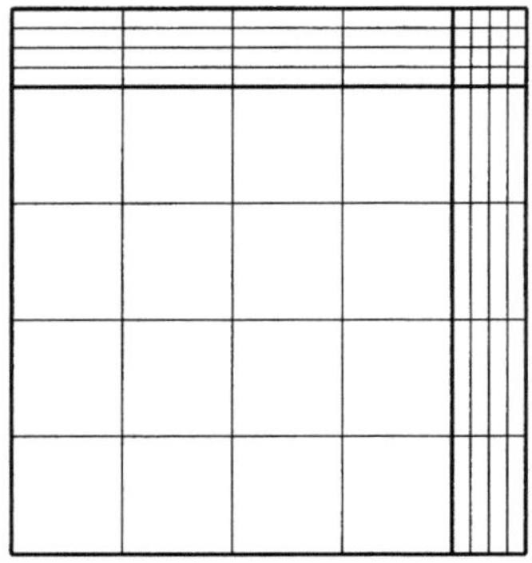

Fig. 4. Shishkin mesh for the convection–diffusion problem ($N = 8$)

6 The Convection–Diffusion Problem

This part of the paper is devoted to the analysis of the case $\mathbf{b} > 0$. We consider the problem (1)–(2) together with the assumptions (4) and (5). Again, a starting point in the mesh construction and the interpolation analysis is a decomposition (6) of the exact solution, which can be found in [8].

The set of mesh points $\Omega_h = \{(x_i, y_j) \in \Omega : i, j = 0, 1, \ldots, N\}$ is constructed similarly to Sect. 5, where the mesh is now dense near the boundary layers $x = 1$ and $y = 1$. The mesh transition parameters are

$$\lambda_x = \min\left\{\frac{1}{2}, 2\frac{\varepsilon}{\beta_1}\ln N\right\}, \qquad \lambda_y = \min\left\{\frac{1}{2}, 2\frac{\varepsilon}{\beta_2}\ln N\right\},$$

where β_1 and β_2 are the lower bounds for the convective part $\mathbf{b} = (b_1, b_2)$; see (4). Now we have

$$x_i = \begin{cases} 2(1-\lambda_x)i/N, & i = 0, 1, \ldots, N/2, \\ 1 - 2\lambda_x(N-i)/N, & i = N/2+1, \ldots, N, \end{cases} \tag{54}$$

$$y_j = \begin{cases} 2(1-\lambda_y)j/N, & j = 0, 1, \ldots, N/2, \\ 1 - 2\lambda_y(N-j)/N, & j = N/2+1, \ldots, N. \end{cases} \tag{55}$$

Again, we construct the mesh by drawing axiparallel lines through the mesh points (see Fig. 4).

The layer region Ω_{f} is defined by $\Omega_{\mathrm{f}} = \overline{\Omega \setminus \Omega_{\mathrm{c}}}$, where $\Omega_{\mathrm{c}} = [0, 1-\lambda_x] \times [0, 1-\lambda_y]$. Furthermore, we shall assume $\lambda_x = 2\varepsilon/\beta_1 \ln N$ and $\lambda_y = 2\varepsilon/\beta_2 \ln N$, which leads to the mesh step sizes

$$h_x = 2\lambda_x/N\,, \quad h_y = 2\lambda_y/N\,, \quad H_x = 2(1-\lambda_x)/N\,, \quad H_y = 2(1-\lambda_y)/N\,.$$

The indices x and y in the previous notation refer to the corresponding directions. It can be easily shown that $\max\{h_x, h_y\} \leq C\varepsilon N^{-1}\ln N$, $\max\{H_x, H_y\} \leq CN^{-1}$ which will often be used in the error analysis.

As already mentioned, the key ingredients in our error analysis are the interpolation error estimates. For the convection–diffusion problem with exponential layers and the Shishkin mesh, they are already proved in [2] and [10]. Furthermore, we frequently use the bounds

$$\|\eta\|_{\infty,\Omega_{\mathrm{f}}} \leq CN^{-2}\ln^2 N\,, \qquad \|\eta\|_{\infty,\Omega_{\mathrm{c}}} \leq CN^{-2}\,, \tag{56}$$

$$\|\eta\|_0 \leq CN^{-2}\,, \tag{57}$$

$$\varepsilon^{1/2}|\eta|_{1,\Omega_{\mathrm{f}}} \leq CN^{-1}\ln N\,, \qquad \varepsilon^{1/2}|\eta|_{1,\Omega_{\mathrm{c}}} \leq CN^{-1}\,, \tag{58}$$

$$\varepsilon\left\|\partial_x\eta\right\|_{\infty,\Omega_{\mathrm{f}}} \leq CN^{-1}\ln N\,, \qquad \left\|\partial_x\eta\right\|_{\infty,\Omega_{\mathrm{c}}^*} \leq CN^{-1}\,, \tag{59}$$

$$\varepsilon\left\|\partial_y\eta\right\|_{\infty,\Omega_{\mathrm{f}}} \leq CN^{-1}\ln N\,, \qquad \left\|\partial_y\eta\right\|_{\infty,\Omega_{\mathrm{c}}^*} \leq CN^{-1}\,. \tag{60}$$

where $\Omega_{\mathrm{c}}^* = [0, 1-\lambda-H]^2$. Before starting with the error analysis, let us recall the definition of the four types of edges for the reaction–diffusion problem from Sect. 5, Fig. 2, right. It is straightforward to re-introduce these types for the convection–diffusion problem and therefore we omit their precise definition.

Now, as well as the terms $I_1, I_2, \ldots, I_7$, when using the bounds (56)–(60) for the interpolation error η, we have to estimate in addition $I_8, I_9, \ldots, I_{14}$. Let us note that for I_2 we have already estimate (43) while for I_1 one has

$$I_1 \leq CN^{-1}\ln N\,. \tag{61}$$

For $I_3, I_4, \ldots, I_7$ we use the same technique as in Sect. 5, again separately analyzing the corresponding terms on the four types of edges. Therefore we have

$$I_{3,\mathrm{I}} + I_{6,\mathrm{I}} \leq C\sqrt{\varepsilon}N^{-3/2}\ln^2 N\,, \quad \sigma_e = N/\ln N\,, \tag{62}$$

$$I_{4,\mathrm{I}} + I_{5,\mathrm{I}} + I_{7,\mathrm{I}} \leq C\sqrt{\varepsilon}N^{-1}\ln^2 N\,, \quad \sigma_e = N/\ln N\,, \tag{63}$$

$$I_{4,\mathrm{II}} + I_{5,\mathrm{II}} + I_{7,\mathrm{II}} \leq C\sqrt{\varepsilon}N^{-1}\ln^2 N\,, \quad \sigma_e = N/\ln N\,, \tag{64}$$

$$I_{3,\mathrm{III}} + I_{6,\mathrm{III}} \leq CN^{-3/2}\ln^{3/2} N\,, \quad \sigma_e = N/\ln N\,, \tag{65}$$

$$I_{4,\mathrm{III}} + I_{5,\mathrm{III}} + I_{7,\mathrm{III}} \leq CN^{-1}\ln^{3/2} N\,, \quad \sigma_e = N/\ln N \text{ on } \Omega_{\mathrm{f}}\,, \tag{66}$$

$$I_{4,\mathrm{III}} + I_{5,\mathrm{III}} + I_{7,\mathrm{III}} \leq CN^{-3/2}\ln^{1/2} N\,, \quad \sigma_e = N\ln N \text{ on } \Omega_{\mathrm{c}} \setminus \Omega_{\mathrm{c}}^*\,, \tag{67}$$

$$I_{4,\mathrm{IV}} + I_{5,\mathrm{IV}} + I_{7,\mathrm{IV}} \leq C\sqrt{\varepsilon}N^{-1}\,, \quad \sigma_e = \varepsilon N\,. \tag{68}$$

Table 3. Different types of edges and corresponding parameters σ_e for the convection–diffusion problem

edge type	parameter σ_e	
I	$N/\ln N$	
II	$N/\ln N$	
III	$N\ln N$	if $e \subset \Omega_c \setminus \Omega_c^*$
	$N/\ln N$	otherwise
IV	εN	

As for the reaction–diffusion problem, we summarize the choices for the discontinuity–penalization parameter σ_e in Table 3.

Let us now analyze the additional terms (30)–(33), beginning first with I_8. For each element $\kappa \in \mathcal{T}$ we estimate

$$\left| \int_\kappa (\mathbf{b} \cdot \nabla\xi)\eta \,\mathrm{d}x \right|$$

assuming $\|\mathbf{b}\|_{\infty,\Omega} = O(1)$. First, if $\kappa \in \Omega_c$, then from the Cauchy–Schwarz inequality, inverse estimates for $\partial_x\xi$, $\partial_y\xi$, and from (57), we have

$$\begin{aligned}\left| \int_\kappa (\mathbf{b} \cdot \nabla\xi)\eta \,\mathrm{d}x \right| &\le \|\eta\|_{0,\kappa} \left(\|b_1\partial_x\xi\|_{0,\kappa} + \|b_2\partial_y\xi|_{0,\kappa}\right) \\ &\le C\|\eta\|_{0,\kappa} \min\{H_x, H_y\}^{-1}\|\xi\|_{0,\kappa} \le CN^{-1}\|\xi\|_{0,\kappa} .\end{aligned}$$

In the case where the element κ is from the layer region Ω_f, we can similarly derive the following estimates:

$$\begin{aligned}\left| \int_\kappa (\mathbf{b} \cdot \nabla\xi)\eta \,\mathrm{d}x \right| &\le C\|\eta\|_{\infty,\kappa} \left(\int_\kappa |\partial_x\xi| \,\mathrm{d}x + \int_\kappa |\partial_y\xi| \,\mathrm{d}x \right) \\ &\le C\|\eta\|_{\infty,\kappa}\sqrt{meas(\kappa)}\|\nabla\xi\|_{0,\kappa} \le CN^{-3}\ln^{5/2} N(\sqrt{\varepsilon}\|\nabla\xi\|_{0,\kappa}) .\end{aligned}$$

Recalling the second way of writing the DG–norm from (9), the bounds in (30) give

$$I_8 \le CN^{-1}\|\xi\|_{\mathrm{DG}} . \tag{69}$$

Let us recall that this estimate is further used in $\|\xi\|_{\mathrm{DG}}^2 = -B(\eta, \xi)$.

Finally, from the inequalities (56) and (57) for the interpolation error, we obtain

$$I_9 \le CN^{-2}, \qquad I_{10} \le CN^{-2}, \tag{70}$$

$$I_{11} \le CN^{-2}\ln^2 N, \qquad I_{12} \le CN^{-3/2}\ln^2 N, \tag{71}$$

$$I_{13} \le CN^{-2}\ln^2 N, \qquad I_{14} \le CN^{-3/2}\ln^2 N. \tag{72}$$

Estimates (61)–(72) imply the error bound for the DGFEM and the convection–diffusion problem that is formulated in the next theorem.

Table 4. The DGFEM on the Shishkin mesh (36) for the problem (73)–(74)

	$\|e_h\|_{DG}$		$\|e_h\|_{d,\infty}$		$\|e_h\|_0$		$\|[u]\|_\infty$	
N	*error*	*rate*	*error*	*rate*	*error*	*rate*	*error*	*rate*
8	3.520(−1)	0.802	3.437(−1)	1.015	5.619(−3)	1.842	3.654(−1)	0.997
16	2.020(−1)	1.083	1.700(−1)	0.922	1.567(−3)	1.928	1.831(−1)	1.399
32	9.536(−2)	1.292	8.970(−2)	1.159	4.119(−4)	1.966	6.942(−2)	1.748
64	3.896(−2)	1.441	4.018(−2)	1.350	1.054(−4)	1.984	2.067(−2)	2.027
128	1.434(−2)	–	1.576(−2)	–	2.667(−5)	–	5.072(−3)	–

Theorem 2. *Let u be a solution of the convection–diffusion problem (1)–(2), and let u_h be a solution of the discrete problem (7) on the Shishkin mesh (54)–(55). Assuming $\varepsilon \le CN^{-1}$, (4) and (5), and choosing the penalty parameter from (62)–(68), we have*

$$\|u - u_h\|_{DG} \le CN^{-1} \ln^{3/2} N .$$

Let us remark that the error estimate from this theorem is obtained without the assumption (10), which is used in [6].

Finally, we present the results of numerical experiments. For the test problem

$$-\varepsilon \Delta u + 2u_x + 3u_y + u = f \quad \text{in} \quad \Omega = (0,1)^2 , \tag{73}$$

$$u = 0 \quad \text{on} \quad \Gamma , \tag{74}$$

with the exact solution

$$u(x,y) = 2 \sin x \left(1 - e^{-2(1-x)/\varepsilon}\right) y^2 \left(1 - e^{-3(1-y)/\varepsilon}\right) ,$$

we apply the discontinuous Galerkin method from Sect. 3 in order to verify the theoretical result from Theorem 2. Again we take $\varepsilon = 10^{-8}$ and use the anisotropic Shishkin mesh (54)–(55).

In the same manner as in Sect. 5, in Table 4 we present the values of the DG–norm of the error $\|e_h\|_{DG}$, the estimate of the maximum norm of the error $\|e_h\|_{d,\infty}$, the L_2–norm of the error $\|e_h\|_0$, and the L_∞–norm of the jumps along interior edges $\|[u]\|_\infty$. The results indicate 3/2 as the order of convergence for $\|e_h\|_{DG}$, and second–order convergence for the L_2–norm of e_h and the maximum norm of the jumps.

We also compare the DGFEM with the bilinear standard Galerkin finite element method (bilFEM) and with the bilinear streamline–diffusion finite element method (bilSDFEM). The estimates of the maximum norm for these methods are presented in Fig. 5. For both bilFEM and bilSDFEM superconvergence techniques for bilinear elements allow one to prove improved estimates for $\|u^I - u_h\|_\varepsilon$, [7], and $\|u^I - u_h\|_{\mathrm{SD}}$, [11], in comparison to estimates for $u - u_h$ itself. These estimates lead also to L_∞–bounds, but because on Ω_c an inverse estimate is used, the theoretically proven order on Ω_c is no better than one.

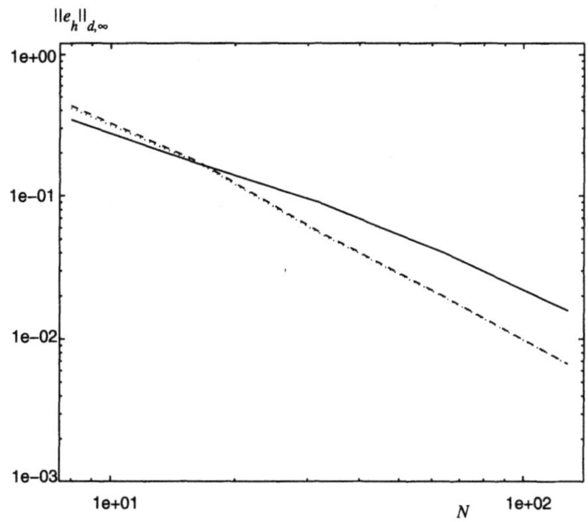

Fig. 5. The discrete maximum norm for the DGFEM (—), bilFEM (- -) and bilSDFEM (···) for the test problem (73)–(74)

Our numerical results indicate that the discontinuous Galerkin method on a layer–adapted mesh does indeed yield robust convergence. It is not surprising that a comparison with bilinear elements (which have poor stability properties) and streamline–diffusion with bilinear elements favours (so far) streamline–diffusion. But we are optimistic that both for problems with characteristic layers — where the tuning of stabilization parameters is difficult — and for still more complicated layer structures where the flexibility of the DGFEM with respect to the mesh is useful, the DGFEM will be a strong competitor for streamline–diffusion. This is a topic for further research.

References

1. Arnold, D.N., Brezzi, F., Cockburn, B., Marini, L.D.: Unified Analysis of Discontinuous Galerkin Methods for Elliptic Problems. SIAM J. Numer. Anal. **39** (2002) 1749–1779
2. Dobrowolski, M., Roos, H.-G.: A Priori Estimates for the Solution of Convection–Diffusion Problems and Interpolation on Shishkin Meshes. J. Anal. Appl. **16** (1997) 1001–1012
3. Georgoulis, E.H., Süli, E.: *hp*–DGFEM on Shape–Irregular Meshes: Reaction–Diffusion Problems. Research Report 01/09, Oxford University Computing Laboratory (2001)

4. Gopalakrishnan, J., Kanschat, G.: A Multilevel Discontinuous Galerkin Method. Numer. Math. (to appear)
5. Han, H., Kellogg, R.B.: Differentiability Properties of Solutions of the Equation $-\epsilon^2 \Delta u + ru = f(x, y)$ in a Square. SIAM J. Math. Anal. **21** (1990) 394–408
6. Houston, P., Schwab, C., Süli, E.: Discontinuous *hp*-Finite Element Methods for Advection–Diffusion–Reaction Problems. SIAM J. Numer. Anal. **39** (2002) 2133–2163
7. Linß, T.: Uniform Superconvergence of a Galerkin Finite Element Method on Shishkin–Type Meshes. Numer.Meth. for PDE **16** (2000) 426–440
8. Linß, T., Stynes, M.: Asymptotic Analysis and Shishkin–Type Decomposition for an Elliptic Convection–Diffusion Problem. J. Math. Anal. Appl. **261** (2001) 604–632
9. Roos, H.-G., Stynes, M., Tobiska, L. (ed.): Numerical Methods for Singularly Perturbed Differential Equations. Computational Mathematics, Vol. 24. Springer-Verlag, Berlin Heidelberg New York (1996)
10. Stynes, M., O'Riordan, E.: A Uniformly Convergent Galerkin Method on a Shishkin Mesh for a Convection–Diffusion Problem. J. Math. Anal. Appl. **214** (1997) 36–54
11. Stynes, M., Tobiska, L.: Analysis of the Streamline Diffusion Finite Element Method on a Piecewise Uniform Mesh for a Convection–Diffusion Problem with Exponential Layers. East–West J. Numer. Anal. **9** (2001) 59–76

Black-Box Preconditioning for Mixed Formulation of Self-Adjoint Elliptic PDEs

Catherine Powell and David Silvester

UMIST, Mathematics Department, P.O. Box 88, Manchester, M60 1QD.

Summary. Mixed finite element approximation of self-adjoint elliptic PDEs leads to symmetric indefinite linear systems of equations. Preconditioning strategies commonly focus on reduced symmetric positive definite systems and require nested iteration. This deficiency is avoided if preconditioned MINRES is applied to the full indefinite system. We outline such a preconditioning strategy, the key building block for which is a fast solver for a scalar diffusion operator based on black-box algebraic multigrid. Numerical results are presented for the Stokes equations arising in incompressible flow modelling and a variable diffusion equation that arises in modelling potential flow. We prove that the eigenvalues of the preconditioned matrices are contained in intervals that are bounded independently of the discretisation parameter and the PDE coefficients.

1 Introduction

We are interested in efficient and robust solution schemes for the general class of saddle-point systems,

$$\begin{pmatrix} A & B^T \\ B & 0 \end{pmatrix} \begin{pmatrix} \underline{u} \\ \underline{p} \end{pmatrix} = \begin{pmatrix} \underline{f}_1 \\ \underline{f}_2 \end{pmatrix}, \tag{1}$$

arising in mixed finite element approximation of second-order, self-adjoint elliptic PDEs. The matrix $A \in I\!R^{n \times n}$ is symmetric and positive definite and we assume that the discretisation is inf-sup stable (see [4],[5]). Without preconditioning, Krylov solution methods are not robust with respect to the discretisation parameter or the PDE coefficients; preconditioners are therefore essential. An *optimal* preconditioner is a matrix operator that accelerates the convergence rate of the chosen iterative solver so that convergence to a fixed tolerance is independent of the problem parameters.

Two distinct types of preconditioning scheme are possible. The first consists of solving the positive definite Schur complement system,

$$S\underline{p} = BA^{-1}\underline{f}_1 - \underline{f}_2, \tag{2}$$

with $S = BA^{-1}B^T$, using the preconditioned conjugate gradient method. This requires matrix-vector multiplication with S and thus computation of the action of A^{-1}. For the problems to be discussed, A is ill-conditioned and also requires a preconditioner. The result is nested iteration, the success of which depends on the selection of inner and outer stopping tolerances or other parameters (see [13]). Alternatively, system (1) can be solved directly using the minimal residual method (MINRES). It is easy to show that an optimal preconditioner for this is,

$$P_* = \begin{pmatrix} A & 0 \\ 0 & S \end{pmatrix}. \tag{3}$$

Clearly, an exact implementation is infeasible. For the Stokes problem, and the variable diffusion problem to be discussed in §2 and §3, however, one of the blocks of P_* corresponds to a discrete Laplacian operator or a scalar diffusion operator. We will show that one V-cycle or W-cycle of black-box algebraic multigrid is an optimal and feasible preconditioner for those blocks.

In §2, we review a preconditioning strategy presented by Silvester and Wathen in [14] and [15]. In §3, we extend it to a variable diffusion problem and prove that algebraic multigrid preconditioning for S combined with diagonal scaling for A yields an optimal preconditioning scheme. Eigenvalue bounds are established. Numerical results are presented for both problems, demonstrating the effectiveness and simplicity of our approach.

First, we review convergence properties of MINRES. Let us denote the symmetric and indefinite system (1) by $C\underline{x} = \underline{b}$ and suppose that a preconditioner with symmetric and positive definite blocks,

$$P = \begin{pmatrix} P_1 & 0 \\ 0 & P_2 \end{pmatrix}, \tag{4}$$

has been chosen. Given a zero initial guess $\underline{x}_0$, the iterative solver MINRES (see [7]) applied to the symmetrically preconditioned system,

$$\begin{pmatrix} P_1^{-\frac{1}{2}}AP_1^{-\frac{1}{2}} & P_1^{-\frac{1}{2}}B^TP_2^{-\frac{1}{2}} \\ P_2^{-\frac{1}{2}}BP_1^{-\frac{1}{2}} & 0 \end{pmatrix}\begin{pmatrix} \underline{y} \\ \underline{z} \end{pmatrix} = \begin{pmatrix} P_1^{-\frac{1}{2}}\underline{f}_1 \\ P_2^{-\frac{1}{2}}\underline{f}_2 \end{pmatrix}, \quad \begin{pmatrix} \underline{y} \\ \underline{z} \end{pmatrix} = \begin{pmatrix} P_1^{\frac{1}{2}}\underline{u} \\ P_2^{\frac{1}{2}}\underline{p} \end{pmatrix}, \tag{5}$$

generates a sequence of iterates $\underline{x}_k$, belonging to the Krylov space,

$$K_k = \underline{x}_0 + \mathbf{span}\left(P^{-1}\underline{r}_0, P^{-1}CP^{-1}\underline{r}_0, \ldots, \left(P^{-1}C\right)^{k-1}P^{-1}\underline{r}_0\right),$$

with the minimisation property,

$$\| \underline{b} - C\underline{x}_k \|_{P^{-1}} = \min_{\underline{x} \in K_k} \| \underline{b} - C\underline{x} \|_{P^{-1}},$$

where $\underline{r}_k = \underline{b} - C\underline{x}_k$ is the kth residual and $\| \underline{v} \|_{P^{-1}}^2 = \underline{v}^T P^{-1} \underline{v}$.

Since P is symmetric, it can be shown that a sharp upper bound for the residual error after k iterations (see [6, p.51]) is given by,

$$\frac{\| \underline{r}^{(k)} \|_{P^{-1}}}{\| \underline{r}^{(0)} \|_{P^{-1}}} \leq \min_{p_k} \max_{i=1:(n+m)} | p_k(\lambda_i) |, \tag{6}$$

where p_k is a polynomial of degree k satisfying $p_k(0) = 1$, and $\{\lambda_i\}_{i=1}^{n+m}$ denotes the eigenvalues of $P^{-1}C$. In summary, the convergence rate of MINRES is completely determined by the eigenvalue spectrum of $P^{-1}C$. For a robust solution scheme, we seek P such that inclusion intervals for the values $\{\lambda_i\}_{i=1}^{n+m}$ are independent of the problem parameters.

1.1 Notation

It is now instructive to recall the definitions of some Sobolev spaces and their associated norms. Let $\Omega \subseteq \mathbb{R}^d$, $(d = 2, 3)$, be a polygon with boundary $\partial\Omega$. $L^2(\Omega)$ is the usual space of square-integrable scalar functions with inner-product $(\cdot, \cdot)$. For the space of vector functions $L^2(\Omega)^d$, the definition is understood to hold componentwise. We write $\| \cdot \|_0$ to denote the Lebesgue measure for both spaces. The subspace,

$$H(div; \Omega) = \left\{ \boldsymbol{v} \in L^2(\Omega)^d \mid \nabla \cdot \boldsymbol{v} \in L^2(\Omega) \right\},$$

contains vectors with square-integrable divergence. The associated inner-product is $(\boldsymbol{u}, \boldsymbol{v})_{div} = (\boldsymbol{u}, \boldsymbol{v}) + (\nabla \cdot \boldsymbol{u}, \nabla \cdot \boldsymbol{v})$ and we denote the induced norm $\| \cdot \|_{div}$. The Sobolev space,

$$H_0^1(\Omega) = \left\{ w \in L^2(\Omega) \mid \nabla w \in L^2(\Omega)^d,\ w = 0 \text{ on } \partial\Omega \right\},$$

is equipped with the inner-product $(w, v)_1 = (w, v) + (\nabla w, \nabla v)$ and the induced norm $\| \cdot \|_1$. The space of vector functions $\left(H_0^1(\Omega)\right)^d$ and the associated norm are analogously defined.

2 The Stokes Problem

Let $\Omega \subset \mathbb{R}^d$, $(d = 2, 3)$, be a specified flow domain, with piecewise smooth boundary $\partial\Omega$. For simplicity in our description, we consider a 'no-flow' boundary condition[1]. The classical Stokes problem is to find a velocity $\boldsymbol{u}$ and a

[1] The generalisation to a non-zero boundary condition is straightforward.

pressure p satisfying,

$$\begin{aligned} \nu\nabla^2 \boldsymbol{u} + \nabla p &= \boldsymbol{f}, \\ \nabla \cdot \boldsymbol{u} &= 0 \quad \text{in } \Omega, \\ \boldsymbol{u} &= 0 \quad \text{on } \partial\Omega. \end{aligned} \tag{7}$$

The parameter ν represents viscosity of the fluid, $\boldsymbol{f}$ is a given forcing term and the second equation is an incompressibility constraint. A standard, conforming mixed finite element approximation is obtained by constructing finite-dimensional subspaces $\boldsymbol{V}_h \subset \left(H_0^1(\Omega)\right)^d$ and $W_h \subset L^2(\Omega)$, consisting of piecewise polynomials defined on a partition of Ω into regular polygons of edge length h. We then seek $\boldsymbol{u}_h \in \boldsymbol{V}_h$, $p_h \in W_h$ satisfying,

$$\nu\left(\nabla \boldsymbol{u}_h, \nabla \boldsymbol{v}\right) - \left(p_h, \nabla \cdot \boldsymbol{v}\right) = (\boldsymbol{f}, \boldsymbol{v}) \quad \forall \boldsymbol{v} \in \boldsymbol{V}_h \tag{8}$$

$$\left(\nabla \cdot \boldsymbol{u}_h, w\right) = 0 \qquad \forall w \in W_h. \tag{9}$$

Choosing basis functions,

$$\boldsymbol{V}_h = \mathbf{span}\left\{\boldsymbol{\psi}_i\right\}_{i=1}^n, \quad W_h = \mathbf{span}\left\{\varphi_j\right\}_{j=1}^m,$$

leads to a linear algebra problem of the form (1). We obtain,

$$\begin{pmatrix} \nu A & B^T \\ B & 0 \end{pmatrix} \begin{pmatrix} \underline{u} \\ \underline{p} \end{pmatrix} = \begin{pmatrix} \underline{f} \\ \underline{0} \end{pmatrix}, \tag{10}$$

where $\boldsymbol{u}_h = \sum_{i=1}^n u_i \boldsymbol{\psi}_i$ and $p_h = \sum_{j=1}^m p_j \varphi_j$ with matrices,

$$\begin{aligned} A(i,j) &= \left(\nabla \boldsymbol{\psi}_i, \nabla \boldsymbol{\psi}_j\right) \quad i,j = 1:n, \\ B(k,j) &= -\left(\nabla \cdot \boldsymbol{\psi}_j, \varphi_k\right) \quad k = 1:m,\ j = 1:n, \end{aligned}$$

and vector $f(i) = (\boldsymbol{f}, \boldsymbol{\psi}_i)$ for $i = 1:n$. Here, A is a vector Laplacian matrix with d diagonal blocks corresponding to scalar Laplacian matrices.

To ensure existence and uniqueness of a solution to the problem (8)–(9), $\boldsymbol{V}_h$ and W_h must be compatible. Specifically, the following discrete inf-sup stability condition must hold:

$$\beta \parallel w \parallel_0 \leq \sup_{\boldsymbol{v} \in \boldsymbol{V}_h} \frac{(w, \nabla \cdot \boldsymbol{v})}{\parallel \nabla \boldsymbol{v} \parallel_0} \quad \forall w \in W_h,\ w \neq \text{ constant} \tag{11}$$

where $\beta \geq \beta_0 > 0$ and β_0 is a constant independent of h. See [5] for a thorough discussion on stability criteria for mixed finite element methods. Defining the pressure mass matrix $Q(k,l) = (\varphi_k, \varphi_l)$, condition (11) can be expressed in matrix form as,

$$\beta^2 \le \frac{\underline{p}^t B A^{-1} B^T \underline{p}}{\underline{p}^t Q \underline{p}} \quad \forall \underline{p} \in I\!R^m \backslash \{\underline{1}\} . \tag{12}$$

That is, β is the square-root of the minimum eigenvalue of $Q^{-1}BA^{-1}B^T$ (see [4], [19]). It can also be shown that for a conforming finite element space,

$$\frac{\underline{p}^t B A^{-1} B^T \underline{p}}{\underline{p}^t Q \underline{p}} \le 1 \quad \forall \underline{p} \in I\!R^m \backslash \{\underline{0}\} . \tag{13}$$

2.1 Preconditioning Strategy

In [14] and [15], Silvester and Wathen proved that for any stable finite element method, an optimal eigenvalue bound is obtained for the Stokes system matrix if it is preconditioned by a matrix of the form,

$$P = \begin{pmatrix} \nu P_A & 0 \\ 0 & \frac{1}{\nu} P_Q \end{pmatrix}, \tag{14}$$

for which there exist constants, λ, Λ, θ and Θ, independent of h such that,

$$\lambda^2 \le \frac{\underline{u}^t A \underline{u}}{\underline{u}^t P_A \underline{u}} \le \Lambda^2 \quad \forall \underline{u} \in I\!R^n \backslash \{\underline{0}\} , \tag{15}$$

$$\theta^2 \le \frac{\underline{p}^t Q \underline{p}}{\underline{p}^t P_Q \underline{p}} \le \Theta^2 \quad \forall \underline{p} \in I\!R^m \backslash \{\underline{0}\} . \tag{16}$$

Specifically, they proved the following theorem.

Theorem 1. *Assume a stable discretisation satisfying (12) and (13) and a preconditioner of the form (14) satisfying (15) and (16), then the eigenvalues of the preconditioned Stokes matrix lie in the union of the intervals,*

$$\begin{aligned} &\left[\tfrac{1}{2}\left(\lambda^2 - \sqrt{\lambda^4 + 4\Theta^2\Lambda^2}\right), \tfrac{1}{2}\left(\lambda^2 - \sqrt{\lambda^4 + 4\beta^2\theta^2\lambda^2}\right)\right], \\ &\quad \cup \left[\lambda^2, \tfrac{1}{2}\left(\Lambda^2 + \sqrt{\Lambda^4 + 4\Theta^2\Lambda^2}\right)\right]. \end{aligned} \tag{17}$$

Proof. See [15]. □

Remark 1. For unstable discretisations, an appropriate stabilisation must be applied. The resulting linear systems can be preconditioned using the same method. See [14] for a discussion of eigenvalue bounds.

Using a result of Wathen [18] it is possible to show that (16) holds with $P_Q = \mathtt{diag}\,(Q)$. In other words, diagonal scaling is an optimal preconditioner

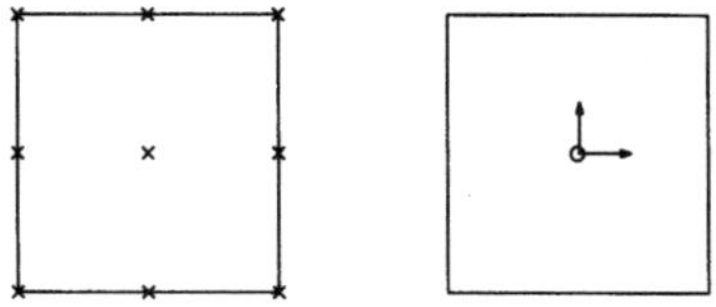

Fig. 1. $Q_2 - P_{-1}$ element; velocity components (left), pressure $\left(p, \frac{\partial p}{\partial x}, \frac{\partial p}{\partial y}\right)$ components (right)

for the Stokes pressure mass matrix. One of the best Stokes approximations in $\mathbb{R}^2$ is the $Q_2 - P_{-1}$ finite element (see Fig. 1), consisting of piecewise biquadratic velocity and discontinuous piecewise linear pressure. In this case, Q is a diagonal matrix. Thus, $P_Q = Q$ is an optimal choice with $\theta^2 = \Theta^2 = 1$.

The task of defining an efficient preconditioner for (10) is thus reduced to finding P_A satisfying (15). Since A is a discrete representation of the Laplacian operator ∇^2, any black-box solver for Poisson problems is a potential candidate. Silvester and Wathen demonstrate numerically in [15] that one V-cycle of *geometric* multigrid (see [2]) is an optimal choice. Recently, there has been a resurgence of interest in *algebraic* multigrid (AMG) schemes. Introduced by Ruge and Stüben (see [11], [12]) in the 1980s, AMG represents the new generation of multigrid (see [16] for a review of current trends and related literature). No geometric information is required; coarsening and interpolation are performed by exploiting only information in the coefficient matrix. In this way, AMG is more generally applicable than traditional multigrid with respect to finite element mesh geometry and PDE coefficients. We choose P_A in (14) to be one V-cycle or W-cycle of the AMG algorithm `amg1r5` (see [11]) applied to the vector Laplacian matrix A. We implement it as a black-box; no parameters are estimated a-priori.

2.2 Numerical Example

Consider the Stokes equations (7), discretised on $[-1,1] \times [-1,1]$ using stable $Q_2 - P_{-1}$ approximation, with $\nu = 1$, $\boldsymbol{f} = \mathbf{0}$ and velocity boundary condition,

$$(u_x, u_y) = \left\{ \left(1 - x^4, 0\right) \mid y = 1, -1 \leq x \leq 1 \right\}. \tag{18}$$

This is the so-called 'regularised driven-cavity' test problem. We apply preconditioned MINRES to the assembled system with a stopping tolerance of 10^{-6} on the relative residual error in the $\|\cdot\|_{P^{-1}}$ norm. The Laplacian blocks of A are preconditioned separately using either one V-cycle ($P_A = A_V$) or one W-cycle ($P_A = A_W$) of `amg1r5` (implemented with symmetric smoothing).

MINRES iteration counts for the preconditioned system are reported in Table 1. The second column contains unpreconditioned counts; the third column lists

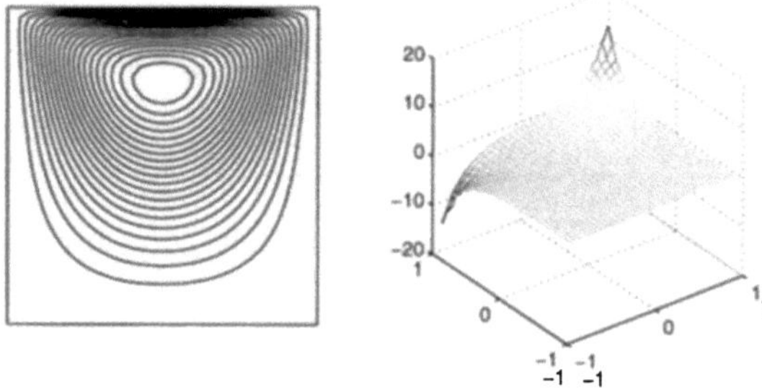

Fig. 2. Contour plot of constant streamlines (left) and pressure solution (right)

Table 1. MINRES iterations and (time), TOL = 10^{-6}

h	$P = I$	$P_A = A$		$P_A = A_V$		$P_A = A_W$	
$\frac{1}{8}$	64	22	(0.2)	25	(0.2)	25	(0.2)
$\frac{1}{16}$	147	24	(1.3)	26	(0.5)	26	(0.6)
$\frac{1}{32}$	274	24	(11.5)	28	(2.2)	28	(2.3)
$\frac{1}{64}$	513	24	(96.8)	29	(9.2)	28	(9.2)

Table 2. Eigenvalues of indefinite preconditioned system; $P_A = A_V$

h	λ^2	Λ^2	β^2	Γ^2	Observed eigenvalues
$\frac{1}{8}$	0.951	1	0.256	0.998	$[-0.615, -0.211] \cup [0] \cup [0.960, 1.613]$
$\frac{1}{16}$	0.917	1	0.235	0.999	$[-0.618, -0.196] \cup [0] \cup [0.939, 1.617]$
$\frac{1}{32}$	0.884	1	0.222	1.000	$[-0.618, -0.186] \cup [0] \cup [0.924, 1.617]$

counts for the exact ($P_A = A$) version of the preconditioner. The time units reported in parentheses are elapsed time in seconds for the total solve using a mex fortran interface in Matlab 6.0 on a SUN ultraSPARC workstation.

Iteration counts do not grow with mesh refinement showing that the AMG preconditioner is optimal. Further, the solve times in the AMG trials grow only linearly with respect to the problem size. The eigenvalues of the preconditioned indefinite system, for the V-cycle case, are reported in Table 2. Observe that λ^2 and Λ^2 are independent of h. Substituting $\theta^2 = 1$, $\Theta^2 = 1$, $\lambda^2 = 0.884$, $\Lambda^2 = 1$ and $\beta^2 = 0.222$ in (17) yields the theoretical bound, $[-0.651, -0.184] \cup [0.884, 1.618]$. Fig. 3 shows the spectral equivalence of the AMG and the exact versions of the preconditioner. Note that in this example the preconditioned system is singular. The nullspace is spanned by the hydrostatic constant pressure solution. This accounts for the zero eigenvalue.

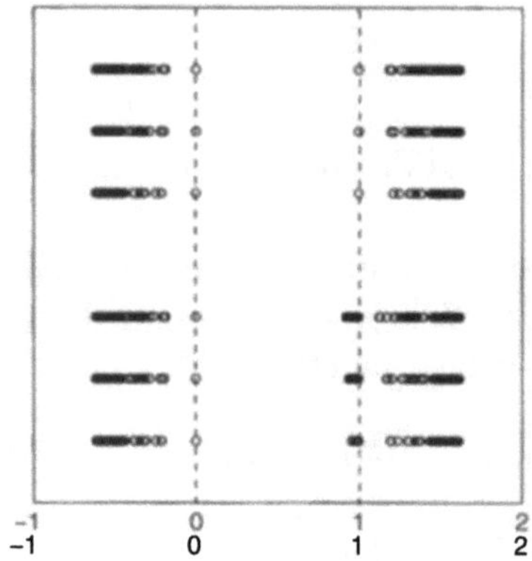

Fig. 3. Eigenvalues of indefinite preconditioned matrix; $P_A = A$ (top), $P_A = A_V$ (bottom), $h = \frac{1}{8}, \frac{1}{16}, \frac{1}{32}$

The experiment shows that we have a simple, optimal preconditioner for the Stokes equations. It is worth pointing out that the preconditioning strategy can also be extended to a generalized (time-dependent) Stokes flow. The corresponding discrete system is given by,

$$\begin{pmatrix} \frac{1}{\Delta_t}G + \nu A & B^T \\ B & 0 \end{pmatrix} \begin{pmatrix} \underline{u} \\ \underline{p} \end{pmatrix} = \begin{pmatrix} \underline{f} \\ \underline{0} \end{pmatrix}, \tag{19}$$

where G is the velocity mass matrix and Δ_t is the time-step. An optimal preconditioner is,

$$P = \begin{pmatrix} P_F & 0 \\ 0 & P_S \end{pmatrix} \tag{20}$$

where P_F is a V-cycle of AMG applied to $F = \frac{1}{\Delta_t}G + \nu A$ and P_S is a preconditioner for the Schur complement $BF^{-1}B^T$. Following the discussion in [3], it is known that an optimal choice is $P_S^{-1} = \frac{1}{\Delta_t}P_A^{-1} + \nu P_Q^{-1}$, where P_A denotes one V-cycle of AMG applied to a discrete Laplacian defined on the pressure space and P_Q is a spectrally equivalent approximation to the mass matrix. We now introduce a second important class of elliptic PDEs.

3 A Variable Diffusion Problem

Let Ω be a bounded domain in $\mathbb{R}^2$. We consider scalar, second-order elliptic problems of the form,

$$
\begin{aligned}
-\nabla \cdot \mathcal{A} \nabla p &= f \quad \text{in } \Omega, \\
p &= g \quad \text{on } \partial\Omega_D, \\
\mathcal{A} \nabla p \cdot \boldsymbol{n} &= 0 \quad \text{on } \partial\Omega_N,
\end{aligned} \tag{21}
$$

where $\partial\Omega_D \neq \emptyset$, $\boldsymbol{n}$ denotes the unit outward normal vector and f and g are given functions. $\mathcal{A} = \sigma(\boldsymbol{x})\mathcal{I}$ is a 2×2 bounded, symmetric and uniformly positive definite matrix-valued function with,

$$
0 < \sigma_{min} \leq \sigma(\boldsymbol{x}) \leq \sigma_{max} < \infty, \quad \forall \boldsymbol{x} \in \Omega. \tag{22}
$$

This is a simple model of fluid flow in porous media; the diffusion coefficient $\sigma(\boldsymbol{x})$ is commonly discontinuous or anisotropic. Substituting $\boldsymbol{u} = \mathcal{A}\nabla p$ yields the first-order system,

$$
\begin{aligned}
\mathcal{A}^{-1}\boldsymbol{u} - \nabla p &= 0, \\
\nabla \cdot \boldsymbol{u} &= -f \quad \text{in } \Omega, \\
p &= g \quad \text{on } \partial\Omega_D, \\
\boldsymbol{u} \cdot \boldsymbol{n} &= 0 \quad \text{on } \partial\Omega_N.
\end{aligned} \tag{23}
$$

This formulation is favoured when the velocity $\boldsymbol{u}$ is the variable of interest. It is known that mixed velocity solutions are more robust than standard finite element methods with respect to any variation in $\sigma(\boldsymbol{x})$ (see [9]). Another crucial advantage in the modelling of groundwater flow ($f = 0$) is local conservation of mass.

To formulate the weak problem, we choose finite-dimensional subspaces, $\boldsymbol{V}_h \subset H_{0,N}(div; \Omega) = \{\boldsymbol{v} \in H(div;\Omega) \mid \boldsymbol{v} \cdot \boldsymbol{n} = 0 \text{ on } \partial\Omega_N\}$ and $W_h \subset L^2(\Omega)$, consisting of piecewise polynomials defined on a partition of Ω into triangles of edge length h. Then we look for $\boldsymbol{u}_h \in \boldsymbol{V}_h$, $p_h \in W_h$ satisfying,

$$
\begin{aligned}
(\mathcal{A}^{-1}\boldsymbol{u}_h, \boldsymbol{v}) + (p_h, \nabla \cdot \boldsymbol{v}) &= < g, \boldsymbol{v} \cdot \boldsymbol{n} > \quad \forall \boldsymbol{v} \in \boldsymbol{V}_h, \\
(w, \nabla \cdot \boldsymbol{u}_h) &= -(f, w) \quad \forall w \in W_h,
\end{aligned} \tag{24}
$$

where $\langle g, \boldsymbol{v} \cdot \boldsymbol{n} \rangle = \int_{\partial\Omega_D} g \boldsymbol{v} \cdot \boldsymbol{n} \, ds$. A unique solution exists (see [5]), if and only if,

$$
\alpha \parallel \boldsymbol{v} \parallel_{div}^2 \leq (\mathcal{A}^{-1}\boldsymbol{v}, \boldsymbol{v}) \qquad \forall \boldsymbol{v} \in \boldsymbol{Z}_h, \tag{25}
$$

with $\boldsymbol{Z}_h = \{\boldsymbol{v} \in \boldsymbol{V}_h \mid (w, \nabla \cdot \boldsymbol{v}) = 0 \quad \forall w \in W_h\}$ and,

$$
\beta \parallel w \parallel_0 \leq \sup_{\boldsymbol{v} \in V_h} \frac{(w, \nabla \cdot \boldsymbol{v})}{\parallel \boldsymbol{v} \parallel_{div}} \qquad \forall w \in W_h, \tag{26}
$$

where $\alpha \geq \alpha_0 > 0$, $\beta \geq \beta_0 > 0$ and α_0 and β_0 are constants independent of h.

We choose the lowest-order Raviart-Thomas finite element spaces. Stability of the method is established in [10]. For a given triangulation T_h of Ω we have,

$$\boldsymbol{V}_h = \{\boldsymbol{v} \in H(div;\Omega) \mid \boldsymbol{v} \mid_T \in (P_0^2(T) + [x,y]P_0(T)) \quad \forall T \in T_h\} \quad (27)$$

and $W_h = \{w \in L^2(\Omega) \mid w \mid_T \in P_0(T) \quad \forall T \in T_h\}$ where $P_0(T)$ denotes the set of polynomials of degree zero defined on triangle T. Velocities are linear in each component; the pressure space consists of piecewise constant functions. For an $H(div;\Omega)$ conforming velocity approximation, the degrees of freedom are normal components at edge midsides; the pressure solution is sampled at element centroids (see Fig. 4).

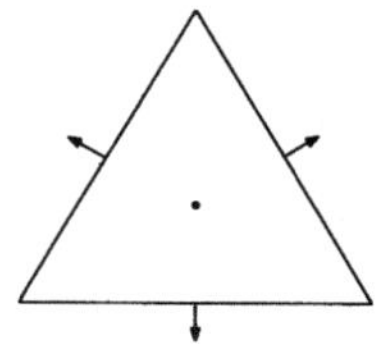

Fig. 4. The lowest order Raviart-Thomas element

Choosing bases, $\boldsymbol{V}_h = \mathbf{span}\,\{\boldsymbol{\psi}_i\}_{i=1}^n$, $W_h = \mathbf{span}\,\{\varphi_j\}_{j=1}^m$, we obtain the linear algebra problem,

$$\begin{pmatrix} A & B^T \\ B & 0 \end{pmatrix} \begin{pmatrix} \underline{u} \\ \underline{p} \end{pmatrix} = \begin{pmatrix} \underline{g} \\ \underline{f} \end{pmatrix}, \quad (28)$$

with matrices A and B defined by,

$$A(i,j) = \left(\mathcal{A}^{-1}\boldsymbol{\varphi}_i, \boldsymbol{\varphi}_j\right) \quad i,j = 1:n,$$
$$B(k,j) = \left(\nabla \cdot \boldsymbol{\varphi}_j, \varphi_k\right) \quad k = 1:m,\ j = 1:n,$$

and vectors $g(i) = \langle g, \boldsymbol{\varphi_i} \cdot \boldsymbol{n}\rangle$, $f(k) = -(f, \varphi_k)$ for $i = 1:n$ and $k = 1:m$.

3.1 Preconditioning Strategy

Preconditioning strategies for the indefinite problem (28) are proposed in [1] and [17]. The former discusses the case $\sigma(\boldsymbol{x}) = 1$; the latter requires the computation of acceleration parameters. A theoretically optimal, parameter-free preconditioner for general coefficients is discussed in [8]. Here, we propose the block-diagonal preconditioner,

$$P = \begin{pmatrix} D_A & 0 \\ 0 & P_S \end{pmatrix}, \quad (29)$$

where $D_A = \mathtt{diag}(A)$ and P_S is black-box AMG preconditioning for the approximate Schur complement $S_D = BD_A^{-1}B^T$. Our choice is motivated by the observation that for $\mathcal{A}$ of the form,

$$\mathcal{A} = \begin{pmatrix} \sigma(\boldsymbol{x}) & 0 \\ 0 & \sigma(\boldsymbol{x}) \end{pmatrix}, \tag{30}$$

diagonal scaling for A is an optimal choice, independent of $\sigma(\boldsymbol{x})$ (see Lemma 1 below). The Schur complement represents the generalised diffusion operator $\nabla \cdot \mathcal{A}\nabla$ and AMG is known to be an efficient *solver* for linear systems arising from discretisations of such operators (see [11]).

We now prove that the preconditioner (29) is optimal. Our starting point is the eigenvalue bound established by Rusten and Winther in [13].

Theorem 2. *Let $0 < \mu_1 \ldots \leq \mu_n$ be the eigenvalues of A and let $0 < \rho_1 \ldots \leq \rho_m$ be the singular values of B, then the eigenvalues of the indefinite system matrix (28) lie in the union of the intervals,*

$$\begin{aligned} &\left[\tfrac{1}{2}\left(\mu_1 - \sqrt{\mu_1^2 + 4\rho_m^2}\right), \tfrac{1}{2}\left(\mu_n - \sqrt{\mu_n^2 + 4\rho_1^2}\right)\right] \\ &\qquad \cup \left[\mu_1, \tfrac{1}{2}\left(\mu_n + \sqrt{\mu_n^2 + 4\rho_m^2}\right)\right]. \end{aligned} \tag{31}$$

Proof. See [13]. □

Consider, first, symmetric preconditioning for (28) with the exact (no AMG) version of the preconditioner (29). We obtain,

$$P^{-\frac{1}{2}} C P^{-\frac{1}{2}} = \begin{pmatrix} D_A^{-\frac{1}{2}} A D_A^{-\frac{1}{2}} & D_A^{-\frac{1}{2}} B^T \left(B D_A^{-1} B^T\right)^{-\frac{1}{2}} \\ \left(B D_A^{-1} B^T\right)^{-\frac{1}{2}} B D_A^{-\frac{1}{2}} & 0 \end{pmatrix} = \begin{pmatrix} \tilde{A} & \tilde{B}^T \\ \tilde{B} & 0 \end{pmatrix}.$$

Applying Theorem 2 to this preconditioned saddle-point system leads, trivially, to the following result.

Corollary 1. *Let $0 < \tilde{\mu}_1 \ldots \leq \tilde{\mu}_n$ be the eigenvalues of $D_A^{-1}A$, then the eigenvalues of,*

$$P^{-\frac{1}{2}} \begin{pmatrix} A & B^T \\ B & 0 \end{pmatrix} P^{-\frac{1}{2}}, \quad \textit{with } P = \begin{pmatrix} D_A & 0 \\ 0 & B D_A^{-1} B^T \end{pmatrix},$$

lie in the union of the intervals,

$$\left[\frac{1}{2}\left(\tilde{\mu}_1 - \sqrt{\tilde{\mu}_1^2 + 4}\right), \frac{1}{2}\left(\tilde{\mu}_n - \sqrt{\tilde{\mu}_n^2 + 4}\right)\right] \cup \left[\tilde{\mu}_1, \frac{1}{2}\left(\tilde{\mu}_n + \sqrt{\tilde{\mu}_n^2 + 4}\right)\right].$$

Proof. Observe that

$$\tilde{B}\tilde{B}^T = \left(B D_A^{-1} B^T\right)^{-\frac{1}{2}} B D_A^{-\frac{1}{2}} D_A^{-\frac{1}{2}} B^T \left(B D_A^{-1} B^T\right)^{-\frac{1}{2}} = I,$$

where I is the identity matrix. The result follows immediately from Theorem 2 since the singular values of $\tilde{B}$ satisfy $\tilde{\rho}_1 = \ldots = \tilde{\rho}_m = 1$ and $\tilde{A}$ has the same eigenvalue spectrum as $D_A^{-1}A$. □

Now, given a preconditioner P_S for $BD_A^{-1}B^T$ satisfying,

$$\theta^2 \leq \frac{\underline{p}^t BD_A^{-1}B^T \underline{p}}{\underline{p}^t P_S \underline{p}} \leq \Theta^2 \quad \forall \underline{p} \in I\!R^m \backslash \{\underline{0}\}.$$

for some constants θ and Θ, Theorem 2 can be extended to obtain the following theoretical eigenvalue bound for the indefinite preconditioned system.

Corollary 2. *Let* $0 < \tilde{\mu}_1 \ldots \leq \tilde{\mu}_n$ *be the eigenvalues of* $D_A^{-1}A$*, then the eigenvalues of,*

$$P^{-\frac{1}{2}} \begin{pmatrix} A & B^T \\ B & 0 \end{pmatrix} P^{-\frac{1}{2}}, \quad \textit{with } P = \begin{pmatrix} D_A & 0 \\ 0 & P_S \end{pmatrix},$$

lie in the union of the intervals,

$$\left[\frac{1}{2}\left(\tilde{\mu}_1 - \sqrt{\tilde{\mu}_1^2 + 4\Theta^2}\right), \frac{1}{2}\left(\tilde{\mu}_n - \sqrt{\tilde{\mu}_n^2 + 4\theta^2}\right)\right] \cup \left[\tilde{\mu}_1, \frac{1}{2}\left(\tilde{\mu}_n + \sqrt{\tilde{\mu}_n^2 + 4\Theta^2}\right)\right].$$

Proof. The result follows from Theorem 2 and Corollary 1 with $\tilde{B} = P_S^{-\frac{1}{2}} BD_A^{-\frac{1}{2}}$. Observe that,

$$\begin{aligned} \underline{p}^T \tilde{B}\tilde{B}^T \underline{p} &= \underline{p}^T P_S^{-\frac{1}{2}} BD_A^{-1}B^T P_S^{-\frac{1}{2}} \underline{p} \\ &\leq \Theta^2 \underline{p}^T P_S^{-\frac{1}{2}} P_S P_S^{-\frac{1}{2}} \underline{p} \\ &= \Theta^2 \underline{p}^T \underline{p}. \end{aligned}$$

That is, the maximum singular value $\tilde{\rho}_m$ of $\tilde{B}$ satisfies, $\tilde{\rho}_m^2 \leq \Theta^2$. Similarly, it can be shown that $\tilde{\rho}_1^2 \geq \theta^2$. □

In section 3.2 we show that black-box AMG is an optimal choice for P_S yielding constants θ and Θ that are independent of h. By Corollary 2, an optimal eigenvalue bound is thus obtained if and only if, the constants $\tilde{\mu}_1$ and $\tilde{\mu}_n$ satisfy,

$$\tilde{\mu}_1 \leq \frac{\underline{u}^t A \underline{u}}{\underline{u}^t D_A \underline{u}} \leq \tilde{\mu}_n \quad \forall \underline{u} \in I\!R^n \backslash \{\underline{0}\},$$

and are independent of the discretisation parameter h and the coefficient function $\sigma(\boldsymbol{x})$. Using a result of Wathen, [18], it is sufficient to consider the *element* matrices. Denoting by λ_{min}^k and λ_{max}^k the minimum and maximum eigenvalues of the diagonally preconditioned element matrix, $\left(D_A^k\right)^{-1} A^k$, associated with triangle k, we have,

$$\min_k \left\{\lambda_{min}^k\right\} \leq \tilde{\mu}_1, \quad \tilde{\mu}_n \leq \max_k \left\{\lambda_{max}^k\right\}.$$

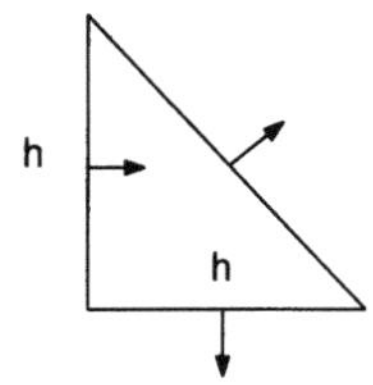

Fig. 5. Uniform triangles with orientated normal vectors

For simple geometries, we can compute explicit bounds for these values. As an illustration, consider uniform triangles as in Fig. 5 and fix normal vectors on each edge as shown.

Lemma 1. *Consider a square domain tiled with uniform triangles of edge length h. Fix normal vectors as shown in Fig. 5. Then,*

$$\frac{1}{2} \leq \min_k \left\{\lambda^k_{min}\right\}, \quad \max_k \left\{\lambda^k_{max}\right\} \leq \frac{3}{2}. \tag{32}$$

Proof. Labelling the Raviart-Thomas velocity degrees of freedom (x_j, y_j) with $j = 1:3$, the element velocity basis functions are,

$$\varphi_1 = \begin{pmatrix} -\frac{x_3}{h} + \frac{x}{h} \\ -1 - \frac{y_1}{h} + \frac{y}{h} \end{pmatrix}, \varphi_2 = \begin{pmatrix} -\frac{x_3\sqrt{2}}{h} + \frac{x\sqrt{2}}{h} \\ -\frac{y_1\sqrt{2}}{h} + \frac{y\sqrt{2}}{h} \end{pmatrix}, \varphi_3 = \begin{pmatrix} 1 + \frac{x_3}{h} - \frac{x}{h} \\ \frac{y_1}{h} - \frac{y}{h} \end{pmatrix}.$$

For element k, let

$$\mathcal{A}^k_c = \begin{pmatrix} \sigma(x^k_c, y^k_c) & 0 \\ 0 & \sigma(x^k_c, y^k_c) \end{pmatrix} = \begin{pmatrix} \sigma & 0 \\ 0 & \sigma \end{pmatrix}$$

denote $\mathcal{A}$ evaluated at the centroid $\left(x^k_c, y^k_c\right)$ of that element. The element velocity mass matrix is then defined by,

$$A^k(i,j) = \int\int_{\triangle} \left(\mathcal{A}^k_c\right)^{-1} \varphi_i \cdot \varphi_j \, dxdy \quad i,j = 1:3.$$

Integrating over triangle k yields,

$$A^k = h^2 \begin{pmatrix} \frac{1}{3\sigma} & 0 & \frac{1}{6\sigma} \\ 0 & \frac{1}{3\sigma} & 0 \\ \frac{1}{6\sigma} & 0 & \frac{1}{3\sigma} \end{pmatrix}, \quad \left(D^k_A\right)^{-1} A^k = \begin{pmatrix} 1 & 0 & \frac{1}{2} \\ 0 & 1 & 0 \\ \frac{1}{2} & 0 & 1 \end{pmatrix}. \tag{33}$$

Hence, $\lambda^k_1 = \frac{1}{2}$, $\lambda^k_2 = 1$, $\lambda^k_3 = \frac{3}{2}$. □

Note that for anisotropic coefficients of the form,

$$\mathcal{A}_c = \begin{pmatrix} \sigma_1(x_c, y_c) & 0 \\ 0 & \sigma_2(x_c, y_c) \end{pmatrix}$$

with $\sigma_1(x_c, y_c) \gg \sigma_2(x_c, y_c)$ or vice versa, diagonal preconditioning for A is not efficient on triangles of this type. The diagonally preconditioned element matrix (33) has a zero eigenvalue in the limit of the anisotropy. Rectangular Raviart-Thomas finite elements should be used in that case.

3.2 Numerical Examples

The following numerical examples illustrate the optimality of the preconditioner (29). We apply one V-cycle ($P_S = S_V$) or one W-cycle ($P_S = S_W$) of `amg1r5` to the sparse matrix $S_D = BD_A^{-1}B^T$.

First, consider (23) with $\sigma(\boldsymbol{x}) = 1$, (unit coefficients), $f = 1, g = 0$ and $\partial\Omega_N = \emptyset$, discretised on $\Omega = [0,1] \times [0,1]$. MINRES iteration counts for the assembled linear system are reported in Table 3. The second column corresponds to unpreconditioned MINRES; the symbol * signifies that more than 500 iterations were needed. We use the same stopping criterion as in the Stokes example.

Table 3. MINRES iterations and (time), TOL = 10^{-6}

h	$P = I$	$P_S = S_D$	$P_S = S_V$	$P_S = S_W$
$\frac{1}{16}$	186	26 (0.5)	26 (0.4)	26 (0.4)
$\frac{1}{32}$	375	26 (2.4)	26 (1.3)	26 (1.6)
$\frac{1}{64}$	*	26 (13.6)	26 (5.3)	26 (6.4)
$\frac{1}{128}$	*	26 (84.0)	26 (23.8)	26 (28.1)

Table 4. Eigenvalues of indefinite preconditioned system; $P_S = S_V$

h	$\tilde{\mu}_1$	$\tilde{\mu}_n$	θ^2	Θ^2	Observed eigenvalues
$\frac{1}{8}$	0.5	1.5	0.957	1	$[-0.777, -0.532] \cup [0.707, 1.896]$
$\frac{1}{16}$	0.5	1.5	0.955	1	$[-0.777, -0.508] \cup [0.707, 1.939]$
$\frac{1}{32}$	0.5	1.5	0.954	1	$[-0.777, -0.496] \cup [0.707, 1.964]$

The iteration counts with V-cycle and W-cycle AMG preconditioning are independent of h and match the iteration counts for the exact version of the preconditioner. Again, we observe that solve times grow almost[2] linearly with respect to the problem size. The eigenvalues of the preconditioned matrix with $P_S = S_V$ are shown in Table 4. We have $\theta^2 \approx 0.954$ and $\Theta^2 = 1$. Substituting these values in Corollary 2 gives the theoretical bound $[-0.781, -0.477] \cup [0.5, 2]$. Note that θ^2 is larger here than in the Stokes example. This can be attributed to the different discretisation method.

Table 5. MINRES iterations and (time), TOL $= 10^{-6}$

h	$P = I$	$P_S = S_D$	$P_S = S_V$	$P_S = S_W$
$\frac{1}{16}$	*	25 (0.5)	25 (0.4)	25 (0.5)
$\frac{1}{32}$	*	25 (2.3)	26 (1.3)	28 (1.5)
$\frac{1}{64}$	*	25 (12.7)	27 (5.6)	25 (6.4)
$\frac{1}{128}$	*	25 (81.1)	27 (25.4)	25 (27.6)

We now consider a discontinuous coefficient example. Take

$$\sigma(\boldsymbol{x}) = \begin{cases} 10^{-3} & \text{in } \Omega_* = [0.25, 0.75] \times [0.25, 1] \\ 1 & \text{in } \Omega \backslash \Omega_* \end{cases}$$

with $f = 0$ and mixed boundary conditions: $\boldsymbol{u} \cdot \boldsymbol{n} = 0$ on $\partial\Omega_N$, $p = 1 - x$ on $\partial\Omega_D$, where $\partial\Omega_N = [0,1] \times 0 \cup \{0,1\} \times [0, 0.75]$ and $\partial\Omega_D = \Omega \backslash \Omega_N$. Iteration counts are given in Table 5.

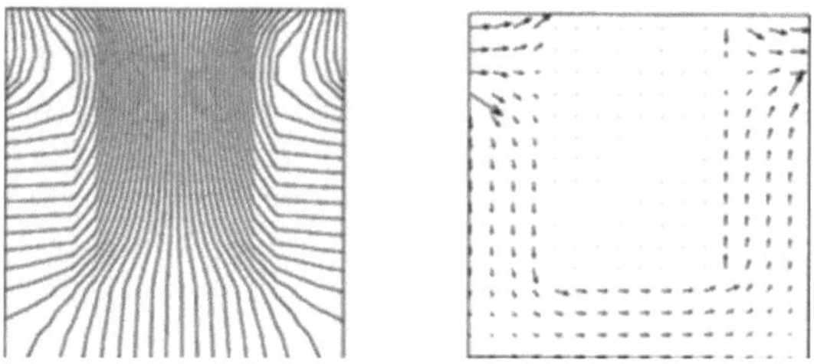

Fig. 6. Pressure contours (left) and velocity field (right)

The AMG preconditioner is optimal with repect to the mesh parameter and the solve times grow linearly. Eigenvalues for the V-cycle case are reported in Table 6. We observe $\theta^2 \approx 0.835$ and $\Theta^2 = 1$, yielding the theoretical bound $[-0.781, -0.433] \cup [0.5, 2]$ for the preconditioned system. Note that θ^2 does change with $\sigma(\boldsymbol{x})$ but not significantly. It is bounded below by a constant that is independent of h and not close to zero. Fig. 7 illustrates the spectral

[2] There is a small overhead associated with the set-up phase of AMG.

equivalence of the exact and inexact versions of the preconditioner. Comparing the iteration counts for the two examples, we see that the AMG preconditioner is completely insensitive to the coefficient $\sigma(\boldsymbol{x})$.

Table 6. Eigenvalues of indefinite preconditioned system; $P_S = S_V$

h	$\tilde{\mu}_1$	$\tilde{\mu}_n$	θ^2	Θ^2	Observed eigenvalues
$\frac{1}{8}$	0.5	1.5	0.918	1	$[-0.774, -0.480] \cup [0.822, 1.965]$
$\frac{1}{16}$	0.5	1.5	0.845	1	$[-0.775, -0.448] \cup [0.822, 1.965]$
$\frac{1}{32}$	0.5	1.5	0.836	1	$[-0.775, -0.452] \cup [0.822, 1.974]$

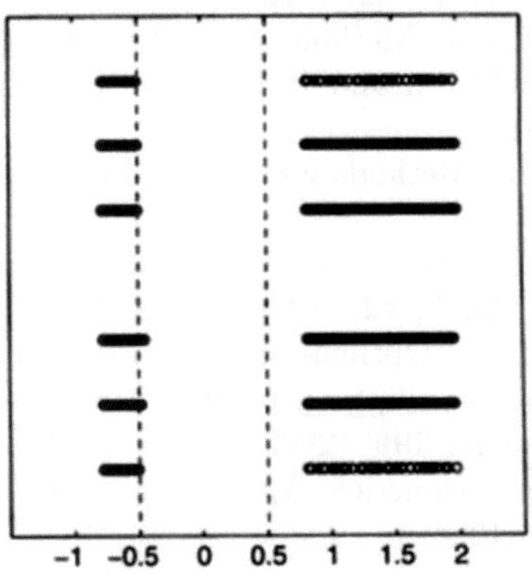

Fig. 7. Eigenvalues of indefinite preconditioned matrix; $P_S = S_D$ (top), $P_S = S_V$ (bottom), $h = \frac{1}{8}, \frac{1}{16}, \frac{1}{32}$

4 Concluding Remarks

In this paper we have shown that the symmetric indefinite linear systems arising from mixed finite element formulation of generalized Stokes and variable diffusion problems can be solved efficiently using a generic black-box preconditioning strategy. The preconditioner discussed essentially consists of diagonal scaling for a mass matrix and one V-cycle or W-cycle of algebraic multigrid applied to discrete Laplacian or diffusion blocks. The resulting method is simple, robust with respect to the PDE coefficients, yields optimal rates of convergence for MINRES and there are no parameters to estimate.

Acknowledgements We would like to thank the Fraunhofer institute SCAI (a former GMD institute) for permission to publish numerical results obtained using the fortran 77 code `amg1r5`.

References

1. Arnold, D.N., Falk, R.S., Winther, R.: Preconditioning in $H(div)$ and applications, Math. Comp., **66**, 219, (1997) 957–984.
2. Bramble, J.H. : Multigrid Methods. Pitman Research Notes in Mathematics Series, **294**, Longman (1993).
3. Bramble, J.H., Pasciak, J.E.: Iterative techniques for time-dependent Stokes problems. Comput. Math. Appl. **33** (1997) 13–30.
4. Brezzi, F., Bathe, K.J.: A discourse on the stability conditions for mixed finite element formulations. Comp. Methods Appl. Mec. Engrg., **82** (1990) 27–57.
5. Brezzi, F., Fortin, M.: Mixed and Hybrid Finite Element Methods. Springer-Verlag, New York (1991).
6. Greenbaum, A.: Iterative Methods for Solving Linear Systems. SIAM, Philidelphia (1997).
7. Paige, C.C., Saunders, A.: Solution of sparse indefinite systems of linear equations. SIAM J. Numer. Anal., **12**, (1975) 617–629.
8. Powell, C.E., Silvester, D.J.: Optimal preconditioning for Raviart-Thomas mixed formulation of second-order elliptic problems. Manchester Centre for Computational Mathematics Report, **399**, (2002).
9. Quarteroni, A., Valli, A.: Numerical Approximation of Partial Differential Equations, Springer-Verlag, (1994).
10. Raviart, P.A., Thomas, J.A.: A mixed finite element method for second order elliptic problems. Mathematical Aspects of the Finite Element Method. Lecture Notes in Mathematics, **606**. Springer-Verlag, New York (1977).
11. Ruge, J.W., Stüben, K.: Efficient solution of finite difference and finite element equations by algebraic multigrid (AMG). In: Paddon, D.J., Holstein, H., (eds.): Multigrid Methods for Integral and Differential Equations. The Institute of Mathematics and its Applications Conference Series. Oxford, (1985) 169–212
12. Ruge, J.W., and Stüben, K.: Algebraic Multigrid. In: McCormick S.F. (ed.): Multigrid Methods. SIAM, Philadelphia, (1987) 73–130.
13. Rusten, T., Winther, R.: A preconditioned iterative method for saddlepoint problems. SIAM J. Matrix. Anal. Appl., **13**, 3, (1992) 887–904.
14. Silvester, D., Wathen, A.: Fast iterative solution of stabilised Stokes systems part II: using general block preconditioners. SIAM J. Numer. Anal., **31**, 5, (1994) 1352–1367.
15. Silvester, D., Wathen, A.: Fast and robust solvers for time-discretised incompressible Navier-Stokes equations. In: Griffiths, D.F., Watson, G.A, (eds.): Numerical Analysis 1995. Pitman Research Notes in Mathematics Series, **344**, Longman (1996).
16. Stüben, K.: A review of algebraic multigrid. J. Comput. Appl. Math., **128**, (2001) 281–309.
17. Vassilevski, P.S., Lazarov, R.D.: Preconditioning mixed finite element saddle-point elliptic problems. Numer. Linear Algebra Appl. **3**, 1, (1996) 1–20.

18. Wathen, A.J.: Realistic eigenvalue bounds for the Galerkin mass matrix. IMA J. Numer. Anal., **7**, 1, (1987) 449–457.
19. Wathen, A., Silvester, D.: Fast iterative solution of stabilised stokes systems part I: using simple diagonal preconditioners. SIAM J. Numer. Anal., **30**, 3, (1993) 630–649.

Author Index

Bastian, Peter 1

Caboussat, Alexandre 69
Clees, Tanja 110

Hübner, Friedrich-Karl 131
Hebermehl, Georg 131
Heinrich, Wolfgang 131
Hron, Jaroslav 87

Klar, Axel 211

Lavrova, Olga 160

Maronnier, Vincent 69
Matthies, Gunar 160
Mitkova, Teodora 160

Neuss, Nicolas 237

Ouazzi, Abderrahim 87

Picasso, Marco 69

Poleviko, Viktor 160
Powell, Catherine 268

Rappaz, Jacques 69
Rivkind, Liudmila S. 37
Roos, Hans–Görg 246

Schlundt, Rainer 131
Seaïd, Mohammed 211
Silvester, David 268
Stüben, Klaus 110

Tischler, Thorsten 131
Tobiska, Lutz 160
Turek, Stefan 37, 87

Walkington, Noel J. 23
Wan, Decheng 37

Zarin, Helena 246
Zscheile, Horst 131
Zunino, Paolo 184

Editorial Policy

§1. Volumes in the following three categories will be published in LNCSE:

i) Research monographs
ii) Lecture and seminar notes
iii) Conference proceedings

Those considering a book which might be suitable for the series are strongly advised to contact the publisher or the series editors at an early stage.

§2. Categories i) and ii). These categories will be emphasized by Lecture Notes in Computational Science and Engineering. **Submissions by interdisciplinary teams of authors are encouraged.** The goal is to report new developments – quickly, informally, and in a way that will make them accessible to non-specialists. In the evaluation of submissions timeliness of the work is an important criterion. Texts should be well-rounded, well-written and reasonably self-contained. In most cases the work will contain results of others as well as those of the author(s). In each case the author(s) should provide sufficient motivation, examples, and applications. In this respect, Ph.D. theses will usually be deemed unsuitable for the Lecture Notes series. Proposals for volumes in these categories should be submitted either to one of the series editors or to Springer-Verlag, Heidelberg, and will be refereed. A provisional judgment on the acceptability of a project can be based on partial information about the work: a detailed outline describing the contents of each chapter, the estimated length, a bibliography, and one or two sample chapters – or a first draft. A final decision whether to accept will rest on an evaluation of the completed work which should include

- at least 100 pages of text;
- a table of contents;
- an informative introduction perhaps with some historical remarks which should be accessible to readers unfamiliar with the topic treated;
- a subject index.

§3. Category iii). Conference proceedings will be considered for publication provided that they are both of exceptional interest and devoted to a single topic. One (or more) expert participants will act as the scientific editor(s) of the volume. They select the papers which are suitable for inclusion and have them individually refereed as for a journal. Papers not closely related to the central topic are to be excluded. Organizers should contact Lecture Notes in Computational Science and Engineering at the planning stage.

In exceptional cases some other multi-author-volumes may be considered in this category.

§4. Format. Only works in English are considered. They should be submitted in camera-ready form according to Springer-Verlag's specifications. Electronic material can be included if appropriate. Please contact the publisher. Technical instructions and/or TEX macros are available via http://www.springer.de/math/authors/help-momu.html. The macros can also be sent on request.

General Remarks

Lecture Notes are printed by photo-offset from the master-copy delivered in camera-ready form by the authors. For this purpose Springer-Verlag provides technical instructions for the preparation of manuscripts. See also *Editorial Policy.*

Careful preparation of manuscripts will help keep production time short and ensure a satisfactory appearance of the finished book.

The following terms and conditions hold:

Categories i), ii), and iii):
Authors receive 50 free copies of their book. No royalty is paid. Commitment to publish is made by letter of intent rather than by signing a formal contract. Springer-Verlag secures the copyright for each volume.

For conference proceedings, editors receive a total of 50 free copies of their volume for distribution to the contributing authors.

All categories:
Authors are entitled to purchase further copies of their book and other Springer mathematics books for their personal use, at a discount of 33,3 % directly from Springer-Verlag.

Addresses:

Timothy J. Barth
NASA Ames Research Center
NAS Division
Moffett Field, CA 94035, USA
e-mail: barth@nas.nasa.gov

Michael Griebel
Institut für Angewandte Mathematik
der Universität Bonn
Wegelerstr. 6
53115 Bonn, Germany
e-mail: griebel@iam.uni-bonn.de

David E. Keyes
Department of Applied Physics
and Applied Mathematics
Columbia University
200 S. W. Mudd Building
500 W. 120th Street
New York, NY 10027, USA
e-mail: david.keyes@columbia.edu

Risto M. Nieminen
Laboratory of Physics
Helsinki University of Technology
02150 Espoo, Finland
e-mail: rni@fyslab.hut.fi

Dirk Roose
Department of Computer Science
Katholieke Universiteit Leuven
Celestijnenlaan 200A
3001 Leuven-Heverlee, Belgium
e-mail: dirk.roose@cs.kuleuven.ac.be

Tamar Schlick
Department of Chemistry
Courant Institute of Mathematical
Sciences
New York University
and Howard Hughes Medical Institute
251 Mercer Street
New York, NY 10012, USA
e-mail: schlick@nyu.edu

Springer-Verlag, Mathematics Editorial IV
Tiergartenstrasse 17
69121 Heidelberg, Germany
Tel.: *49 (6221) 487-8185
e-mail: peters@springer.de
http://www.springer.de/math/
peters.html

Lecture Notes in Computational Science and Engineering

Vol. 1 D. Funaro, *Spectral Elements for Transport-Dominated Equations.* 1997. X, 211 pp. Softcover. ISBN 3-540-62649-2

Vol. 2 H. P. Langtangen, *Computational Partial Differential Equations.* Numerical Methods and Diffpack Programming. 1999. XXIII, 682 pp. Hardcover. ISBN 3-540-65274-4

Vol. 3 W. Hackbusch, G. Wittum (eds.), *Multigrid Methods V.* Proceedings of the Fifth European Multigrid Conference held in Stuttgart, Germany, October 1-4, 1996. 1998. VIII, 334 pp. Softcover. ISBN 3-540-63133-X

Vol. 4 P. Deuflhard, J. Hermans, B. Leimkuhler, A. E. Mark, S. Reich, R. D. Skeel (eds.), *Computational Molecular Dynamics: Challenges, Methods, Ideas.* Proceedings of the 2nd International Symposium on Algorithms for Macromolecular Modelling, Berlin, May 21-24, 1997. 1998. XI, 489 pp. Softcover. ISBN 3-540-63242-5

Vol. 5 D. Kröner, M. Ohlberger, C. Rohde (eds.), *An Introduction to Recent Developments in Theory and Numerics for Conservation Laws.* Proceedings of the International School on Theory and Numerics for Conservation Laws, Freiburg / Littenweiler, October 20-24, 1997. 1998. VII, 285 pp. Softcover. ISBN 3-540-65081-4

Vol. 6 S. Turek, *Efficient Solvers for Incompressible Flow Problems.* An Algorithmic and Computational Approach. 1999. XVII, 352 pp, with CD-ROM. Hardcover. ISBN 3-540-65433-X

Vol. 7 R. von Schwerin, ***M**ulti **B**ody **S**ystem **SIM**ulation.* Numerical Methods, Algorithms, and Software. 1999. XX, 338 pp. Softcover. ISBN 3-540-65662-6

Vol. 8 H.-J. Bungartz, F. Durst, C. Zenger (eds.), *High Performance Scientific and Engineering Computing.* Proceedings of the International FORTWIHR Conference on HPSEC, Munich, March 16-18, 1998. 1999. X, 471 pp. Softcover. 3-540-65730-4

Vol. 9 T. J. Barth, H. Deconinck (eds.), *High-Order Methods for Computational Physics.* 1999. VII, 582 pp. Hardcover. 3-540-65893-9

Vol. 10 H. P. Langtangen, A. M. Bruaset, E. Quak (eds.), *Advances in Software Tools for Scientific Computing.* 2000. X, 357 pp. Softcover. 3-540-66557-9

Vol. 11 B. Cockburn, G. E. Karniadakis, C.-W. Shu (eds.), *Discontinuous Galerkin Methods.* Theory, Computation and Applications. 2000. XI, 470 pp. Hardcover. 3-540-66787-3

Vol. 12 U. van Rienen, *Numerical Methods in Computational Electrodynamics.* Linear Systems in Practical Applications. 2000. XIII, 375 pp. Softcover. 3-540-67629-5

Vol. 13 B. Engquist, L. Johnsson, M. Hammill, F. Short (eds.), *Simulation and Visualization on the Grid.* Parallelldatorcentrum Seventh Annual Conference, Stockholm, December 1999, Proceedings. 2000. XIII, 301 pp. Softcover. 3-540-67264-8

Vol. 14 E. Dick, K. Riemslagh, J. Vierendeels (eds.), *Multigrid Methods VI.* Proceedings of the Sixth European Multigrid Conference Held in Gent, Belgium, September 27-30, 1999. 2000. IX, 293 pp. Softcover. 3-540-67157-9

Vol. 15 A. Frommer, T. Lippert, B. Medeke, K. Schilling (eds.), *Numerical Challenges in Lattice Quantum Chromodynamics.* Joint Interdisciplinary Workshop of John von Neumann Institute for Computing, Jülich and Institute of Applied Computer Science, Wuppertal University, August 1999. 2000. VIII, 184 pp. Softcover. 3-540-67732-1

Vol. 16 J. Lang, *Adaptive Multilevel Solution of Nonlinear Parabolic PDE Systems.* Theory, Algorithm, and Applications. 2001. XII, 157 pp. Softcover. 3-540-67900-6

Vol. 17 B. I. Wohlmuth, *Discretization Methods and Iterative Solvers Based on Domain Decomposition.* 2001. X, 197 pp. Softcover. 3-540-41083-X

Vol. 18 U. van Rienen, M. Günther, D. Hecht (eds.), *Scientific Computing in Electrical Engineering.* Proceedings of the 3rd International Workshop, August 20-23, 2000, Warnemünde, Germany. 2001. XII, 428 pp. Softcover. 3-540-42173-4

Vol. 19 I. Babuška, P. G. Ciarlet, T. Miyoshi (eds.), *Mathematical Modeling and Numerical Simulation in Continuum Mechanics.* Proceedings of the International Symposium on Mathematical Modeling and Numerical Simulation in Continuum Mechanics, September 29 - October 3, 2000, Yamaguchi, Japan. 2002. VIII, 301 pp. Softcover. 3-540-42399-0

Vol. 20 T. J. Barth, T. Chan, R. Haimes (eds.), *Multiscale and Multiresolution Methods.* Theory and Applications. 2002. X, 389 pp. Softcover. 3-540-42420-2

Vol. 21 M. Breuer, F. Durst, C. Zenger (eds.), *High Performance Scientific and Engineering Computing.* Proceedings of the 3rd International FORTWIHR Conference on HPSEC, Erlangen, March 12-14, 2001. 2002. XIII, 408 pp. Softcover. 3-540-42946-8

Vol. 22 K. Urban, *Wavelets in Numerical Simulation.* Problem Adapted Construction and Applications. 2002. XV, 181 pp. Softcover. 3-540-43055-5

Vol. 23 L. F. Pavarino, A. Toselli (eds.), *Recent Developments in Domain Decomposition Methods.* 2002. XII, 243 pp. Softcover. 3-540-43413-5

Vol. 24 T. Schlick, H. H. Gan (eds.), *Computational Methods for Macromolecules: Challenges and Applications.* Proceedings of the 3rd International Workshop on Algorithms for Macromolecular Modeling, New York, October 12-14, 2000. 2002. IX, 504 pp. Softcover. 3-540-43756-8

Vol. 25 T. J. Barth, H. Deconinck (eds.), *Error Estimation and Adaptive Discretization Methods in Computational Fluid Dynamics.* 2003. VII, 344 pp. Hardcover. 3-540-43758-4

Vol. 26 M. Griebel, M. A. Schweitzer (eds.), *Meshfree Methods for Partial Differential Equations.* 2003. IX, 466 pp. Softcover. 3-540-43891-2

Vol. 27 S. Müller, *Adaptive Multiscale Schemes for Conservation Laws.* 2003. XIV, 181 pp. Softcover. 3-540-44325-8

Vol. 28 C. Carstensen, S. Funken, W. Hackbusch, R. H. W. Hoppe, P. Monk (eds.), *Computational Electromagnetics.* Proceedings of the GAMM Workshop on "Computational Electromagnetics", Kiel, Germany, January 26-28, 2001. 2003. X, 209 pp. Softcover. 3-540-44392-4

Vol. 29 M. A. Schweitzer, *A Parallel Multilevel Partition of Unity Method for Elliptic Partial Differential Equations.* 2003. V, 194 pp. Softcover. 3-540-00351-7

Vol. 30 T. Biegler, O. Ghattas, M. Heinkenschloss, B. van Bloemen Waanders (eds.), *Large-Scale PDE-Constrained Optimization.* 2003. VI, 349 pp. Softcover. 3-540-05045-0

Vol. 31 M. Ainsworth, P. Davies, D. Duncan, P. Martin, B. Rynne (eds.) *Topics in Computational Wave Propagation.* Direct and Inverse Problems. 2003. VIII, 399 pp. Softcover. 3-540-00744-X

Vol. 32 H. Emmerich, B. Nestler, M. Schreckenberg (eds.) *Interface and Transport Dynamics.* Computational Modelling. 2003. XV, 432 pp. Hardcover. 3-540-40367-1

Vol. 33 H. P. Langtangen, A. Tveito (eds.) *Advanced Topics in Computational Partial Differential Equations.* Numerical Methods and Diffpack Programming. 2003. XIX, 658 pp. Softcover. 3-540-01438-1

Vol. 34 V. John, *Large Eddy Simulation of Turbulent Incompressible Flows.* Analytical and Numerical Results for a Class of LES Models. 2004. XII, 261 pp. Softcover. 3-540-40643-3

Vol. 35 E. Bänsch, *Challenges in Scientific Computing - CISC 2002.* Proceedings of the Conference *Challenges in Scientific Computing*, Berlin, October 2-5, 2002. 2003. VIII, 287 pp. Hardcover. 3-540-40887-8

Texts in Computational Science and Engineering

Vol. 1 H. P. Langtangen, *Computational Partial Differential Equations.* Numerical Methods and Diffpack Programming. 2nd Edition 2003. XXVI, 855 pp. Hardcover. ISBN 3-540-43416-X

Vol. 2 A. Quarteroni, F. Saleri, *Scientific Computing with MATLAB.* 2003. IX, 257 pp. Hardcover. ISBN 3-540-44363-0

For further information on these books please have a look at our mathematics catalogue at the following URL: `http://www.springer.de/math/index.html`

MIX
Papier aus verantwortungsvollen Quellen
Paper from responsible sources
FSC® C105338

If you have any concerns about our products,
you can contact us on
ProductSafety@springernature.com

In case Publisher is established outside the EU,
the EU authorized representative is:
Springer Nature Customer Service Center GmbH
Europaplatz 3, 69115 Heidelberg, Germany

Printed by Libri Plureos GmbH
in Hamburg, Germany